Climate Change and Natural Resources

About the Author

Prof. H.S. Sharma is a senior geographer with long established career spanning over 40 years. He holds post-graduate and doctoral degrees in Geography from University of Saugar, Madhya Pradesh. Prof. Sharma is a former Professor and Head, Department of Geography, University of Rajasthan, Jaipur. He is also a former Director of Indira Gandhi Centre for Human Ecology, Environment and Population Studies, Academic Staff College and Dean, Faculty of Science, University of Rajasthan.

Prof. Sharma has over 100 research papers in National and International Journals including NATURE and 18 books to his credit. He has special inclination about environmental geomorphology, desertification, climate change and human impact on environment. He has supervised more than three dozen Ph.D's. He attended and chaired various International and National Conferences. Prof. Sharma has been President of IIG, IGI, NAGI, Sectional President of Section of the Earth System Sciences in Indian Science Congress and RGA. Prof. Sharma is a widely travelled geographer in U.K., Germany, Norway, Canada, U.S.A., Malaysia, Seychelles and Netherlands under exchange programme and attending International conferences.

CLIMATE CHANGE AND NATURAL RESOURCES

A Study of Indian Desert

H.S. Sharma
Formerly Professor and Head, Department of Geography
University of Rajasthan, Jaipur-302004

CONCEPT PUBLISHING COMPANY PVT. LTD.
NEW DELHI-110059

ISBN-13: 978-93-5125-280-1

First Published 2018

Published and Printed by

Since 1974

Concept Publishing Company Pvt. Ltd.
A/15-16, Commercial Block, Mohan Garden,
New Delhi-110059 (INDIA)
T : +91 11 25351460, +91 11 25351794
E : publishing@conceptpub.com *W* : www.conceptpub.com

Cataloging in Publication Data--Courtesy: D.K. Agencies (P) Ltd. <docinfo@dkagencies.com>

Sharma, H. S. (Hari Shanker), 1945- **author.**
Climate change and natural resources : a study of Indian desert / H.S. Sharma.
pages cm
Includes bibliographical references and index.
ISBN 9789351252801

1. Climatic changes--India--Rajasthan. 2. Natural resources--India--Rajasthan. I. Title.

LCC QC903.2.I4S53 2018 DDC 551.609544 23

Dedicated

to My Wife

Late Smt. Maya Sharma

Foreword

Seldom does one come across a book that evaluates the veracity of the widely accepted phenomenon of climatic change on the basis of a thorough investigation of the climatic parameters spread over a period of 100 years in a regional frame. Professor H.S. Sharma has accomplished this arduous task in the present book "**Climate Change and Natural Resources: A Study of Indian Desert**". The analysis and evaluation of periodic variation in temperature, rainfall and the frequency of extreme events, together with an assessment of the desertification trend, over a large area, spread over 32 districts of Rajasthan, form the basis of his conclusive observations.

In the spirit of a true scholar, Dr. Sharma has mined and critically reviewed the available literature on climate change, especially the research findings related to Rajasthan, outlined his own methodology and sources of data and proceeded to systematically process the secondary as well the primary data to reach a conclusion that may concur or disagree with the observations of previous researchers. A fascinating part of his study is to give a comparative picture of .water and other biological resources that seem, apparently, to correspond with the trends in temperature and rainfall. The most critical part of his study is 'Climate and Climate Change' as observed in Rajasthan, that evaluates most of the climate variables, like, temperature, rainfall, numbers of rainy days, rainfall variability, extremes of rainfall conditions, droughts and trends of variation in temperature and rainfall, the former over a period of 50 years and the latter over a hundred-year period.

To get a clearer picture of climate change, if and as perceived by the people, and has a grass-root impact, Dr. Sharma has taken a stratified sample of 20 villages in the five agricultural regions of Rajasthan, covering over 400 households, and, as expected, finds the farmers perpetually complaining about drought and water scarcity. More to the point is his finding that "there is a clear warming trend in the past temperature records that support the hypothesis of an increasing trend of mean annual temperature over time" especially after 1971. The analysis of rainfall data for a hundred-year period, on the other hand, is marked by periodic fluctuations, with conflicting trends of increase and decrease in alternative decades, thus suggesting a cyclic trend, with no definite trend.

Dr. Sharma has, very aptly, looked at the possible consequences of climate change on water resources and the biological landscape of the region. Depletion of natural resources like groundwater or the shrinking area of forests or changes in the land-use, though statistically established, could be attributed as much to anthropogenic causes, a burgeoning population and growing demand for food, fodder and shelter, as to change in climate. It is difficult not to talk of desert and desertification while discussing Rajasthan in the context of climate change. The fourteen north-western districts of Rajasthan, grouped under arid and semi-arid regions based on aridity and moisture index, experience variable stresses, depending on their location, especially the distance from Indira Gandhi canal. The seepages from the canal and irrigated areas have raised the water table in some areas where a high degree of evaporation has caused accumulation of salts close to the surface, rendering the soils alkaline. Desertification of the semi-arid region often perceived as the result of recurring droughts could as well be because of overgrazing and faulty land use. Desertification is a challenging theme and the author has succeeded in producing a convincing picture of the contemporary reality of Rajasthan desert. Dr. Sharma suggests some measures that could be adopted to face the possible adverse scenarios..

A book, based on region-based study of climate-change, though not a certainty, and the existing state of natural resources of Rajasthan, presents a contemporary theme and topical addition to the existing literature on the subject. In fact, Dr Sharma's book may prove a trendsetter, with many such research based studies and publications to follow.

I am sure, the book will receive wide appreciation for the approach adopted, a rigorous analysis of data and independent observations.

K.R. Dikshit

Preface

Climate change is a global challenge with diverse implication at the national and regional levels, through impacts on various sectors such as agriculture, water resources, forestry and biodiversity, human health, energy and infrastructure. The long term perspective is needed in debating and communicating climate change and for converging the importance of sustainability. The constant dream beats of Gloom and Doom have often led to wait and see attitudes and decision-paralysis in the developed countries and a sense of hopelessness for the youth in the developing countries (Misra, 2013).

The scientific studies indicate that climate change is inevitable. The climate has transformed in the past, is varying in the present and is bound to change in the future. The studies indicate that before global warming between 18,000-20,000 years ago, most of the earth was frozen and was an arid wasteland (U.B. Mathur, 2010). Over half of the earth's surface was covered by glaciers or extreme deserts, and forest covers were rare. Hence, it was not a very congenial place to live. 'Global Warming' over the last 15,000-16,000 years has changed our earth from an ice box to a garden. Today, we are erroneously worried about the global warming and the impacts it is likely to bring. Contrary to it, some scientists forecast that 'global cooling' would be more disastrous in future.

The scientists are debating on the causes of climate change. The natural scientists believe that climate changes are due to several natural forces like changes in the obliquity of earth's axis, its rotational speed, solar variability or sunspot and volcanic activities. On the contrary, some scientists who believe in global warming say that increase in the emission of green house gases (GHGs) after industrial revolution has been mainly responsible for global warming and climate changes. It is beyond doubt that since 1850 A.D. the temperature is rising at such a rate that it was never experienced during last 8,00,000 years during geological past. The 'pro-warming lobby' advocates that the earth's warming occurs not only in response to external forcing, but also due to release of GHG's in the atmosphere by human beings especially the burning fossil fuels. The global warming leads to increase in both average and extreme temperature and rainfall and consequent melting of ice caps, retreat of glaciers, rise in sea level, increasing coastal erosion, lengthening growing seasons and increasing range of many infectious diseases. There are more hot days and less cool days. Rainfall is more in the high latitudes and less in tropical regions. The duration of rainfall gets shorter but the amount increases. Further, rain is accompanied by massive storms and dry spells. The frequency of drought increases. While some scientists subscribe to the point that global warming phase is likely to end in the end of 2015 and the cooling phase may hamper to life system. Accordingly, they state that symptoms of cooling are visible in the Northern Hemisphere especially in Europe and North America which have been experiencing heavy snowfall during 2012 and 2013 winters. Peru in March 2013 experienced heavy flooding, and temperature in March 2013 at Bombay was touching 40°C due to anti-cyclonic system.

Scientific understanding is now remarkably coherent on the following fundamental conclusions about climate change:

1. The Earth system has clearly moved far outside the range experienced over the last 7,00,000 years and hence is operating in a 'No Analogue' State.
2. The earth is warming due to increased concentration of GHG in atmosphere. Snowy winters in some parts of the world do not alter this fact.
3. Most of the increase in emission of GHG over last century is due to anthropogenic factors and have become dominant force, that a new geological epoch the Anthropocene has been proposed.
4. Natural causes always play a role in changing earth's climate, but are now being out competed by anthropogenic changes.
5. Warming of the planet will cause many other climatic pattern to change at speeds unprecedented in modern times including increasing rates of sea level rise and alternations in hydrological cycle.
6. The combination of these complex climate changes threatens coastal communities, urban centres and rural systems.
7. A prophesy has been made by several scientists that in future cooling phase of climate would be more disastrous than global warming.

The climate change is emerging as a major area of interest for civic organizations, universities, research institutes and public bodies in India. Many foreign NGO's and others are setting-up base in India convinced that the Indian subcontinent could play a major role in addressing global climate change. A fairly good amount of work has been done by Indian scientists on climate changes. India has witnessed extreme weather events during last three decades such as drought in Orissa, cloudburst in Leh, unprecedented rainfall in Uttarakhand, Barmer and Jammu & Kashmir which show that various parts of country are already reeling under the impacts of climate change.

Why Research on Climate Change is Crucial for Rajasthan?

Rajasthan is the largest state in India in terms of size comprising 10.4 per cent of the country's total area with 2/3 of its area is Thar Desert. The total population of Rajasthan is 686.21 lakh (2011) of which nearly 75.10 per cent of the states population resides in rural areas. Why Rajasthan warrants priority in adaptation for climate change research? Rajasthan is a unique geographical region for the study of societal adaptation to climate change (V.S. Singh, 2010). Antiquity of human occupation of the desert region goes back of the late Pleistocene (1,60,000 years) or even earlier as indicated by archaeological studies. During the mid-Holocene, the regions around the Indian Ocean witnessed rise of Three Great Civilization of the World (i.e. Mesopotamia, Egyptian and Indus Valley). Indus-Saraswati Civilization is the earlier urban civilization in South Asia that flourished and fell in the region that includes parts of present day Rajasthan.

In the present book climate change and its impacts on natural resources particularly land and water resources have been examined. Climatically, Rajasthan is a very sensitive state having scanty rainfall, high air temperature, intense solar radiation and high wind velocity. In an era of climate change Rajasthan is likely to suffer further more shortage due to uncertainty of rainfall and increasing number of weather related hazards and disasters. The analysis of excessive rainfall, droughts, changing nature of number of rainy days and trend of temperature and rainfall during last 112 years or so make it clear that Rajasthan has been witnessing climate change during past and present centuries. The frequency of weather related hazards have increased both in western arid region as well as in the eastern semi-arid region. As a consequence, it may be stated that the impacts of climate change will be felt across most sections of economy of Rajasthan. Drought and floods could mean higher evaporation which will result in low river flows and the quantity of available runoff may decrease. Crop yields are likely to decrease with rising temperatures. Forest character will change over because of pressing demand for fuel. However, more severe droughts could lead to greater desertification if land resources are not carefully managed. Strategies for drought are also needed. Water supplies have to be maintained.

In view of changing climatic conditions everyone has to contribute slowing climate change and reducing its impacts. People have to take individual responsibility for themselves and families. They have to support the green economy, lower their green house gas emissions, live sustainably and adopt the way of life to the changing climate. At the community level efforts should be made to ensure green economy to make use of products in public buildings such as solar power, lighting control and rain water harvesting.

I take this opportunity to express my sincere thanks to the Indian Council of Social Science Research, New Delhi for awarding Senior Fellowship to me.

I express my profound sense of gratitude to Prof. Sukhadeo Thorat, Chairman and Prof. Ramesh Dadhich, Member Secretary for their generous support.

I warmly appreciate the sincere support provided by Prof. Somdeo, Senior Academic Consultant and Dr. Sanchita Dutta, Incharge (RFS) of ICSSR in completion this academic assignment. I owe my sincere thanks to them.

Prof. R. N. Mishra and Prof. R.D. Doi, former Chairpersons, Department of Geography, University of Rajasthan, Jaipur, deserve my thanks for providing infrastructural facilities and administrative support.

I am grateful to Prof. H. N. Mishra of Allahabad University and his research scholar Mr. Ashutosh Mishra (JRF) for providing and analyzing temperature and rainfall data. Dr. P. R. Binda, Senior Cartographer (Retd.) and Miss Garvita Kaushik, Research Assistant in the SAC project deserve my sincere thanks and appreciation for their help throughout the project work.

Dr. Ganesh Sharma, Assistant Professor, Department of Geography, Government College, Jaipur deserves my appreciation for his assistant in this work. I warmly appreciate the research and administrative assistance provided by several of my colleagues during different stages of the study.

Dr. S.S. Singh of India Meteorology Department, Government of India, Shri P.K. Parchure, Regional Director and Dr. D.D. Sharma, Scientist 'G', Central Ground Water Board, Jaipur also deserve my heartfelt thanks for their advice and providing data related to climate and ground water resources.

I am also thankful to Shri H. L. Maheswari for typing the manuscript and to Shree Printing Center, Jaipur for designing and formatting the report in a publishable form.

In the last, I am thankful to the villagers of sample villages who spared their time to share their views and perception on the issue of climate change and other aspects.

Jaipur **H.S. Sharma**

Contents

List of Tables

List of Figures

1
Introduction

Conceptual Framework

Earth, life and climate have always changed and climate change itself may have been crucial for the evolution of human being. The climate system is a complex interactive system consisting of the atmosphere, hydrosphere, lithosphere, cryosphere and outer space. Climate change is a global challenge with diverse implication at the national and regional levels, through impacts on various sectors such as agriculture, water resources, forestry and biodiversity, human health, energy and infrastructure. The long-term perspective is needed in debating and communicating climate change and for converging the importance of sustainability. The constant dream beats of Gloom and Doom have often led to wait and see attitudes and decision-paralysis in the developed countries and a sense of hopelessness for the youth in the developing countries (Misra, 2013).

The scientific studies indicate that climate change is inevitable. The climate has transformed in the past, is varying in the present and is bound to change in the future. The studies indicate that before global warming between 18,000-20,000 years ago, most of the earth was frozen and was an arid wasteland (U.B. Mathur, 2010). Over half of the earth's surface was covered by glaciers or extreme deserts, and forest covers were rare. Hence, it was not a very congenial place to live. 'Global Warming' over the last 15,000-16,000 years has changed our earth from an ice box to a garden. Today, we are erroneously worried about the global warming and the impacts it is likely to bring. Contrary to it, some scientists forecast that 'global cooling' would be more disastrous in future.

The scientists are debating on the causes of climate change. The natural scientists believe that climate changes are due to several natural forces like changes in the obliquity of earth's axis, its rotational speed, solar variability or sunspot and volcanic activities. On the contrary, some scientists who believe in global warming say that increase in the emission of green house gases (GHGs) after industrial revolution has been mainly responsible for global warming and climate changes. It is beyond doubt that since 1850 A.D. the temperature is rising at such a rate that it was never experienced during last 8,00,000 years during geological past. The 'pro-warming lobby' advocates that the earth's warming occurs not only in response to external forcing but also due to release of GHGs in the atmosphere by human beings especially the burning fossil fuels. The global warming leads to increase in both average and extreme temperature and rainfall and consequent melting of ice caps, retreat of glaciers, rise in sea level, increasing coastal erosion, lenthening growing seasons and increasing range of many infectious diseases. There are more hot days and less cool days. Rainfall is more in the high latitudes and less in tropical regions. The duration of rainfall gets shorter but the amount increases. Further, rain is accompanied by massive storms and dry spells. The frequency of drought increases. While some scientists subscribe to the point that global warming phase is likely to end in the end of 2015 and the cooling phase may hamper to life system. Accordingly, they state that symptoms of cooling are visible in the Northern Hemisphere especially in Europe and North America which have been experiencing heavy snowfall during 2012 and 2013 winters. Peru in March 2013 experienced heavy flooding, and temperature in March 2013 at Bombay was touching 40°C due to anti-cyclonic system.

Expected Global Cooling

On April 2006, a group of 60 scientists wrote to the Canadian Prime Minister questioning the so-called consensus that the Earth faces a '*climate emergency*'. '*Climate Change is Nothing New*' the article appeared in *Washington Post* on October 16, 2006 referred to the findings of Simon Brassell, an Indiana University geologist asserting that climate change occurred during the Mesozoic Era when the earth inhabited by dinosaurs and there were no human beings on the earth (http://www.publicaffairs.noaa.gov/realses2006/Jul06/noaa06-065.html). There are some scientists who think that the prospect of global cooling is more rather than global warning, which should concern us. All these unusual freak weather phenomena are warming signals of climate uncertainties in future. It may, therefore, be safely said that climate change is happening in which natural forces have been predominant, but during last 150 years the process of climate change has been accelerated due to anthropogenic forcing. A very significant study on global cooling has been brought out by Joseph D'Aleo and Don Easterbrook (2010) in the paper on Multidecadal tendencies in ENSO and Global Temperatures related to Multidecadal Oscillations. The widespread snowfall in Europe and North America in November-December 2013, Srinagar (J & K) 2013 and unprecedented snowfall in United States of America in November 2014 and February 2015 are warning signals of global cooling in near future.

The recently published IPCC Fifth Assessment Synthesis report (Climate Change, 2014) provides an integrated view of climate change. The report advocated that human influence on climate system is clear, and recent anthropogenic emission of green house gases are the highest in history. Recent climate changes have had widespread impacts on human and natural systems. Warming of climate system is unequivocal and since 1950's many of the observed changes are unprecedented over decades to millennia. The atmosphere and oceans have warmed, the amount of snow and ice have diminished, and sea level has risen (1901 to 2010 rises 0.17 to 0.21 metre). The report further indicates that changes in many extreme weather and climate event have been observed since 1950. Some of these changes have been linked to human influences, including a decrease in cold temperature extreme, in increase in extreme high sea levels and an increase in the number of heavy precipitation events in a number of regions.

The Inter-governmental Panel on climate change (IPCC, 2007) and other scientific assessments of climate change, which involve thousands of scientists producing massive and comprehensive reports, have quite expectedly and normally made some mistakes. When errors are pointed out, they are corrected. But there is nothing remotely identified in recent events that changes the fundamental conclusions about climate change. Indeed, there is compelling, comprehensive, constant and objective evidences that humans are altering the climate in ways that threaten our societies and the ecosystems during last 150 years. Scientific understanding is now remarkably coherent on the following fundamental conclusions about climate change:

1. The Earth system has clearly moved far outside the range experienced over the last 7,00,000 years and hence is operating in a **'No Analogue' State.**
2. The earth is warming due to increased concentration of GHG in atmosphere. Snowy winters in some parts of the world do not alter this fact.
3. Most of the increase in emission of GHG over last century is due to anthropogenic factors and have become dominant force, that a new geological epoch the ***Anthropocene*** has been proposed.
4. Natural causes always play a role in changing earth's climate, but are now being out competed by anthropogenic changes.
5. Warming of the planet will cause many other climatic pattern to change at speeds unprecedented in modern times including increasing rates of sea level rise and alternations in hydrological cycle.
6. The combination of these complex climate changes threatens coastal communities, urban centres and rural systems.
7. A prophesy has been made by several scientists that in future cooling phase of climate would be more disastrous than global warming.

Review of Literature

A fairly good amount of work has been done by Indian scientists on climate changes. The climate change is emerging as a major area of interest for civic organizations, universities, research institutes and public bodies in India. Many foreign NGOs and others are setting-up base in India convinced that the Indian subcontinent could play a major role in addressing global climate change. India has witnessed extreme weather events during last three decades such as drought in Orissa, cloudburst in Leh, unprecedented rainfall in Uttarakhand, Barmer and Jammu & Kashmir which show that various parts of country are already reeling under the impacts of climate change.

The study of Anderson, D.M., *et al.* (2002) reveal in their paper an increase in the Asian south-west monsoon during the past four centuries. Climate reconstruction reveal unprecedented warming in the past century. Bandhopadhyaya, A., *et al.* (2009) stated that a significant decreasing trend was found in evapo-transpiration all over India during 1971-2002 due to global warming. However, a general increase in rainfall was not found in recent years. A focus in climate change and Malaria in India by Bhattacharya, S.C., *et al.*, (2006) has been to explain the likely impact of climate change on vector production and malaria transmission in India. Bhuiyan, C., *et al.* (2009) in their paper attempted to assess relative influence of different controlling factors on groundwater recharge. A very interesting study on "*Dividing Climate Change: Global Warming in Indian Mass Media*" by Billett, S. (2010) reveals that the communication of climate change from scientists and policy-makers to the public *via* the mass media has been a subject of interest because of its implications for creating national variation in public understanding of a global environmental issue.

The paper by Bokil (2000) on drought in Rajasthan develops a systematic perspective on drought to guide both policy and practice. The paper argues that drought in arid and semi-arid regions is not a calamity, but a regular climatic feature. Instead of restoring to calamity relief, government should consider drought mitigation as a principal strategy of agricultural and rural development.

The vulnerability of India and Indian states to climate change was assured using Vulnerability-Resilience Indicator Prototype (VRIP) in the paper by Brenkert, A., *et al.*, (2005). A case study by Chatterjee, K., *et al.*, (2005) on traditional adaptation practices by vulnerable communities in drought prone district Tonk, Rajasthan reveal that communities here already bear the brunt of drought and have learnt to cope.

A study on *Impact of Heat Waves over India* by Choudhary, S.K., *et al.*, (2000) states that a severe heat wave conditions exists when the day temperature exceeds the normal maximum temperature over the place by 6° C. During 1978-1999 heat waves over Bihar, Rajasthan and Uttar Pradesh have been witnessed during the month of June compared to others. The loss of human lives was found to be more in Rajasthan.

Daily gridded (1° × 1°) rainfall data prepared by IMD for the period 1951-2004 have been used by Dash, S.K., *et al.* (2009) in the study to examine the frequency of rain events in India in terms of their duration and intensities per day. The results show the weakening of the summer monsoon circulation over India.

Further, Dash, S.K., *et al.* (2007) in their paper on some evidences of climate change in twentieth century in India indicate decreasing tendency in summer monsoon rainfall over Indian landmass and increasing trend in the rainfall during pre-monsoon and post-monsoon months.

The study by Gosain, A.K., *et al.* (2006) reveals that under GHG scenario the conditions may deteriorate in terms of severity of droughts in some parts of India and enhanced intensity of floods in other parts of the country. However, there is a general over all reduction in the quantity of the available runoff under the GHG scenario. The Luni with west flowing rivers shall face water scare condition. So, shall be the case of the Mahi, Pennar, Sabarmati and Tapi.

Increasing trend of extreme rain events over India in a warming environment has been analyzed by Goswami, B.N., *et al.* (2006). Against a backdrop of rising global surface temperature, the stability of the Indian monsoon rainfall over the past century has been a puzzle. The paper reveals that substantial increase in hazards related to heavy rains is expected over Central India in future. The study by Goswami, P. and K.V. Ramesh (2008) based on weighted epochal ensemble approach shows a sharp increase in the size of the Indian desert in the next hundred years.

The paper on sensitivity of evapo-transpiration to global warming with reference to arid zone of Rajasthan studied by Goyal, P.K. (2004). The paper brings out an increase of 14.8 per cent of the total ET demand with increase in temperature by 20 per cent (maximum 8°C). The analysis of 100 years rainfall data of the arid region of Rajasthan indicates an increasing trend of 0.5 mm/year, but increased evapo-transpiration demand due to global warming can put tremendous pressure on existing overstressed water resources of this region. R.K. Gupta (1968) in his paper on anthropogenic influence on the vegetation of western Rajasthan asserted that Rajasthan desert, if not a 'man made desert' is surely a 'man maintained desert'.

Water for India in 2050 first order assessment of available options by Gupta, S.K. and R.D. Deshpande (2004) examined the water demand in the context of growing population. Jain, S.K., *et al.* (2009) identified drought—vulnerable areas using NOAA AVHRR data and found that in 2002 all the areas in Rajasthan was affected by drought with greater intensity, mostly classed as extreme and severe drought conditions.

Kaul, M., *et al.* (2009) in their paper on Land-use change and net C flux in Indian forests indicate that the conversion of land to forest land led to a net uptake of 0.86. By contrast, the net CO emissions from the forest land conversion to another land-use (deforestation) resulted in annual emissions of 9.9.

Agriculture is highly dependent on the spatial and temporal distribution of monsoon rainfall. The paper by Kumar, K.K., *et al.* (2004) presents an analysis of crop-climate relationship for India, using historic production statistics for major crops.

High resolution climate change scenario for India for the twenty-first century has been studied by Kumar, K.R., *et al.* (2006) using PRECIS (Providing Regional Climates for Impact Studies) developed by the Hadley Centre for Climate Prediction and Research. Precis simulations under scenario of increasing green house gas concentrations and sulphate aerosols indicate marked increase in both rainfall and temperature towards the end of the twenty-first century. PRECIS estimates 20 per cent rise in all India summer monsoon rainfall in future scenario as compared to present. Rise of rainfall is seen over all states except Punjab, Rajasthan and Tamil Nadu which show slight decrease in precipitation in the future scenario.

Spatial and sub-seasonal patterns of the long-term trends of Indian summer monsoon rainfall have been analyzed by Kumar, K.R., *et al.*, (1992) using over 114 years data (1871-1984). Areas of increasing trend in the monsoon seasonal rainfall are found along the west coast, north Andhra Pradesh and north-west India and those of decreasing trend over east Madhya Pradesh and adjoining areas, north-east India and part of Gujarat and Kerala.

In recent times several studies around the globe show that climate change is likely to impact significantly upon fresh water resources availability. The paper by Mall, R.K., *et al.* (2006) examines the potential for sustainable development of surface water and groundwater resources within the constraints imposed by climate change. In 124 years, probability of occurrence of drought was found maximum in west Rajasthan (25%), Saurashtra and Kutch (23%) followed by Jammu & Kashmir (21%) and Gujarat (21%).

Further, Impact of climate change on Indian agriculture: A Review has been viewed by Mall, R.K. *et al.* (2006). The paper presents an overview of the State of the knowledge of possible effect of the climate variability and change on foodgrain production in India.

Impact of climate change on the lives and livelihood of people in India is now widely recognized. The study by Panda, A. (2009) points out that the states of Bihar, Rajasthan, Gujarat, Punjab, Haryana, Madhya Pradesh, Maharashtra, Andhra Pradesh and Karnataka have the lowest adaptive capacity. The areas of greatest climate sensitivity are Rajasthan, Madhya Pradesh and Uttar Pradesh.

Rainfall and temperature trend during 1901-1982 have been studied for the north-west Indian region consisting of the meteorological sub-divisions of Punjab, Haryana, west Rajasthan, east Rajasthan and west Madhya Pradesh by Pant, G.B. and Hingane (1988). The results indicate a decreasing trend in the mean annual surface temperature, which is mainly contributed by the south-west monsoon season. The mean annual and south-west monsoon season rainfall series over most parts of the region indicate a conspicuous increasing trend. The increasing trend is significantly marked for the peripheral areas of the Rajasthan desert. Earlier findings reporting the decreasing trend in rainfall is found to be unacceptable due to the unrealistic assumptions in his study.

Evidences are growing to suggest that climate change clearly has implications for drought vulnerable India with studies projecting possible reductions in monsoon related rainfall in India (Prabhakar, S., *et al.*, 2008). A paper by Ragab, R. and C. Prudhomme (2002) estimates that the annual average increase in temperature in Thar desert ranges from 2° to 2.5°C in summer. Average annual rainfall is shown to decrease by 5.25 per cent in the region.

The impact of future climate change on the Indian summer monsoon has been investigated by Rajendran, K. and A. Kitoh (2008) using a super high resolution global general circulation model. The study shows that widespread, but spatially varying increase in rainfall over interior regions of India and significant reduction over west coasts of Kerala and Karnataka and eastern hilly regions around Assam. The study further suggests comparatively weaker enhancement seen over regions with reduced future rainfall such as Rajasthan, J & K and parts of Assam, Arunachal Pradesh, Manipur, Meghalaya and Mizoram.

Station level hourly precipitation data from 1980 to 2002 spread across the Indian subcontinent were analyzed for trends in extreme hourly precipitation events by Roy, S.S. (2009). The results indicate rising trends in extreme heavy precipitation events in Himalayan region extending south into the Indo-Gangetic basin. The highest annual temperatures for the whole year are experienced during this season across the entire subcontinent, led by north-western India.

Climate change and its impact on drought and floods in Luni river basin of north-west arid India has been studied by Singh, R.S., *et al.* (1999). The paper analyzed long-term fluctuations and trends in rainfall within and adjoining areas for the period 1901-96. A significant rising trend in annual mean air temperature of 0.015°C per year in Jodhpur and 0.008°C year in Pali was observed. An increasing linear trend for the rainfall at 8 stations with highest (1.6 mm year^{1}) in Ajmer was observed during the study period.

A state of art regional climate modelling system known as PRECIS and Meteorological Research Institute-Japan Meteorology Agency (MRI-JMA) Atmosphere General Circulation Model (AGCM) have been applied to develop high resolution climate change scenario for north-west India winter season by Yadav, R.K., Rupa Kumar and M. Rajeevan (2010). In the present study simulation (1961-1990) with PRECIS and with MRI-JMA was evaluated. The study suggests increasing trend of green house gas concentrations and sulphate aerosols and marked increase in both precipitation and surface temperature towards the end of the twenty-first century.

In order to substantiate the results of Inter-governmental Panel on Climate Change (IPCC) and the PRECIS model that the hotter days and warm nights and reduction in rainfall in the Thar desert are ahead by twenty-first century. Surendra

Poonia and A.S. Rao (2013) in their paper on climate change and it impact on Thar Desert ecosystem long-term trends in annual rainfall and temperatures for Thar region were analyzed. The study shows that the air temperatures by the end of twenty-first century are likely to increase by +3.3°C at Bikaner, +3.4°C at Jaisalmer, +2.9°C at Jodhpur and + 2.5°C at Pali. Similarly, the annual rainfall of the region is likely to be increased by +100 mm at Bikaner, +124 mm at Jaisalmer, –40 mm at Jodhpur and +21 mm at Pali. The spatial and temporal variation in potential evapo-transpiration requirement of Thar region ranged from 2.1 mm/day to 12.2 mm/day and on an annual basis between 1500 mm to 2220 mm. This range would reduce the water and food resources of the region.

Trends in rainfall pattern over India has been studied by Guhatha Kurta, P. and M. Rajeevan (2006). Monthly, seasonal and annual rainfall time series of 36 meteorological sub-divisions of India were constructed using fixed, but a large network of about 147 rain-gauge stations. Trend analysis was carried out to examine the long-term trends in rainfall over different sub-divisions. The study has pointed out the contribution of June, July and September rainfall to annual rainfall is decreasing for few sub-divisions, while contribution of August rainfall is increasing in few other sub-divisions. Though Indian monsoon rainfall as a whole does not show any significant trend. However, significant rainfall trends are observed over some specific areas. The study brings out that the pattern of rainfall has been interesting and significant changes are noticed.

Impact of climate change on land degradation over India has been studied by Gore, P.G., B.A. Roy and H.R. Hatwar (2011). In this study land degradation has been examined with the help of moisture status in two different periods (1901 to 1950 and 1941-1990). Land degradation would lead to evolve a climate change, if any.

Influence of land-use/land cover (LULC) changes on atmospheric dynamics over the arid region of Rajasthan state, India has been examined by Shailesh Kumar Kharol, *et al.* (2013). The study examines the long-term effects of land-use/land cover changes in Rajasthan. The multi-decadal variability (1951-2007) in annual rainfall and total number of rainy days reveals significant year-to-year fluctuations exhibiting a slight increasing trend over arid western Rajasthan.

A report on Climate Change: *Vulnerability and Adaptation, Experiences from Rajasthan and Andhra Pradesh* has been brought out by Swiss Agency for Development and Cooperation (SDC) in 2009. The climate change vulnerability and adaptation have been studied taking Udaipur district of Rajasthan and Mahabub Nagar of Andhra Pradesh.

Several studies on climate change in India (Hingani, L.S., *et al.* (1985); Rupa Kumar, K., *et al.* (2003-05); Moolay, D.A. (1984), Thappiyal, V., *et al.* (1991); Rupa Kumar, K. (2003); Kriplani, R.H., *et al.* (2007); Koteshwaram, P., *et al.* (1958); Singh, R.S., *et al.* (2001); Doi, R.D. (2000); Bhaskaran, B, *et al.* (1995); Longergan, S. (1998); Lal, M., *et al.* (2001); Alka Gadgil and Amit Dhorde (2005); Amit Dhorde, *et al.*

(2009); and Bhardwaj, V. and Geeta Sharma (2008) further confirm the increase in temperature and change in rainfall pattern during the last three decades.

In the study facing environmental challenges Misra, R.P. (2013) has covered a broad canvas highlighting major environmental threats like climate change, gaia and environment principles of nature, instability of modern civilization, increasing population, industrialization, urbanization and dehumanized technology. The paper further addressing the climate change, discussed contradicting views and advocated Gandhian way for attaining sustainability.

Ian Douglas (2013) in his paper on *Climate Change and Green Economy with Reference to India* has discussed that the rising of global population and perhaps more significantly rising affluent middle class with high rates per capita consumption, the emission of green house gases continues to rise.

Sharma, H.S. (2013) in his study on climate change, desertification and water resource management in India has discussed the issue of climate change, its natural and anthropogenetic factors and no analogue state of the earth and its impact on land degradation and water resources in India.

Recently C.A. Rama Rao, *et al.* (2016) have mode a district-level assessment of vulnerability of Indian Agriculture to climate change. This study presents the analysis of vulnerability of agriculture to climate change and variability of a district-level. Out of 33 districts of Rajasthan 25 districts fall in very high, 6 in High, 1 in medium category of Vulnerability.

Rathore, M.S. (2004) while analyzing the state level analysis of drought policies and impacts in Rajasthan, India advocated that "the problem is further compounded by the fact that government perceives drought as a *'crisis situation'* and a short term problem and manages as an *'isolated event'*. At the household level, individuals perceive drought as a *natural hazard*, beyond *human control*." The present policy of the state to deal with drought is of a 'reactive' (*post facto*) action rather than 'pro-active', by framing a long-term drought mitigation policy.

Groundwater exploration and augmentation efforts in Rajasthan: A Review has been studied by Rathore, M.S. (IDS, 2005). He pointed out that one of the largest challenges in evaluating the viability of groundwater harvesting for recharge is the lack of accessible technical information on over all groundwater context in Rajasthan. Feasibility studies on groundwater recharge techniques their relative merits and economics under different terrain rainfall and socio-economic conditions are essential.

Droughts and Rainfall Pattern, 1987-1999 has been studied by Sivasamy, K.S. He observed that rainfall data from 1985 to 1999 shows a cyclical pattern. Severe and widespread drought has followed years of good monsoons. An extrapolation of these observations is not valid for prediction. But considering that India has had good monsoons since 1987, it must plan for droughts now.

Observed Climate Change in India

The studies related to temperature and rainfall trend in India have shown an

Table 1.1: Observed climate change during the Past Century over India

Region	*Temperature*	*Rainfall*	*Reference*
All-India	Increase in 0.4°C, 100 yrs in mean annual temperature		Hingane, L.S., *et al.*
All-India	Increase in 0.4°C, 100 yrs in mean annual temperature		Rupakumar, K., *et al.*
Western Himalayas	Winter season—Srinagar. Mussoorie and Mukteswar show increasing trend (0.5°C 100 yrs) Monsoon season—Srinagar, which is beyond the monsoon regime, shows significant increasing trend, whereas Mussoorie and Dehradun, which are at the foothills of the Himalayas, show decreasing trend	No increasing or decreasing trend for the last 100 years	Pant, G.B., *et al.*
Indo-Gangetic Plain Region (IGPR)	Annual surface air temperature of the IGPR shows rising trend (0.53°C/100 yrs during 1875-1958). Decreasing trend (–0.93°C 100 yrs during 1958-1997) Decreasing trend in temperature minimum	Summer monsoon rainfall over western IGPR shows increasing trend (170 mm, 100 yrs) from 1900, while over central IGPR it shows decreasing trend (5 mm 100 yrs) from 1939 and over eastern IGPR a decreasing trend (50 mm 100 yrs) during 1900-1984 and increasing trend (480 mm 100 yrs) during 1984-1999 Westward shift in rainfall activities over the IGPR	Singh, N., *et al.*, Amit Dhorde, *et al.*
All-India	Significant decrease in mean annual and mean maximum temperature	Monsoon rainfall is trendless and is mainly random in nature over a long period Increase in mosoon raifall	Moolay, D.A., *et al.*, Thapliyal, V., *et al.*, Alka Gadgil and A. Dhorade
NE Peninsula. ME India. NW Peninsula, west coast and central Peninsula	Mean annual temperature increasing at the rate 0.15 and 0.30 degree centigrade/decade	NE Peninsula. NE India and NW Peninsula show decreasing trend in the Indian summer monsoon rainfall (–6 to 8% of normal 100 yrs), while increasing trend was noticed along the west coast and over central Peninsula (–10 to 12% of normal 100 yrs)	Rupakumar, K., *et al.*, Amit Dhorde, *et al.*
Western and eastern Himalayas		Western Himalayas gets more snowfall than eastern Himalayas during winter. More rainfall in the eastern Himalayas than in the western Himalayas during monsoon season	Kriplani, R.H., *et al.*
All-India		Decadal departures in summer monsoon rainfall are found above and below the long-time average alternatively for three consecutive decades	Kothyari, U.C. and Singh, V.P.
Rajasthan desert		Slight increases in monsoon rainfall in spite of large inter-annual variations	Pant, G.B., *et al.*
In different locations		Pockets of significant long-term rainfall changes	Kotswaram, P., *et al.* Raghavendra, V.K.
Luni River Basin (arid west Rajasthan)	Rising trend at Barmer, Jodhpur. Ajmer and Pali Decreasing trend at Udaipur and Jawaibandh	Annual rainfall indicating increasing tendency at 19 stations (around Ajmer in upper part of the Luni basin) Decreasing trend at the remaining nine stations in lower Luni basin, i.e. Barmer	Singh, R.S. *et al.*, Doi, R.D. Bhardwaj, V., *et al.*

Source: Adopted from R.K. Mall *et al.*, *Current Science*, Vol. 90, No. 12, 25 June, 2006 and modified with latest references.

increasing trend of temperature and year to year variability in rainfall pattern during last 100 years or so. Table 1.1 presents details of the various studies on temperature and rainfall pattern over India.

It has been observed that during the last four decades the monsoon rainfall has been trendless, particularly on all India scale (Table 1.1), but trends in regional monsoon rainfall in the past century have been observed. Trend analysis of 1 day extreme rainfall series based on the period 1951-2007 shows that extreme rainfall amounts are increasing at many places over India. In Rajasthan also some pockets show an increasing trend of extreme rainfall events (Lower Luni River, July 1979, Jaipur, July 1981 and Barmer, August 2006).

Future Projections over India

High resolution climate change scenario for Indian region generated by PRECIS indicates a rise in annual mean surface air temperature for all parts of India. According to this model temperatures are likely to rise by 2.5° C and 2.5-4° C with warming more pronounced over the northern parts of India. The warming is expected to be relatively greater in winter and post-monsoon reasons than in summer monsoon rainfall for the future compared to present. The model projections for mean annual surface temperature in Rajasthan indicates an increase by 1.4° C for the 1971-2100 period. Mean annual rainfall is predicted to decrease slightly, where as the extreme rainfall is expected to increase in frequency and intensity.

Kothawala, *et al.* (2010) have pointed out that the annual mean temperature for the period 1901-2007 over India has shown a significant warming trend of 0.51° per 100 years. In the recent period 1971-2007 accelerated warming has been observed, which has mainly been contributed by the warming in winter and post-monsoon seasons, which have increased by 0.80° C and 0.82° C in the last 100 years respectively. The all India maximum temperature also have shown an increasing trend in temperature by 0.71° C per 100 years. The all India minimum temperature also shows an increase in temperature by 0.27° C per 100 years during 1901-2007 (Rajasthan State Action Plan, Teri Report 2010). Which discussing long-term temperature trends at four largest cities of India during the twentieth century (Alka Gadgil and Amit Dhorde, 2005) and Amit Dhorde and Anargha Wakhare (2009) have pointed out that some of the metropolitan cities in India have recorded significant increase in minimum temperature during winter. For all India as a whole show a significant increasing trend in frequency of hot days as well as decreasing trend in frequency of cold days during the premonsoon season over the period 1970-2005 (Kothawala, *et al.* 2005). The ever increasing urbanization has led to increase in temperature (Table 1.2).

Table 1.2: Projected climate change during present century over India

Region	*Temperature*	*Rainfall*	*Reference*
All-India	Increase in winter temperature by 1-4°C with increased CO; concentration	Precipitation increase of approximately 20 per cent Increase in heavy rainfall days during the summer monsoon period and an increased inter-annual variability	Bhaskaran, B., *et al.*
All-India	Average temperature change is predicted to be in the range of 2.33 to 4.78°C. with a doubling in CO; concentration	Increase in frequency of heavy rainfall events	Lonergan, S.
All-India	Area-averaged annual mean surface temperature rise is projected to range between 3.5 and 5.5°C by the end of the century. More warminc in winter season.	Increase of about 7 to 10 per cent in annual mean precipitation Decline of 5-25 per cent in winter precipitation Increase in monsoon precipitation is 10-15 per cent Monsoon season over northwest India — increase of 30 per cent or more in rainfall by 2050. Western semi-arid regions of India could receive higher than normal rainfall in a warmer atmosphere. Decrease in winter precipitation between 10 and 20 per cent over central India by 2050.	Lal, M., *et al.*
All-India	Over the region south of 25°N (south of cities such as Udaipur. Khajuraho and Varanasi), maximum temperature will increase by 2-4°C during 2050. In the northern region, increase in maximum temperature may exceed 4°C. A general increase in minimum temperature up to 4°C all over the country.	Decrease in number of rainy days over a major part of the country. This decrease is more in the western and central parts (by more than 15 days), while near the foothills of the Himalayas (Uttarakhand) and in northeast India, the number of rainy days may increase by 5-10 days. Increase in rainy days intensity by 1-4 mm/day, except for small areas in northwest India, where rainfall intensities may decrease by 1 mm/day.	Rupakumar, K., *et al.*

Source: Adopted from R.K. Mall, *et al., Current Science*, Vol. 90, No. 12, 25 June, 2006.

In view of the above environmental harshness and challenges the state falls within the areas of greatest climatic sensitivity (O. Brien, K., *et al.*, 2004). In an era of climate change, Rajasthan is likely to suffer further more shortage due to over reduction in rainfall. In addition, the state has maximum vulnerability and lowest adaptive capacity to climate change challenges. Rajasthan has the maximum probability of occurrence of droughts in India (Mall, R.K., *et al.*, 2006).

Condition may further deteriorate in terms of severity of drought in the state. Even though 20 per cent rise in all India summer monsoon rainfall is projected in Rajasthan over all rainfall is projected to decrease, and evapo-transpiration to increase, due to global warming (Kumar, K.R., *et al.*, 2006). Although analysis of 100 years rainfall data for the desert district of Rajasthan indicates an increasing trend of 0.5 mm/year, but increased evapo-transpiration demand due to global warming can put tremendous pressure on existing overstressed water resources of Rajasthan (Goyal, R.K., 2004). As the total available surface water resources of the arid zone are to the tune of 1361.21 MCM, a robust policy intervention is required for sustainable water management.

Several studies have shown a rising trend in temperature in western Rajasthan particularly in Barmer, Jodhpur, Ajmer and Pali. In the same region, annual rainfall has shown increasing trend in the arid region. Decreasing temperature trends have been observed at Udaipur and Jawaibandh, and decreasing rainfall trend at the other nine stations in Lower Luni Basin, i.e. at Barmer has been reported by Singh, R.S., *et al.* (2001). It is pertinent to note that Barmer received unprecedented rainfall in the year 2006 from 18th August to 5th September, 2006. Over all, there is a prediction of an increase in the rainy day intensity by 1-4 mm/day in India, but some areas of north-west India including Rajasthan the rainfall intensities are predicted to decrease by 1 mm/day (Mall, R.K., *et al.*, 2007).

Widespread land degradation is a persistent challenge in Rajasthan (Ajai, *et al.*, 2009). Recent studies by Goswami, P. and K.V. Ramesh (2008) have further predicted that due to climate change there may be significant increase in the desert areas over India in the next 100 years with potential disproportionate impact of global warming on coupled human and natural systems. Model studies on the wind erosion potentials in the Thar Region for A.D. 1951 to 2100 suggest that larger efforts in land conservation practices would be required than at present to stabilize the aeolian bed forms in the Thar desert (Amal Kar, 2009, 2012 and 2014).

An analysis of the present and future climate of Arid western India and its implications for land degradation has been brought out by Kar, Amal (2012). Based on General Circulation Model (GCM) data for AZ scenario revealed that only few GCM could be used with greater confidence. The study suggests that strong possibility of a prolonged drought phase during 2021-2025, which may continue with some respite till mid-century. This study also shows the likelihood of aridity spreading beyond the present eastern boundary of the region. Since wind strength is likely of increase gradually and soils dry, the wind erosivity is expected to increase manifold across Thar desert and Megathar region, destabilizing the sandy landscape and contributing more to the atmospheric dust load. The study further reveals higher monsoon rainfall then 1951-2000, which may encourage higher fluvial erosion along the Aravalli foothills due to increasing rainfall intensity that was predicted earlier by PRECIS models (Amal Kar, 2012).

In Rajasthan plenty of past-desiccation of climate that led the spread of the desert surface in the north-eastern part of the present desert boundary (Allchin, B., 1982). Is there evidence of a similar desiccation in progress today? Is the widespread desertification now in progress due to such a downward trend of rainfall? There is strong evidence to suspect that more and more humid and sub-humid zone of the country, which had earlier experienced no significant moisture stress, are now under the intensification of man induced desertification process.

The unprecedented drought situation that has occurred within last four five decades in Rajasthan is not a new thing. There are plenty of references of such a situation in the Vedic literature and in the recorded history of India. The droughts of 1877, 1900, 1918, 1940, 1965, 1972, 1979, 1984-85, 1986-87, 2002-03 and 2009 are well known because of their serious consequences. The most notable characteristics of such droughts are a random distribution in time, spatial whence over large areas and in some regions, persistence for years at an end.

In contrast to drought heavy rainfall in desert region is rare. Whether the Barmer floods in 2006 were produced by freak weather, or by global warming is matter of further research?

Despite of harsh geographical conditions and climate sensitivity, Rajasthan during last few decades has witnessed rapid growth of population, modernization in agriculture, accelerated urbanization, changes in land-use, industrialization and enhanced mining activities. This rapid transformation in nature by anthropogenic activities has contributed in damaging physical environment leading to land degradation, droughts, deforestation, depletion of water resources which together resulted in an ecological imbalance with catastrophic effects.

The Study Region

Rajasthan is the largest state in India in terms of size comprising 10.4 per cent of the country's total area with 2/3 of its area is Thar Desert. The total population of Rajasthan is 686.21 lakh (2011) of which nearly 75.10 per cent of the state's population reside in rural areas. The forest cover of the state contributes 9.51 per cent as per India forest report (7% as per the Government of Rajasthan). As much as 7909.927 lakh hectares of land is irrigated in Rajasthan, which is about 11 per cent of the net irrigated area of India and 39 per cent of net sown area of Rajasthan. The state has 566.63 lakh livestock—mainly cows, buffaloes and goats comprising 10.50 per cent of the country's livestock population. Administratively, Rajasthan comprises 33 districts, 44672 revenue villages (2011), census, 249 Panchayat Samities and 9175 Gram panchayats. On the basis of climate conditions and agricultural practices, Rajasthan has been divided into five major and five meso-agro-climatic zones.

Table 1.3: An Overview of Rajasthan

Sl. No.	*Indicators*	*Rajasthan*	*% of India*
1.	Area	3,42.239 Sq. km.	10.41
2.	Population	68,621,012	5.67
3.	Rural Population	51,540,236	—
4.	Urban Population	17,080,776	—
5.	Decadal growth rate of population (2001-11)	21.44	17.64
6.	Total Forest Cover (India Forest Report-2011) as per IRS data 2008	9.51% 4.50%	
7.	Net irrigated area (2009-10) showing decrease of 6.33 per cent over 2008-09	58.50 lakh hectares	11.00
8.	Livestock (2007)	566.63 lakh	10.50
9.	Index of Agricultural Production	185.09	—
10.	Rainfall	575 mm (annual average)	—
11.	Total State Water Resources	50.96 BCM	1.16

Source: Directorate of Economics and Statistics, Rajasthan, 2011 and Census of India, 2011.

The desert part of Rajasthan is one of the world's most thickly populated arid region with a density of human population of 110 km.^{2} and livestock population of 120 km.^{2} while livestock population has doubled over the last five decades, human population has registered a fourfold increase (Kar, Amal, 2014).

The entire state receives scanty rainfall. The desert part of western Rajasthan is characterized by low and erratic rainfall, high air and soil temperature, intense solar radiation and high wind velocity. The interactions of these factors not only give rise to frequent droughts and famines, they also make local livelihood highly vulnerable.

Physiographically, the state may be divided into four major physiographic divisions, namely the Aravalli range, South-Eastern plateau and Eastern undulating riverine plain of the Chambal and Banas and north-western sandy desert, the Thar. The Aravalli range traverses the state in south-west to north-eastern direction dividing into two halves. It is a life line of the state and acts as orographic barrier for the monsoon system in Rajasthan.

Broadly, the following drainage systems can be identified in Rajasthan (Fig. 1.1):

A. The disorganized Ghaggar drainage system,
B. The Chambal-Banas drainage system,
C. The Mahi drainage system,
D. The Luni-Jawai-Sukri drainage system,
E. The Sabarmati drainage system,
F. The Western Banas-Sipu drainage system,
G. The Gambhir drainage system, and
H. The inland drainage systems, which consist of:
   - the Banganga system,
   - the Katli-Dohan-Sabi and Berah system,
   - the Sambhar Didwana-Kuchaman drainage system, and
   - the inland drainage of Jaisalmer-Barmer-Like river and Gughari system.

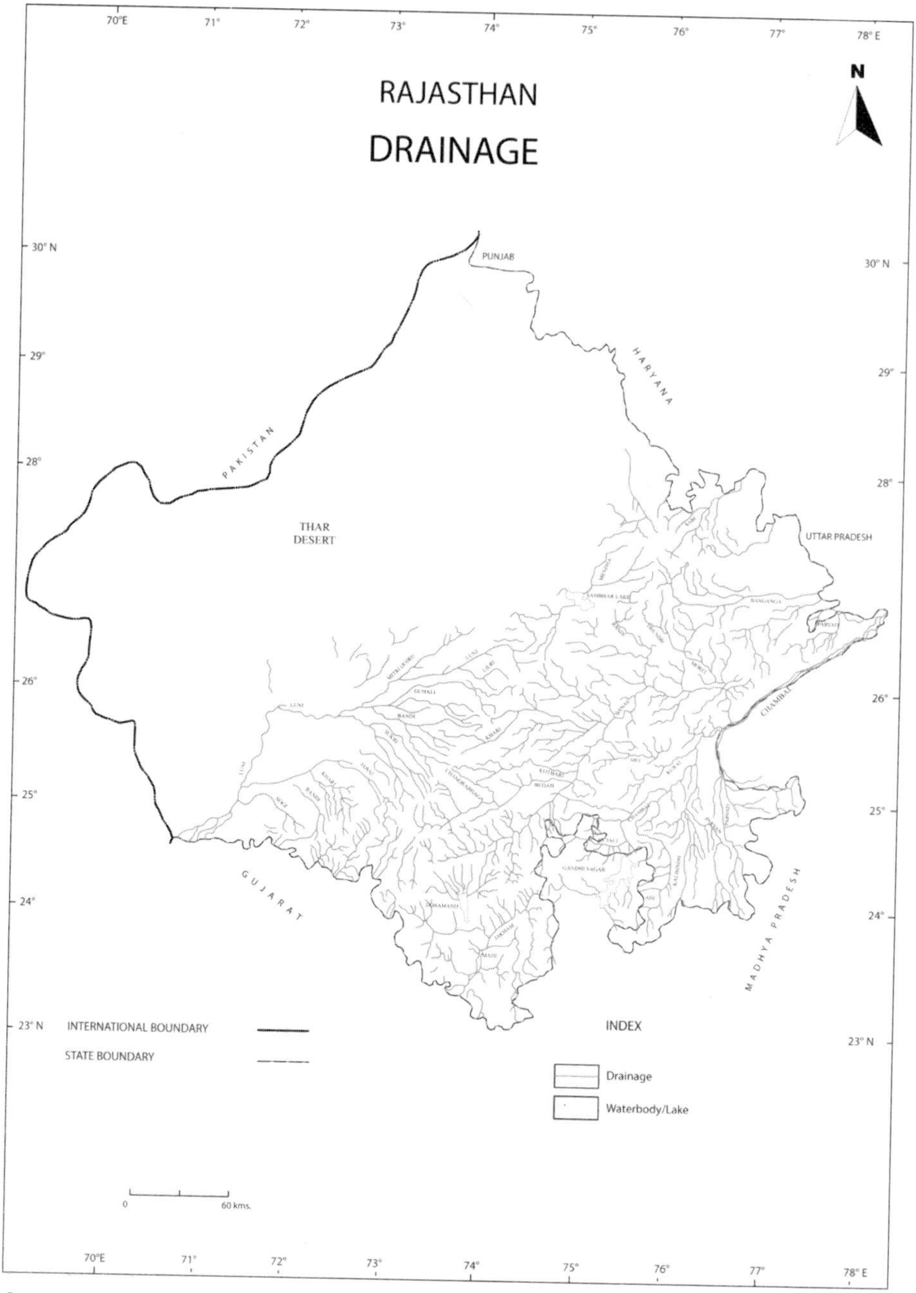

Source: Survey of India

Fig. 1.1: Drainage of Rajasthan

Why Research on Climate Change is Crucial for Rajasthan?

Why Rajasthan warrants priority in adaptation for climate change research? Rajasthan is a unique geographical region for the study of societal adaptation to climate change (V.S. Singh, 2010). Antiquity of human occupation of the desert region goes back of the late Pleistocene (1,60,000 years) or even earlier as indicated by archaeological studies. During the mid-Holocene, the regions around the Indian Ocean witnessed rise of Three Great Civilizations of the World (i.e. Mesopotamia, Egyptian and Indus Valley). Indus-Saraswati Civilization is the earlier urban civilization in South Asia that flourished and fell in the region that includes parts of present day Rajasthan.

There are increasing evidences in favour of climate change as driver for societal disruption and collapse of Indus Valley Civilization (Gupta, A.K., 2009 and Gupta, A.K., *et al.*, 2009). Some evidences have indicated the climate change as the reason for the collapse of Indus Valley Civilization (Allchin & Allchin, B., 1982 and 1997). Although controversy still remains. The climate change theory was supported by others (Enzel, Y., *et al.*, 1999), rather they advocated that chronology indicates that there is no relation between the proposed drought that caused the desiccation of the lakes and the collapse of Indus Valley Civilization, as the lakes in the region dried out >1500 years earlier. As per their conclusions north-western India during the period of Indus Valley Civilization experienced semi-arid climatic conditions that are similar to those of present. However, the recent study by Staubwasser, M., *et al.* (2003) based on planktonic oxygen isotope ratio of the Indus delta reveals climate changes with a multi-centennial pacing during the last 6,000 years with the most prominent change recorded at 4.2 Ka BP. This study suggests that solar variation is one fundamental cause behind Holocene rainfall changes over South Asia.

Past Climatic Changes in Rajasthan

Rajasthan not only witnessing climate change in the present century but has been witnessing climate changes in the past 20,000 to 28,000 years. The State bears incomparable geomorphological, archaeological, palynological and pedological evidences of climate change. The landforms created by various geomorphic processes under different climatic phases, have been used to construct the chronological sequences of the climatic changes in different parts of the world during the Quaternary period. In Rajasthan especially in desert part, several climate-morphogenetic studies have been conducted by Singh, G. (1971), Singh, *et al.* (1974), Singh (1977), Ghose (1964, 1965), Abhichandani and Ray (1966) and other workers from Central Arid Zone Research Institute, Jodhpur: Stein (1942), Sood *et al.* (1982), and Yash Pal, *et al.* (1980), Allchin, Goudie and Hegde (1978), Meher-Homiji (1973), Brysen and Barlis (1967), Ahmad (1969), Verstappan (1970), Agrawal, *et al.* (1975, 1976, 1980), Mishra and Rajaguru (1973, 1980), Whilhelmy (1969), Bakliwal (1999), Dhir (2010), Raghav, K.S. (1992), Singhvi, A. (1995), Wadhawan (1995), and Sharma, H.S. (1968, 1983, 1987), Kale, V.S., *et al.* (2000).

The tentative chronology of the climatic changes in Rajasthan and adjoining parts has been presented in Table 1.4.

Table 1.4: Geomorphic evidences of desertification/dedesertification during the late Quaternary period in Western Rajasthan

Period in Years B.P. (After Singh, et al., 1974)*	*Major gemorphic events*	*Geomorphic processes*	*Climatic phases*	*Desertification/ dedesertification*
0-Early Centuries A.D.	Increasing Fluvial activities but increasing biotic stress on the land has resulted as salinity hazard along priod drainage channels and more deflation and deposition of sediments in the sandy plains and duny tracts.	Biotice activities	Slight amelioration of climate	Small scale amelioration of climate but biotic desertification is present
Early centuries A.D. 3,800	Formation of low active dunes like barchans. Shrubscoppice etc. reactivation of the flanks and crests of the stabilised dunes. Inland basins turned into saline depression.	Aeolian	Arid	Climatic desertification
3,800-10,000	Major dunes stabilised, drainage channels revived, sandy inland basins filed with fresh water and fluvial sediments.	Fluvial	Humid	Amelioration of climate leading to dedesertification
10,000 - ?	Formation of major dunes, disintegration of drainage systems of aeolian sand, creation of sandy inland basins.	Aeolian	Arid	Climatic desertifcation
?	Well integrated drainage systems, Vast alluvial plains.	Fluvial	Humid	No evidence of desertification

*B.P. means before present.
Source: Compiled by H.S. Sharma (1987).

A fairly large amount of work has been done at the Central Arid Zone Research Institute, Jodhpur, Cambridge-Baroda-Poona team of Archaeologists and Scientists of Physical Research Laboratory, Ahmedabad, and Geomorphologists on the changes in climate of Rajasthan desert in Quaternary and Recent times. While all agree that there has been a change from arid to semi-arid to sub-humid climate and a repetition of this cycle, contradiction are to be found in the duration assigned to these climatic changes. Broadly speaking dry river beds are supposed to be indicative of aridity. There have also been attempts to correlate geomorphic features with the glacial and inter-glacial periods in the Himalaya. It is summarized that sand shields formed during the glacial period were washed out during the much longer fluvial (raining) inter-glacial periods. During these periods in Rajasthan desert, the climate is supposed to have been arid and humid respectively fossil parabola seem to have shaped into longitudinal dunes and later berchans predominated. Major dunes are supposed to have been formed during the arid phase and stabilized during the humid phase. One interesting aspect of these studies is that a chronology could be established because of the occurrence of human implements with these landform. Great emphasis has been laid on the degeneration of drainage in these parts though different authors attribute different reasons for this (tectonic, climate, river piracy, human interference, etc.). Now, there is greater need of determination of TL ages of different representative samples from various parts of Rajasthan. However, a beginning has already been made by Singhvi and his team at RPL, Ahmedabad.

In recent times during last 112 years or so, Rajasthan has experienced severe and frequent spells of droughts than any other region in India. Climate change, thus, presents a serious risk to poverty eradication and sustainable livelihood. The adverse impacts of climate change is more severely felt by poor people who are more vulnerable than rich.

Hypotheses and Research Questions

Based on the analysis of earlier review of literature on climate change the following hypotheses and research questions have been formulated to understand cause effects of climate change in Rajasthan:

- Frequency of extreme weather events including droughts and floods have increased in Rajasthan during last 112 years.
- Rainfall pattern has been trendless.
- Temperatures are showing increasing trend.
- Climate changes coupled with anthropogenic pressures, biophysical factors and poor management practices may lead to desertification/ land degradation and reduce productivity adversely impacting ecological and socio-economic benefits.
- Rajasthan is reeling under the impact of climate change and land and water resources are negatively correlated with climate change.

Objectives

- To understand the changing pattern of climate during last more than 112 years particularly rainfall and temperature.
- To study potential impact of climate change on land and water resources.
- To address the vulnerability to current water scarcity and climate risks.

- To analyze the decadal land-use changes (1985-86, 1995-96, 2005-06 and 2009-10) and agricultural productivity in relation to climate.
- To study the land degradation/desertification and its relationship with climate change.
- To suggest adaptive strategies to mitigate the effects of climate change.

Methodology

The study has been undertaken under a broad framework linking the objectives to derive at a conclusion. The work has been done at three levels. As has been indicated earlier that Rajasthan is the area of study. The analysis of the general land-use, water resources and agricultural productivity has been done at macro-level. A meso-level analysis has been done to analyze land-use changes at the district-level and finally in micro-level analysis has been undertaken at a village-level (households form the core of analysis) to understand the perception of the people on climate change, land-use practices, agricultural productivity, land degradation and groundwater depletion.

Time Frame and Spatial Units

The temporal frame chosen under study for showing land-use changes is taken for four points of time, i.e. 1985-86, 1995-96, 2005-06 and 2009-10. The 33 districts of Rajasthan are considered as individual spatial unit, which means that it is a district-level analysis for land-use and land-use changes.

The trend analysis of rainfall and temperature to ascertain the climate change in Rajasthan has been done in the time frame of more than hundred years (i.e. 1871-2012 and 1901-2012). The simple linear regression technique was applied to develop a model between the temperature/rainfall as dependent variable and years as an independent variable. Total number of years for this model were taken 112 years. ARIMA model has been used to predict the trend of temperature and rainfall in future. The spatial unit for temperature and rainfall is also a district.

For analysis of land-use and trend of rainfall, secondary data were obtained from Directorate of Statistics and Economics, Government of Rajasthan, Jaipur and India Meteorology Department, New Delhi respectively. Data related to groundwater resources have been procured from Central Ground Water Board, Jaipur.

Primary Data

In order to get the inner view of the problem of climate change and its impacts on land and water resources, a stratified proportionate sampling was adopted.

To have a spatial coverage of selection of sample villages have been made from different 5 agro-climatic region of the state. Based on this 0.05 per cent of the villages from different agro-climatic regions have been selected using the random numbers. The total 20 villages were selected as per Table 1.5 and Fig. 1.2).

Table 1.5: Sample Villages

Sl. No.	*Agro-Climatic Regions*	*Name of Villages*
1.	Arid Region Ganganagar, Hanumangarh, Bikaner, Jodhpur, Jaisalmer, Barmer	1. Karuo (Jodhpur) 2. Dabla (Jaisalmer) 3. Sanavada (Barmer) 4. Takhatpura (Bikaner) 5. Chak 4C Chhoti (Ganganagar)
2.	Transitional Zone Jalor, Sirohi, Pali, Nagaur, Churu, Jhunjhunu, Sikar	6. Daman aka Golia (Jalore) 7. Gharat (Pindwara, Sirohi) 8. Kitalsar (Nagaur)
3.	Semi-Arid Region Rajsamand, Ajmer, Bhilwara, Tonk, Jaipur, Alwar, Dholpur, Bharatpur, Dausa, Karauli, Sawai Madhopur	9. Chapli (Rajsamand) 10. Kardala (Ajmer) 11. Dhodhakari (Behror, Alwar) 12. Bhandana (Dausa) 13. Chainpura (Tonk) 14. Beri (Bharatpur) 15. Lilee (Laxmangarh, Alwar)
4.	Sub-Humid Zone Bundi, Kota, Jhalawar, Baran, Chittorgarh	16. Kanwas (Kota) 17. Bagdar (Jhalawar)
5.	Humid Zone Udaipur, Dungarpur, Banswara, Pratapgarh	18. Ghasiyar (Udaipur) 19. Sadari (Banswara) 20. Nandore (Sangwara, Dungarpur)

Source: Compiled by Author.

The randomly selected households in each village (about 20 households from each village) were interviewed with the help of structured schedules. Total 407 households from 20 villages have been interviewed and primary data generated for perceptional analysis.

Two districts were selected for analyzing status of desertification, i.e. Ajmer and Dausa districts. For this IRS LISS III multidate geocoded FCC paper prints on 1:50,000 scale for two seasons data, viz. *Kharif* and *Rabi* (2002-03) and LISS III data on the same scale and seasons for 2012-13 have been used. The methodology adopted for mapping has been at three levels, i.e. Level 1 for land-use/land cover, Level 2 for processes of desertification and Level 3 severity of degradation. All the IRS data have been provided by Space Application Centre, Ahmedabad under the project DSM second cycle.

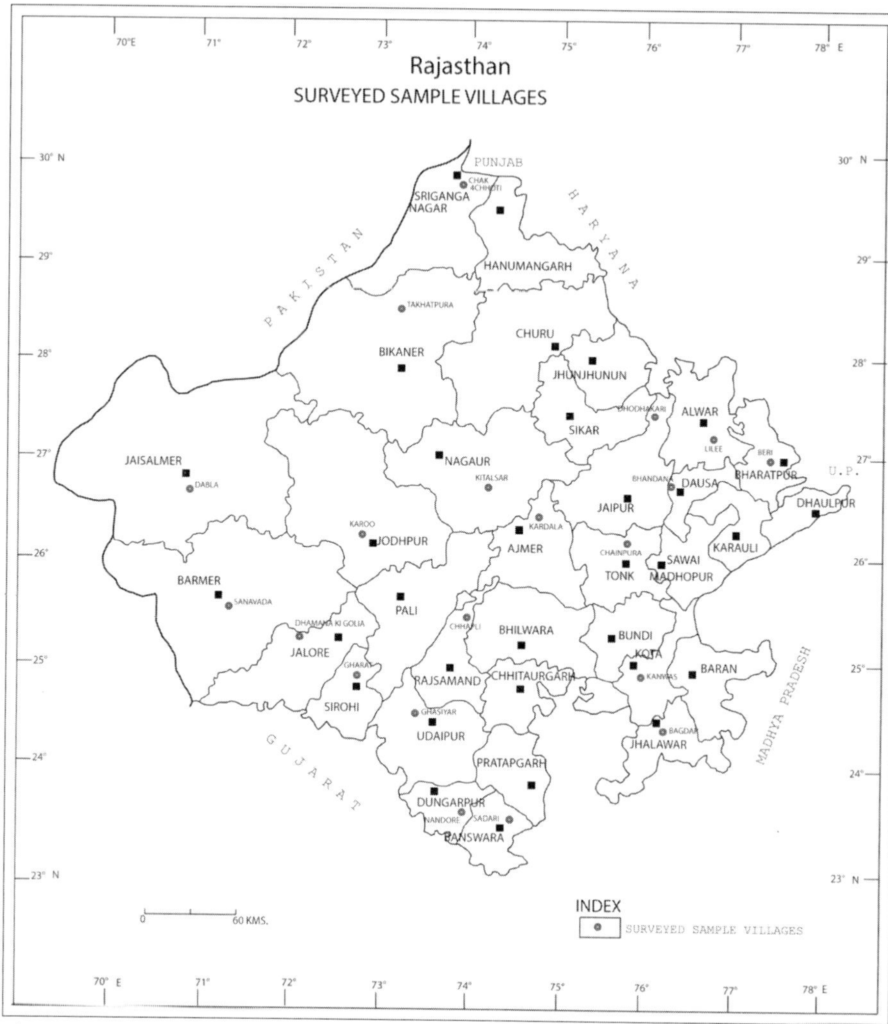

Fig. 1.2: Rajasthan: Surveyed Sample Villages

Chapterization Scheme

The study has been divided into seven chapters. The first chapter deals with the introduction, conceptual framework, review of literature on climate change, relevance of climate change study for Rajasthan, hypotheses and research questions, objectives, methodology and chapterization scheme. The second chapter deals with climatic elements such as temperature, rainfall, rainfall variability, excessive and deficient rainfall years, droughts, rainfall departures, decadal changes of rainfall and temperature, district-level temperature and rainfall trends in Rajasthan and causes of climate change.

The third chapter deals with the status of water resources in Rajasthan, groundwater resources, surface water resources, groundwater quality, water demand-supply gap, industrial demand, status of drinking water in rural and urban areas and depletion of groundwater resources in relation to climate change.

Land-use, land-use changes and agricultural productivity have been analyzed in chapter fourth. The fifth chapter deals with desertification in relation to climate change. In this chapter two districts *viz*. Dausa and Ajmer have been selected to find out the changes in the status of desertification.

In sixth chapter the perception and response of villagers of 20 sampled villages have been analyzed and presented in a systematic format. In the last chapter main findings and conclusions have been presented. The detailed adaptive strategies have also been highlighted in this chapter. Field survey questionnaire and photographs have also been included along with bibliography and index.

2
Climate and Climate Change

Climatically, Rajasthan lies within a region which according to the classification of Koppen (1931), Thornthwaite (1931) and Meigs (1953) may be described as having arid (Ed), semiarid (Dd), transnational between Ed and Dd, and *sub-humid* (Cw) zones. The area under study receives limited rainfall. The rainfall variability is high and evaporation losses are also very high. Almost 91 per cent of the rainfall occurs between the months of June and September. During the remaining months of the year rainfall is light and sporadic and several consecutive rainless months and perhaps years are not uncommon in the arid part of Rajasthan. Annual and daily ranges of temperature are high. Marked seasonal contrasts in humidity, cloud, wind speed and velocity and radiation are significant.

Rajasthan occupies the north-west portion of India and hence, the pattern of atmospheric circulation with slight variations is the same as in north-west India. The principal seasons can be divided into four seasons like –

- Winter (January and February)
- Pre-monsoon (hot weather—March, April and May)
- South-West monsoon (rainy season—June, July, August and September)
- Post-monsoon (retreating monsoon—October, November and December)

Atmospheric Sea Level Pressure and Winds

The seasonal variation of atmosphere pressure over the state takes place in a systematic manner with maximum in the winter (January) and a minimum in the monsoon season (July). The pressure gradient over the state generally remains weak expect during the late summer and monsoon season. During winter the higher pressure is to the north. In July, the pressure decreases from south to north in Rajasthan.

The winds are light and mainly from northwest-north-northeast over the state in January, turn gradually anti-clockwise and are replaced by light north-westerly to westerly or south-westerly winds in April. With the advance of the summer, the pressure gradient increases and correspondingly the winds from north-west to south-west also strengthen reaching their maximum strength in July. In July, the pressure decreases from west-southwest to east-northeast over the state. October is the month of transition, with weakest pressure gradient. From October onwards, the change over of the pressure and wind pattern commences and north-northeasterly winds appear.

Temperature

Day temperatures are more or less uniform over the plains except during the winter and monsoon season, when temperatures increase southwards and northwards respectively. In general, the night minimum temperatures are lower in higher latitudes except during the southwest monsoon when they are more or less uniform. Both day and night temperatures are lower over the plateau and at high on the Aravalli (Mt. Abu). May is the hottest month with the mean maximum temperature at about 41°C - 42°C in the plains and 2°C to 4°C lower over the plateau regions. The highest maximum temperature ever recorded at an individual station in the plains is 50.6°C at Alwar on 10th May, 1956, which is about 10°C higher than the normal of the warmest month. Mount Abu, a hill station registered the highest maximum temperature at 39.2°C on 15 May, 1985, which was 7°C higher than the normal for the warmest month.

January is the coldest month in the state when the mean minimum temperature for the state as a whole is 7.4°C varying from 4°C in the north to 12°C in the south. During winter, much lower temperature may be experienced in the wake of western disturbances. On such occasions, minimum temperature 2°C to 5°C below the freezing point can be registered at few stations in the northern parts of Rajasthan. The lowest minimum temperature ever recorded at an individual plain station was -5.9°C at Jaisalmer on 12 January, 1967, while the hill station of Abu had recorded the lowest temperature of —7.4°C on 12 December, 1994. These were 12.8°C and 13.1°C below the respective normals for the coldest months.

The maximum temperature rise rapidly from February onwards till May and minimum temperature from February onwards till June. The increase in maximum in the period from January to May ranges from 13°C to 20°C at individual stations as we proceed from south to north of the state. From the beginning of June to the end of July, the maximum temperature falls by about 3°C to 7°C whereas the minimum temperature falls only by about 3°C to 5°C from June to September. A slight rise in the maximum temperature is experienced in the month of September due to increased insolation. The night temperatures start falling rapidly after September while day temperatures follow this trend after October and both attain lowest values by January. The fall in minimum temperature and maximum temperature is about 8°C to 15°C and 5°C to 7°C respectively during these periods in both cases, the fall increases from southern parts of the state to the northern parts.

July and August have the smallest diurnal range of temperature of about 9°C in the state. The diurnal range increases rapidly after the withdrawal of the monsoon. During the period from November to May, the diurnal range is of the order of 15°C to 18°C, being greatest in November.

Humidity

The relative humidity is generally high during the period from July to September. It is about 45 per cent-47 per cent in June rising to a little less than 70 per cent during August in western Rajasthan and to about 76 per cent-77 per cent in eastern Rajasthan. The diurnal variation is relative humidity is least during monsoon being higher in western Rajasthan. The relative humidity is least during the summer afternoons when it becomes about 20 to 30 per cent in major parts of the state making the summer very dry and host. The diurnal variation is highest during the period January and February.

Rainfall

The total annual rainfall in state varies from 140 mm over the extreme northwestern parts to 1000 mm over the southeastern parts. The southwest monsoon is the principal rainy season when the state receives 91 per cent of its annual rainfall. Rainfall in the winter season (January-February) is about 2 per cent of the annual total rainfall in the summer season (March-May) is about 3 per cent and the post-monsoon season (October-December) it is about 4 per cent.

Southern/southeastern districts adjacent to Madhya Pradesh constitute the area of maximum rainfall in the state. Pali and Jalore districts on the west of Aravalli hills receive maximum amount of rain of 500 mm and 430 mm in western Rajasthan. In the north/northwest districts like Bikaner, Ganganagar, Jaisalmer receive annual rainfall of 260 mm, 240 mm and 170 mm respectively, annual rainfall over Jaisalmer district being the lowest. These districts and the adjoining areas constitute the driest zone of the state. The mean annual rainfall in the eastern and western Rajasthan is about 649 mm and 327 mm respectively with the districts of eastern Rajasthan receiving more rainfall than those of western Rajasthan.

The southwest monsoon sets in over the eastern part of the state by about the last week of June and extends over the entire state by the first week of July. July and August are the rainiest months, each accounting individually to about 34 per cent of the annual rainfall. In each of these months, there are about 5 rainy days (with daily rainfall of about at least 2.5 mm) in western Rajasthan and about 9-10 rainy days in eastern Rajasthan.

Rainfall Variability

Co-efficient of Variation (CV) of annual rainfall is more than 60 per cent in extreme western parts of the state. In southeastern and extreme eastern parts of the state, it is less than 30 per cent. In other parts of the state it ranges between 30 per cent to 60 per cent. In the months of January and February, the CV is extremely high and it ranges from 100 to 200 per cent. It increases from northeast to southwest. During the summer months, i.e. March to May, it is the lowest in the northeastern parts of the state, adjacent to Haryana state, which is less than 100 per cent and increases to its southwestern and extreme southern parts of the state which is greater than 200 per cent. During the monsoon season CV is more than 60 per cent in the extreme western parts of the state. In the extreme southern parts of the state, it is less than 35 per cent and in the other parts of the state, it ranges from 30 to 60 per cent. During post-monsoon season CV is less than 150 in eastern parts of the state while in the extreme northern-northwestern parts of the state CV is greater than 200 per cent. In other region of the state it varies in between 150 and 200 per cent.

Excessive Rainfall

Rainfall sufficiently in excess of the normal is a predominant factor for occurrence of floods, particularly in high rainfall regions. For the purpose of the present description, annual rainfall of 125 per cent or more of the normal is considered as excessive rain (IMD. 2010).

Western Rajasthan

Table 2.1 gives the district-wise excessive rainfall years and highest annual rainfall (expressed as percentage of normal) with the year of occurrence.

From Table 2.1 it is evident that during the 50 years period from 1951-2000, there were 27 years in which some districts or other in the western Rajasthan recorded excessive rainfall, the maximum and minimum amount being 299 per cent and 207 per cent of the normal annual rainfall in the year 1990 and 1975 for the district Barmer and Jodhpur respectively.

The districts Barmer and Jaisalmer have got maximum number (140 years with excessive rainfall while Nagaur has got minimum number (8) years with such rainfall. In the year 1975, 1976 and 1983 eight districts out of nine districts of the western part received excessive rainfall. Successive years of excessive rainfall are shown against each district in Table 2.2.

All the districts of the arid Rajasthan except Ganganagar and Pali experienced excessive rainfall in successive years 1975-76. Districts Barmer and Jaisalmer experienced excessive rainfall in four consecutive years 1992-

Table 2.1: District-wise years of excessive rainfall with highest percentage of annual rainfall and year

District	Years of Excessive Rainfall (Rainfall > 125%)	Highest amount of rainfall (expressed as % of annual normal) with year
Barmer	1955, 1959, 1961, 1967, 1973, 1975, 1976, 1990, 1992, 1993, 1994, 1995, 1997, 1998	753 mm in 1990 (299%)
Bikaner	1959, 1964, 1975, 1976, 1978, 1982, 1983, 1992, 1995, 1997, 1998	538 mm in 1997 (208%)
Churu	1964, 1970, 1975, 1976, 1978, 1983, 1988, 1995, 1997, 1998	794 mm in 1978 (214%)
Ganganagar	1956, 1966, 1967, 1976, 1978, 1983, 1990, 1992, 1995, 1997	557 mm in 1997 (229%)
Jalore	1952, 1955, 1956, 1959, 1961, 1973, 1975, 1976, 1983, 1992, 1994, 1997	950 mm in 1956 (219%)
Jodhpur	1956, 1967, 1970, 1973, 1975, 1976, 1983, 1990, 1992	664 mm in 1975 (207%)
Jaisalmer	1953, 1955, 1956, 1961, 1973, 1975, 1976, 1983, 1992, 1993, 1994, 1995, 1998, 1999	484 mm in 1973 (280%)
Nagpur	1955, 1964, 1970, 1971, 1975, 1976, 1983, 1995	1131 mm in 1975 (289%)
Pali	1955, 1959, 1961, 1967, 1970, 1973, 1975, 1979, 1983, 1990, 1992, 1994, 1997	1093 mm in 1973 (219%)

Source: *Climate of Rajasthan, Climatological Summaries of States Series* No. 16, India Meteorology Department, 2010.

Table 2.2: District-wise Successive Years of Excessive Rainfall

District	Successive Years of Excessive Rainfall
Barmer	1975-1976, 1992-1993-1993-1994-1995, 1997-1998
Bikaner	1975-1976, 1982-1983, 1997-1998
Churu	1975-1976, 1997-1998
Ganganagar	1966-1967
Jaisalmer	1955-1956, 1975-1976, 1992-1993-1994-1995, 1998-1999
Jalore	1955-1956, 1975-1976
Jodhpur	1975-1976
Nagaur	1970-1971, 1975-1976

Source: *Climate of Rajasthan, Climatological Summaries of States Series* No. 16, India Meteorology Department, 2010.

1995. Only Pali district had no occasion when district experienced such rainfall in two consecutive years.

The heaviest rainfall recorded at any station for one day in the sub-division was 780.3 mm at Parbatsar.

Eastern Rajasthan

Table 2.3 gives the district-wise excessive rainfall years and highest annual rainfall (expressed as percentage of normal) with the year of occurrence.

Table 2.3: Years of Excessive Rainfall in Eastern Rajasthan

District	*Years of Excessive Rainfall*	*Highest amount of rainfall (expressed as % of annual normal) with year*
Ajmer	1955, 1956, 1973, 1975, 1976, 1979, 1983, 1994, 1997, 1998	905 mm in 1975 (199%)
Alwar	1955, 1958, 1967, 1976, 1983, 1985, 1995	1093 mm in 1975 (175%)
Baran	1956, 1961, 1971, 1975, 1982, 1994	1314 mm in 1975 (153%)
Banswara	1952, 1954, 1959, 1961, 1973, 1976, 1978, 1981, 1984, 1990, 1994, 1997	1632 mm in 1973 (185%)
Bharatpur	1955, 1958, 1961, 1967, 1976, 1977, 1982, 1983, 1995, 1996, 1998, 1999	1052 mm in 1958 (169%)
Bhilwara	1952, 1955, 1956, 1958, 1961, 1970, 1973, 1976, 1982, 1990, 1994	966 mm in 1994 (163%)
Bundi	1955, 1956, 1958, 1959, 1961, 1969, 1971, 1973, 1974, 1975, 1976, 1979	1174 mm in 1956 (168%)
Chittorgarh	1952, 1955, 1956, 1959, 1961, 1973, 1976, 1994	1413 mm in 1973 (178%)
Dausa	1955, 1958, 1959, 1961, 1975, 1978, 1981, 1995	1156 mm in 1995 (170%)
Dholpur	1955, 1957, 1958, 1961, 1967, 1977, 1992	1198 mm in 1977 (177%)
Dungarpur	1959, 1973, 1976, 1977, 1984, 1990, 1994, 1996, 1998	1301 mm in 1977 (186%)
Hanumangarh	1966, 1970, 1971, 1975, 1976, 1978, 1982, 1983, 1992, 1994, 1995	600 mm in 1978 (205%)
Jaipur	1956, 1957, 1959, 1971, 1975, 1977, 1981, 1983, 1990, 1995, 1997	1259 mm in 1977 (211%)
Jhalawar	19952, 1961, 1969, 1971, 1973, 1985, 1986	1446 mm in 1961 (157%)
Jhunjhunu	1955, 1956, 1964, 1967, 1975, 1976, 1977, 1978, 1988, 1993, 1995, 1997	843 mm in 1978 (177%)
Karauli	1956, 1961, 1971, 1975, 1983, 1992, 1995	1160 mm in 1961 (169%)
Rajsamand	1959, 1967, 1973, 1975, 1976, 1983, 1990, 1992, 1994	1113 mm in 1973 (200%)
Kota	1955, 1956, 1961, 1971, 1974, 1975, 1977, 1986	1275 mm in 1971 (168%)
Sawai Madhopur	1956, 1958, 1959, 1961, 1971, 1975, 1981, 1983, 1994, 1995, 1996, 1997, 1998, 1999	1122 mm in 1956 (154%)
Sikar	1955, 1956, 1964, 1967, 1970, 1975, 1977, 1978, 1983, 1995	1034 mm in 1977 (225%)
Sirohi	1952, 1956, 1959, 1961, 1967, 1970, 1973, 1975, 1976, 1973, 1990, 1992, 1994	1525 mm in 1973 (274%)
Tonk	1956, 1961, 1971, 1973, 1975, 1976, 1977, 1983	945 mm in 1971 (155%)
Udaipur	1954, 1959, 1973, 1975, 1976, 1994	1436 mm in 1973 (206%)

Source: *Climate of Rajasthan, Climatological Summaries of States Series* No. 16, India Meteorology Department, 2010.

From Table 2.3, it may be observed that during the period under consideration, there were 37 years in which some districts or the other in the eastern Rajasthan recorded excessive rainfall, the maximum and the minimum amount being 274 per cent and 153 per cent of the normal annual rainfall in the year 1973 and 1975 for the Sirohi and Baran districts respectively. Excessive rain occurred in the each and every district of the eastern Rajasthan, viz. Ajmer (10), Alwar (7), Banswara (12), Baran (6), Bharatpur (12), Bhilwara (12), Bundi (12), Chittorgarh (8), Dausa (8), Dholpur (7), Dungarpur (9), Hanumangarh

(11), Jaipur (11), Jhalawar (7), Jhunjhunu (12), Karauli (7), Kota (8), Rajsamand (9), Sawai Madhopur (14), Sikar (10), Sirohi (13), Tonk (8), and Udaipur (6), in 6 to 14 years as shown within the brackets against each district. In the years 1975 and 1976, 15 and 14 districts out of 23 districts of eastern Rajasthan received excessive rainfall.

Districts Sawai Madhopur (1994-1999), Bundi (1973-1976), Jhunjhunu (1975-1978) and Tonk (1975-1977) had experienced excessive rainfall in 6, 4, 4 and 3 successive years respectively in the years mentioned in brackets against each district. Four districts in the sub-division, viz. Alwar, Banswara, Baran and Karauli had no consecutive years of excessive rainfall in the period under consideration. Eight districts of sub-division registered excessive rainfall in the consecutive years 1975-76. The heaviest rainfall in 24 hours recorded at any station in the eastern Rajasthan was 802.9 mm at Chittor in Chittorgarh district on 14 July, 1982.

Analysis of Number of Rainy Days

It has been noted that there is a spatial and temporal variations in the number of rainy days which shows possible linkage with rainfall and land-use/land cover changes. Table 2.4 gives a spatial distribution of number of rainy days in Rajasthan. Table 2.4 indicates that the number of rainy days in western desert districts is comparatively less than eastern Rajasthan. The days in western districts range from 7.7 days at Jaisalmer to 30.2 days at Sikar with an average of 16.8 days. In eastern districts the number of rainy days vary from 26.4 at Ajmer to 43.9 at Jhalawar with an average of 34.3 days. The average number of rainy days in Rajasthan is 28.9.

Some studies have been made to examine the number of rainy days observed over eastern (a) and western (b) during 1951-2007 by Kharol, S.K. *et al.* (2013). Kharol S.K., *et al.* (2013) examined that eastern part presents significantly larger number of rainy days (64.03 ±11.27) compared to the western (39.83 ±10.24). Despite of the significant year to year fluctuation and the similarity in peaks and gaps, the two regions reveal contrasting trends, i.e. an increasing over western and a decreasing over eastern Rajasthan. This indicates that the rainfall pattern over the area although exhibiting high spatial heterogeneity is mainly controlled by similar meteorological dynamics defining the intensity and duration of the monsoon period (Kaskaoutis, *et al.*, 2012). The significance of the trends was checked by applying the two-sample paired t-test. The results show that both trends are not significant at 95 per cent confidence level and are influenced by extreme values (high and low) in specific years like 1997, 2002, 2006 etc. Except of this, a pronounced increasing trend in number of rainy days is observed over both regions from mid-1980s and more specifically, after drought years in the beginning of 2000s. Earlier studies (Basistha *et al.*, 2007; Gupta, 1996; Pant and Hingane, 1988; Rao *et al.*, 1997) also reported increasing rainfall trends over arid western part of Rajasthan with remarkable reduction in wind erosion and frequency of dust storms probably attributed to the land-use/land cover changes and increased moisture due to extensive canal irrigation by IGC (Indira Gandhi Canal) (Jaglan and Qureshi, 1996).

The multilayer decadal variation of number of rainy days having rainfall amount <10.10–25 and 25 mm over eastern and western Rajasthan during 1951-2007 has been discussed by S.K. Kharol *et al.*, 2013.

Table 2.4: District-wise Normal Number of Rainy Days

Districts	Jan.	Feb.	March	April	May	June	July	Aug.	Sept.	Oct.	Nov.	Dec.	Annual
Ajmer	0.5	0.6	0.4	0.4	1.0	2.8	8.0	8.1	3.6	0.6	0.2	0.5	26.4
Alwar	1.0	1.2	0.9	0.9	1.1	3.2	9.1	8.7	4.7	0.6	0.2	0.3	32.1
Banswara	0.4	0.2	0.1	0.1	0.3	4.4	13.6	12.2	6.3	0.8	0.4	0.1	38.9
Baran	1.0	0.5	0.3	0.3	1.0	4.7	13.6	12.3	6.3	0.9	0.6	0.5	42.1
Barmer	0.2	0.3	0.2	0.2	0.5	1.1	4.0	4.4	1.6	0.1	0.1	0.1	12.7
Bharatpur	1.1	0.9	0.6	0.6	0.9	3.3	10.3	10.5	5.5	0.7	0.2	0.6	35.1
Bhilwara	0.5	0.3	0.4	0.4	0.6	3.2	11.0	10.3	4.2	0.4	0.1	0.3	31.6
Bikaner	0.4	0.7	0.5	0.5	0.9	2.0	4.6	4.6	1.9	0.3	0.0	0.3	16.6
Bundi	0.6	0.5	0.4	0.4	0.7	4.0	11.8	10.9	5.3	0.5	0.2	0.2	35.4
Chittorgarh	0.5	0.3	0.2	0.2	0.5	3.9	11.4	10.5	5.1	0.5	0.3	0.2	33.6
Churu	0.8	0.7	0.6	0.6	1.0	2.3	5.3	5.0	2.5	0.3	0.1	0.4	19.4
Dausa	1.0	0.9	0.5	0.5	1.1	3.2	10.3	9.6	4.8	0.7	0.2	0.6	33.3
Dholpur	1.3	0.9	0.6	0.6	1.0	3.4	11.5	11.5	6.3	0.9	0.2	0.6	38.8
Dungarpur	0.3	0.1	0.1	0.1	0.6	4.3	12.4	10.7	5.1	0.6	0.3	0.1	34.7
Ganganagar	1.1	0.9	0.7	0.7	0.6	1.7	4.2	3.7	1.6	0.3	0.0	0.4	15.7
Jaipur	1.0	0.9	0.6	0.6	1.1	3.3	9.5	9.4	4.4	0.6	0.2	0.4	31.8
Jaisalmer	0.2	0.2	0.2	0.2	0.3	0.6	2.6	2.7	0.7	0.0	0.0	0.1	7.7
Jalore	0.2	0.4	0.2	0.2	0.4	1.7	6.3	6.4	2.5	0.3	0.1	0.1	18.7
Jhalawar	1.1	0.5	0.5	0.5	0.8	5.6	13.5	11.9	7.2	1.0	0.9	0.6	43.9
Jhunjhunu	1.1	1.0	0.9	0.9	1.5	3.1	7.2	6.9	3.4	0.6	0.2	0.5	27.0
Jodhpur	0.4	0.5	0.3	0.3	0.8	1.8	5.1	5.7	2.1	0.3	0.1	0.2	17.6
Kota	1.0	0.6	0.4	0.4	0.8	4.5	12.8	11.8	6.3	0.9	0.6	0.5	40.6
Nagaur	0.6	0.6	0.5	0.5	1.2	2.5	6.4	6.4	2.8	0.4	0.1	0.3	22.1
Pali	0.4	0.5	0.3	0.3	0.8	2.5	6.9	7.0	3.3	0.5	0.1	0.1	22.6
Rajsamand	0.4	0.1	0.4	0.4	0.2	2.0	7.8	5.6	4.0	0.3	0.1	0.1	21.0
Sawaimadhopur	1.0	0.8	0.5	0.5	0.8	3.5	11.0	10.9	5.2	0.7	0.2	0.5	35.5
Sikar	1.4	1.0	0.8	0.8	1.1	2.7	8.8	8.5	4.3	0.4	0.2	0.5	30.2
Sirohi	0.5	0.5	0.2	0.2	0.8	2.9	9.9	9.2	3.6	0.7	0.2	0.2	28.9
Tonk	0.9	0.6	0.5	0.5	0.8	3.3	10.4	10.4	4.6	0.6	0.2	0.3	33.0
Udaipur	0.4	0.3	0.3	0.3	0.7	3.1	9.6	8.1	4.6	0.6	0.2	0.2	28.3
Hanumangarh	1.0	1.0	0.6	0.6	0.7	1.9	4.4	3.9	2.0	0.2	0.0	0.4	16.7
Rajasthan East	0.8	0.6	0.5	0.5	0.8	3.7	10.8	10.2	5.1	0.7	0.3	0.4	34.3
Rajasthan West	0.5	0.6	0.4	0.4	0.7	1.8	4.9	4.9	2.1	0.3	0.1	0.2	16.8
Rajasthan State	**0.7**	**0.6**	**0.5**	**0.5**	**0.8**	**3.1**	**9.0**	**8.6**	**4.2**	**0.6**	**0.2**	**0.3**	**28.9**

Note: Rainy day is a day on which 2.5 milimetres or more rain is recorded. Figures for District Head Quarter only.

Source: Department of Agriculture, Government of Rajasthan, Jaipur.

Rainfall Departures

Rainfall departures from the monsoon rainfall of eastern and western Rajasthan have been examined from 1871 to 2011-2012 (Figs. 2.1 and 2.2). Fig. 2.1 indicates that the lowest amount of rainfall (194.5 mm) was received in 1877 and the highest amount of rainfall (1118.5 mm) in 1917. Figure 2.1 also reveals that in eastern Rajasthan out of 141 years, 65 years were of deficient rainfall. However, after 1987, 2002 and 2009 there has been a slight increasing trend of rainfall in eastern Rajasthan. Fig. 2.2 indicates that the highest amount of rainfall in western Rajasthan was recorded in 1908 amounting 573.2 mm or 123 per cent departure from normal-monsoon rainfall. Contrary to it, the lowest amount of rainfall was recorded in 1918, i.e. 33.6 mm or –87 per cent departure from normal. It further reveals that the frequency of deficient rainfall year is more in western Rajasthan numbering 72 years out of 140 years.

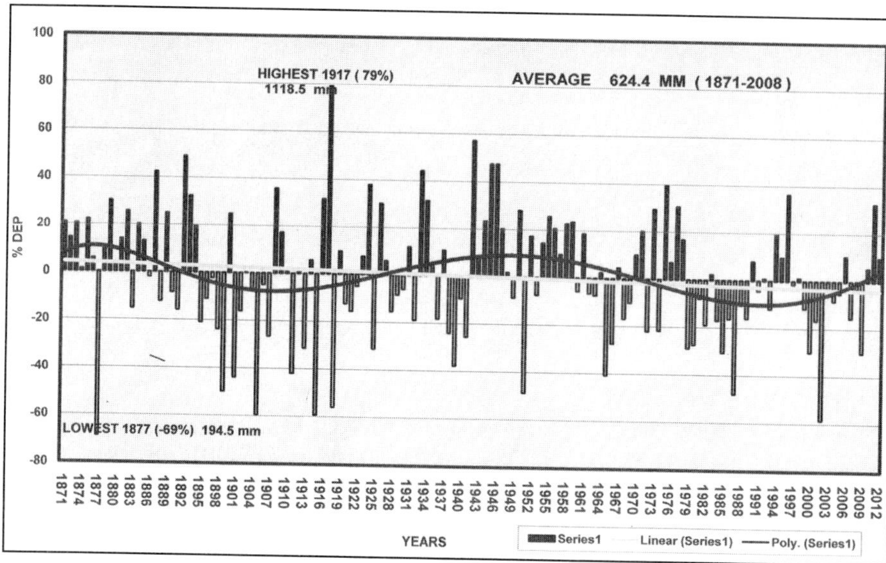

Fig. 2.1: Departure of Rainfall in Eastern Rajasthan
Source: IMD.

The analysis of rainfall data from 1961 to 2012 also reveals interesting results (Fig. 2.3). Over all monsoonal rainfall in the states shows increasing and decreasing trend of rainfall. The decadal change in rainfall in Rajasthan may be interpreted with the help of Fig. 2.3. The decade from 1962-1972 was the decade of declining trend of rainfall. The period 1972 to 1978 showed an increasing trend of rainfall. Again the declining trend may be noticed from 1979 to 1996 with a severe drought in 1987 when –55 per cent departure was noticed. After 1987 there was period of 5 years when there has been slight

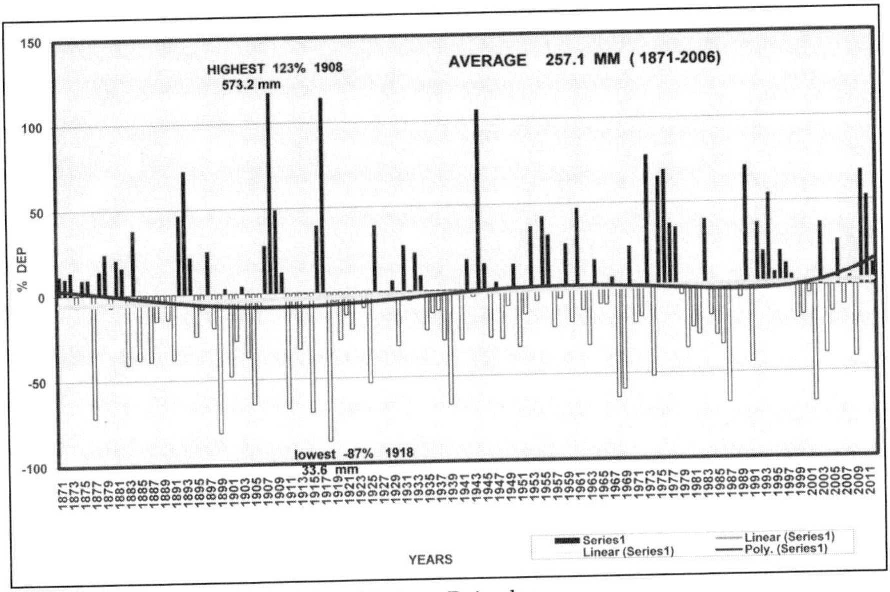

Fig. 2.2: Departure of Rainfall in Western Rajasthan
Source: IMD.

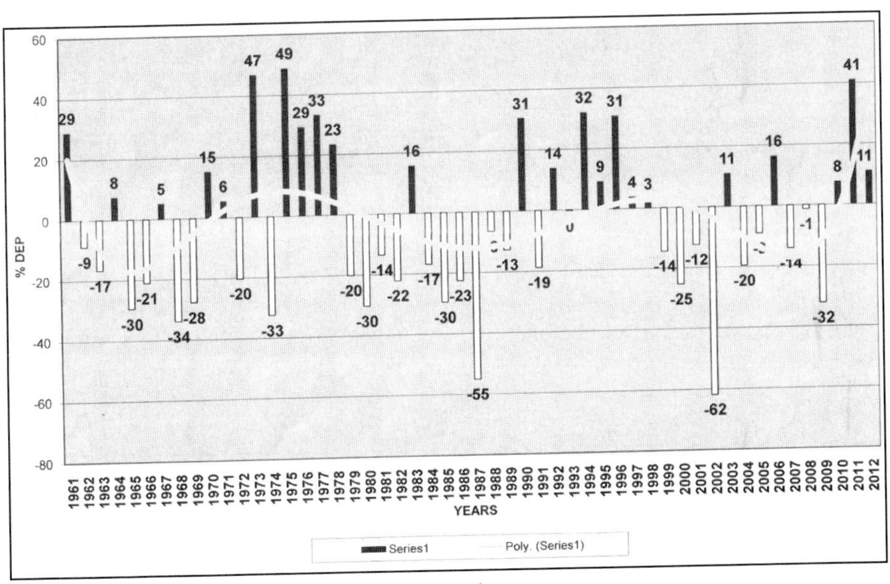

Fig. 2.3: Decadal change in Rainfall of Rajasthan
Source: IMD

increase in rainfall. From 2000 onward till 2009 (except 2003 and 2006) there has been deficiency of rainfall in the state. However, from 2009 onward the current decade is indicating an increasing trend of rainfall. The same trend of rainfall pattern in eastern Rajasthan may be noticed (Fig. 2.4). But, it may be inferred from Figure 2.4 that after 1995-96 the eastern Rajasthan has been witnessing declining trend of rainfall except in 2011 when this region received +33 per cent of rainfall and showing increasing trend of rainfall.

The district-wise analysis of average rainfall from 1901 to 2012 further indicates distribution of rainfall pattern (Fig. 2.5) and departure from the normal rainfall.

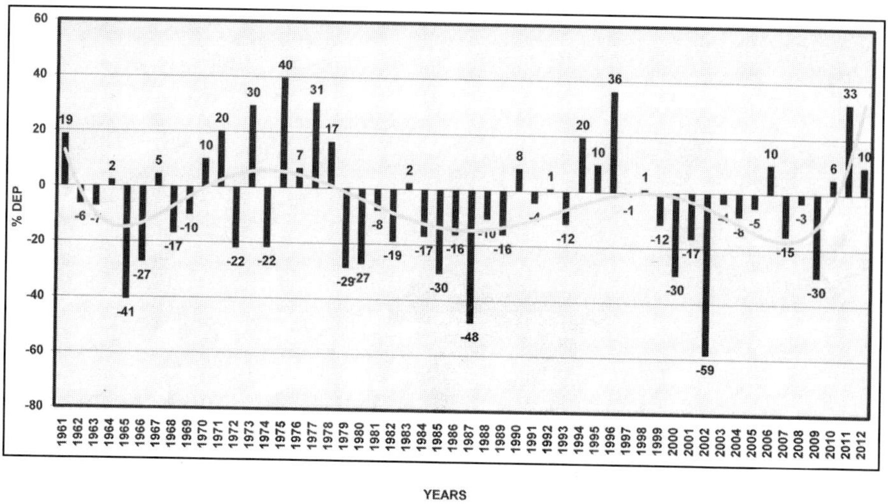

Fig. 2.4: Decadal change in Rainfall departure in East Rajasthan
Source: IMD.

Decadal Analysis of Temperature Trend

The analysis of mean temperature of Rajasthan from 1901-2010 reveals that temperature in Rajasthan since 1971 onward is showing an increasing trend (Fig. 2.6).

Decade	*1901-10*	*1911-20*	*1921-30*	*1931-40*	*1941-50*	*1951-60*	*1961-70*	*1971-80*	*1981-90*	*1991-2000*	*2001-10**
Avg. Temp in °C	25.62	25.60	25.60	25.54	25.71	25.82	25.76	25.54	25.79	25.94	26.57

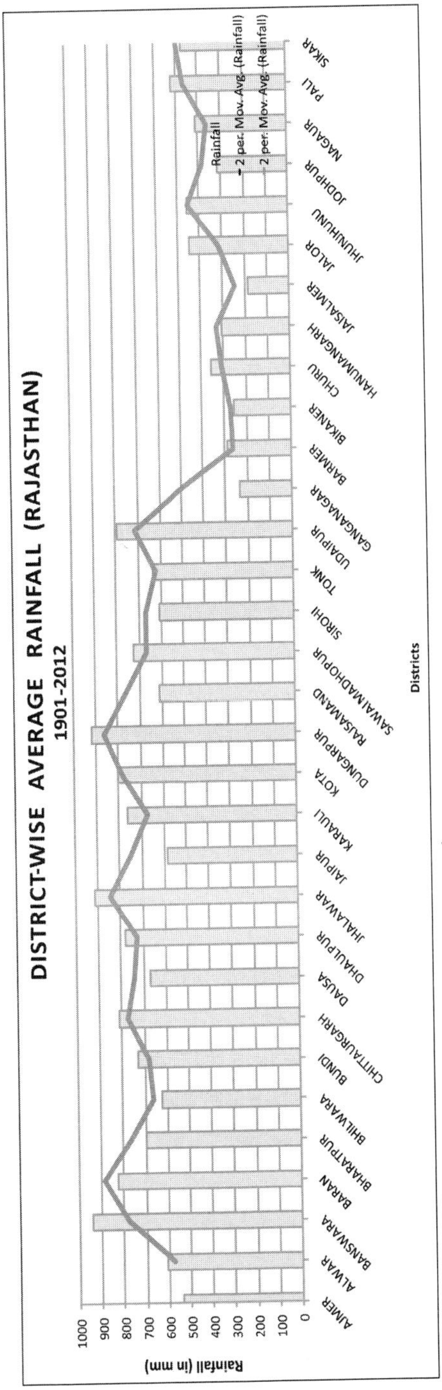

Fig. 2.5: District-wise rainfall departure in Rajasthan

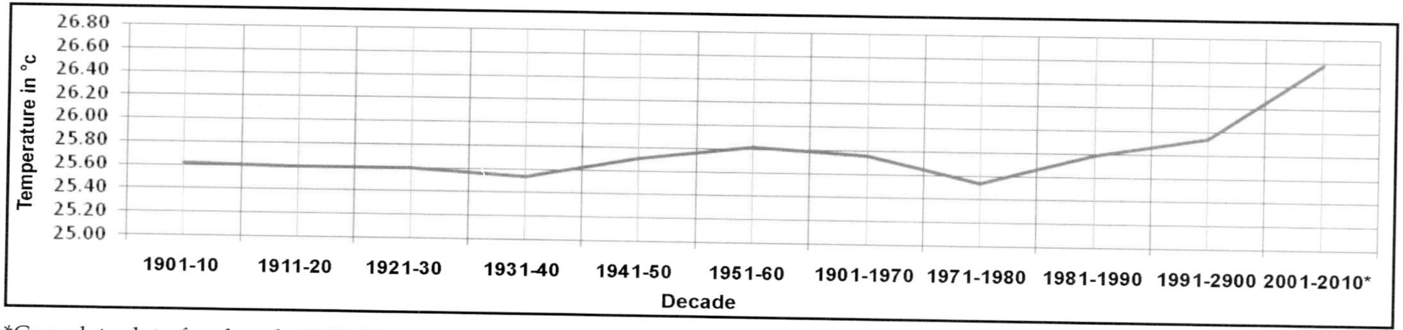

*Complete data for decade 2001-10 was not available, hence average of 7 years has been taken into consideration.

Fig. 2.6: Mean Temperature from 1901-2010 in Rajasthan

The plotting of temperature pattern from 1901 to 2012 also reveals an increasing trend (27°C) except 1996-97 when average temperature recorded up to 25°C (Fig. 2.7).

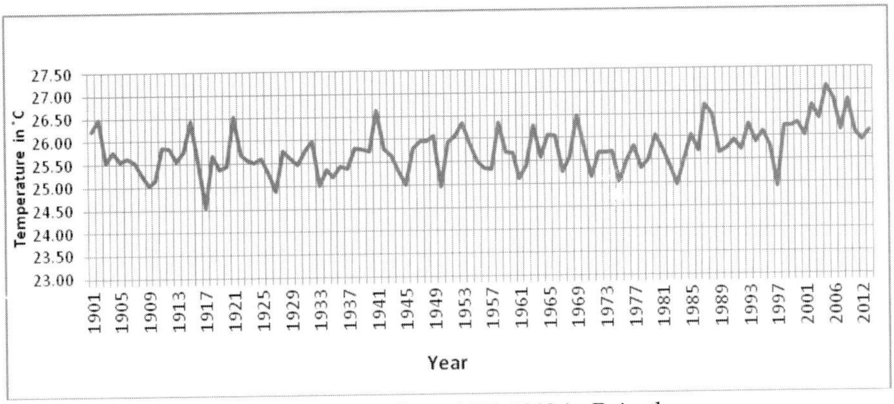

Fig. 2.7: Decadal Mean Temperature from 1901-2012 in Rajasthan

The average temperature district-wise shows that Banswara, Sawai Madhopur, Sirohi, Barmer, Nagaur, Jalore, Bikaner, Jaisalmer and Jodhpur recorded high mean temperature while Rajsamand and Jhunjhunu recorded low average temperature (Fig. 2.8).

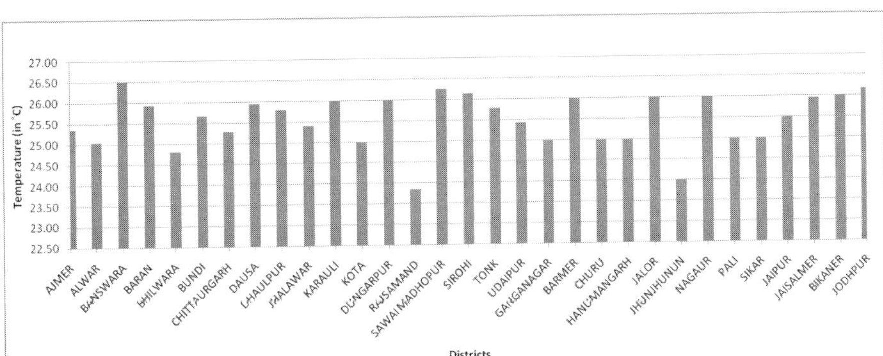

Fig. 2.8: District-wise Average Temperature in Rajasthan

When the pattern of temperature plotted indicting per cent deviation from mean for eastern Rajasthan, it was found that the deviation from average temperature after 1981 has been showing higher up to 2 per cent to 3 per cent. The highest mean temperature departure of 4 per cent has been recorded in 1987 and in 2002. Over all the deviation has been higher after 1981 onward (Fig. 2.9). The analysis of average temperature deviation for western Rajasthan indicates that there has been continuous rise in temperature after 1981 with highest deviation of 8 per cent in 2002-03 (Fig. 2.10).

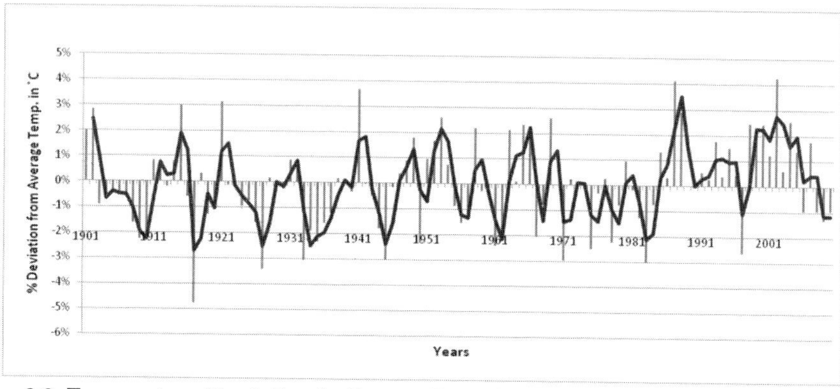

Fig. 2.9: Temperature Deviation in Eastern Rajasthan

The temperature deviation from average of Rajasthan from 1901 to 2012 indicates (Fig. 2.11) the same trend of temperature increase in Rajasthan. It may be noted that the deviation of more than 4 per cent has been seen after 2001 onwards. In the earlier years from 1901 till 1981 the deviation from average was between −2 to −3 per cent. The continuous rise in temperature after 1981 is well indicated in Fig. 2.11.

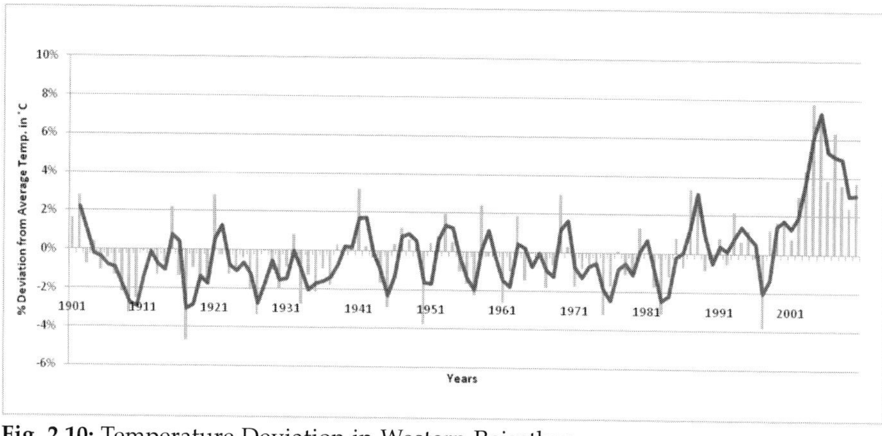

Fig. 2.10: Temperature Deviation in Western Rajasthan

Droughts

Rainfall is the ultimate source of water, affecting production of crops and other

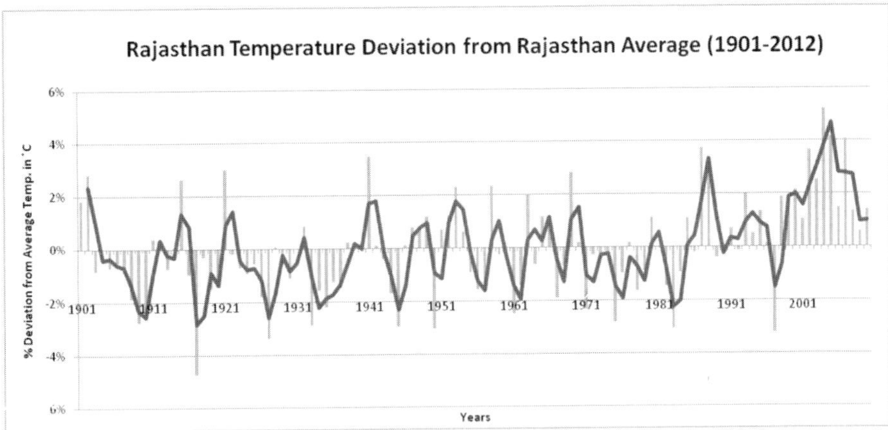

Fig. 2.11: Average Temperature Deviation in Rajasthan

biomass by direct falling on the fields as well as supporting surface and ground for irrigation. However, the possibility of drought occurrence in India varies from once in two years in arid Rajasthan to once in 15 years in Assam. It has been noted that frequency of extreme events like drought, floods, heat waves/ cold was, cyclones, delayed and early onset of monsoon, long dry spells etc. have increased in last 2-3 decades due to global warming and resultant climate change.

In Rajasthan, there have been 50 drought years of varied intensity in the period from 1901-2012, which means that the chance of occurrence of a meteorological drought in the state is about 48 per cent. A detailed analysis has revealed in only after 2009 out of these years none of the districts in Rajasthan were affected by droughts, because of increasing trend of monsoon rainfall.

The frequency of drought from 1963-2010 as indicated in Fig. 2.12 brings out clearly that the frequency of drought is more in western desert districts in comparison to eastern districts. The frequency of drought is more than 30. The southern districts such as Sirohi, Dungarpur, Banswara, Chittorgarh fall in the frequency range of 22-26, despite of high amount average rainfall. The occurrence of droughts in Jalore, Nagaur, Pali, Ajmer, Rajsamand, Udaipur, Bhilwara and Jhunjhunu are in the frequency range from 27 to 30. The north-eastern districts of Rajasthan also fall in the range of 22-26. Minimum occurrence of frequency of droughts during 1963-2010 are in Bharatpur, Dholpur, Kota, Baran and Jhalawar districts. The frequency of drought in these districts is less than 21.

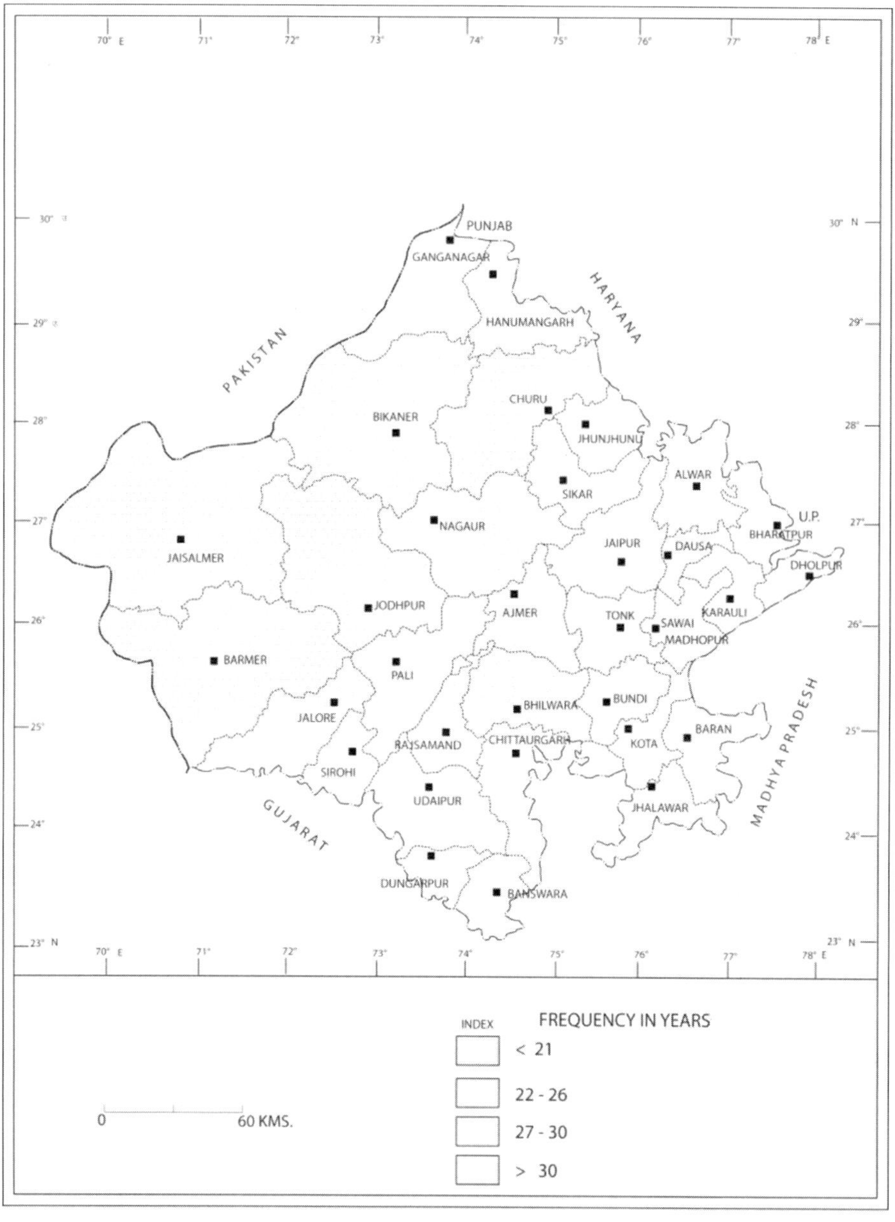

Fig. 2.12: Frequency of Drought in Rajasthan 1963-2010

Western Rajasthan

The drought conditions which prevailed over western Rajasthan during last 58 years period from 1951 to 2010 are described in Table 2.5.

Table 2.5: Years of Successive Droughts and the Affected Districts of Western Rajasthan

Years of Successive Drought	*Affected Districts*
1951-1952	Bikaner, Churu
1962-1963	Barmer, Pali
1965-1966	Nagaur
1968-1969	Barmer, Bikaner, Churu, Ganganagar, Jaisalmer, Jalore, Jodhpur, Pali
1971-1972	Pali
1980-1981	Jalore
1980-1981-1982	Pali
1984-1985-1986-1987	Bikaner, Jaisalmer, Jalore, Jodhpur, Nagaur, Pali
1985-1986-1987-1988	Barmer
1996-1997-1998-1999	Jodhpur

Source: *Climate of Rajasthan, Climatological Summaries of States Series* No. 16, India Meteorology Department, 2010.

Table 2.5 reveals that all the districts of western arid Rajasthan have been under drought conditions. The table clearly brings out the area which was simultaneously affected by drought conditions. Churu and Bikaner experienced drought in 1951-52 while Bikaner, Jaisalmer, Jalore, Jodhpur, Nagaur and Pali districts witnessed continuous drought conditions from 1984 to 1987. So, is the case of Barmer which experienced drought continuously from 1985 to 1988. Again the Jodhpur witnessed drought conditions from 1996 to 1999. Pali and Jalore affected by drought conditions in 1971-72 and 1980-81 respectively.

It has further been observed that the rainfall of less than 50 per cent of the annual normal representing severe drought conditions occurred in various districts of western Rajasthan indicated in Table 2.6.

Table 2.6: District-wise years of severe drought with lowest percentage of annual rainfall and year (Western Rajasthan)

Districts	*Years of Severe Drought (Rainfall < 50%)*	*Lowest amount of rainfall in cms (expressed as % of annual normal) with year*
Barmer	1966, 1968, 1969, 1974, 1987	45 mm in 1969 (18%)
Bikaner	1963, 1968, 1991	75 mm in 1968 (29%)
Churu	1952	171 mm in 1952 (46%)
Ganganagar	1968	112 mm in 1968 (46%)
Jaisalmer	1963, 1968, 1969, 1974, 1987, 1991	22 mm in 1969 (13%)
Jalore	1968, 1969, 1974, 1980, 1986, 1987, 1991	91 mm in 1968 (21%)
Jodhpur	1968, 1969, 1987, 1996, 1997, 1998, 1999	16 mm in 1998 (5%)
Nagaur	1951, 1963, 1987	164 mm in 1951 (42%)
Pali	1969, 1974, 1981, 1987	160 mm in 1969 (32%)

Source: *Climate of Rajasthan, Climatological Summaries of States Series* No. 16, India Meteorology Department, 2010.

Table 2.6 reveals that the percentage of rainfall delficiency ranges from

5 per cent at Jodhpur in the years 1968, 1969, 1987, 1996, 1997, 1998 and 1999 to 46 per cent at Churu in 1952. It is interesting to note that the year 1969 witnessed drought condition in all the western districts of Rajasthan.

Eastern Rajasthan

All the districts of eastern Rajasthan experienced drought conditions ranging from 7 years in Ajmer, Baran and Dausa to highest 13 years in Jaipur, Jhunjhunu during the period from 1951 to 2010. The data reveals that the districts of eastern Rajasthan such as Ajmer (23%), Banswara (20%), Bundi (25%), Hanumangarh (23%), Dungarpur (21%), Jaipur (27%), Jhalawar (24%), Jhunjhunu (27%), Karauli (29%), Kota (2%), Sawai Madhopur (21%), Sikar (21%), Sirohi (23%) and Udaipur (23%) had drought conditions for more than 20 per cent of the years (1951-2010).

It has been noted that the occasions of occurrence of drought conditions have been very frequent in eastern districts of Rajasthan. Jaipur, Jhunjhunu, Sawai Madhopur districts experienced drought condition continuously from 1951-52 and 1953. The southern districts (Banswara and Dungarpur) experienced occurrence of successive drought in 1964 and 1966. However, the Dungarpur experienced successive drought in 1968-69. The other districts like Karauli (1953-54), Ajmer (1962-63), Dholpur (1993-94) and Kota (1998-99) witnessed drought continuously twice during 1950-2010.

The rainfall of less than 50 per cent of the annual rainfall occurred in eastern districts is indicated in Table 2.7.

Table 2.7: District-wise years of severe drought with lowest percentage of annual rainfall and year (Eastern Rajasthan)

District	*Years of Severe Drought (Rainfall < 50%)*	*Lowest amount of rainfall in mm (expressed s % of annual normal) with year*
Ajmer	1972, 1987	214 mm in 1972 (27%)
Alwar	1987	285 mm in 1987 (46%)
Banswara	1966	388 mm in 1966 (44%)
Baran	1965	378 mm in 1965 (44%)
Bharatpur	1979	199 mm in 1979 (32%)
Bhilwara	1951	279 mm in 1951 (47%)
Chittorgarh	1951	286 mm in 1951 (36%)
Dungarpur	1966	315 mm in 1966 (45%)
Hanumangarh	1951, 1969, 1974	99 mm in 1969 (34%)
Jhunjhunu	1951, 1968, 1999	190 mm in 1951 (40%)
Karauli	1989	316 mm in 1989 (46%)
Sikar	1972	216 mm in 1972 (47%)
Sirohi	1974, 1987	178 mm in 1987 (32%)
Tonk	1951	299 mm in 1951 (49%)

Source: *Climate of Rajasthan, Climatological Summaries of States Series* No. 16, India Meteorology Department, 2010.

To sum up, it may be noted that in 112 years period of rainfall trend analysis there are 3 years of disastrous droughts, 7 years of famine, 66 years of moderate drought and 36 years are good years. The decades of 1901-10, 1911-20, 1961-70, 1981-90 and 2000-2010 have been marked for concentration of drought years. During this period there have been more than 66 drought years of different intensity in Rajasthan.

District-Level Temperature and Rainfall Trends of Rajasthan

Correlation analysis of annual average temperature with time series shows that fifteen districts of the State—Jaisalmer, Bikaner, Jodhpur, Banswara, Dungarpur, Barmer, Udaipur, Jalore, Jhalawar, Sirohi, Chittorgarh, Rajsamand, Baran, Pali and Kota, have recorded very significant warming, while two districts—Bhilwara and Bundi, have witnessed significant warming trend at centennial time scale (Fig. 2.13). The districts Sikar and Dausa have shown a cooling trend although this trend is insignificant.

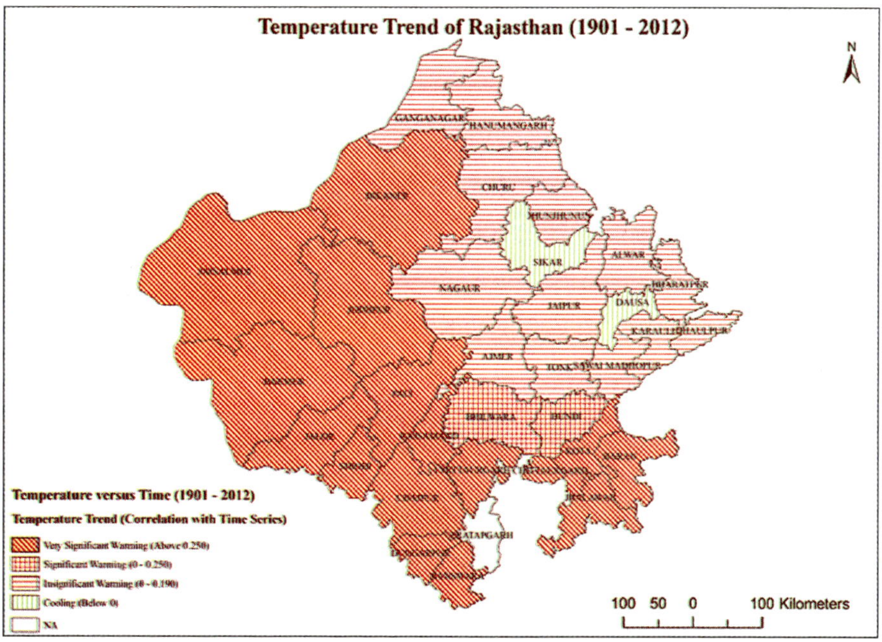

Fig. 2.13: Centennial Temperature Trend in Rajasthan

Annual total rainfall pattern shows that Banswara, Dungarpur, Jhalawar, Sirohi and Udaipur have recorded less rainfall than normal while rest 27 districts of the State have received more rainfall, although all correlation values are insignificant (Fig. 2.14). The data of Pratapgarh district was not available.

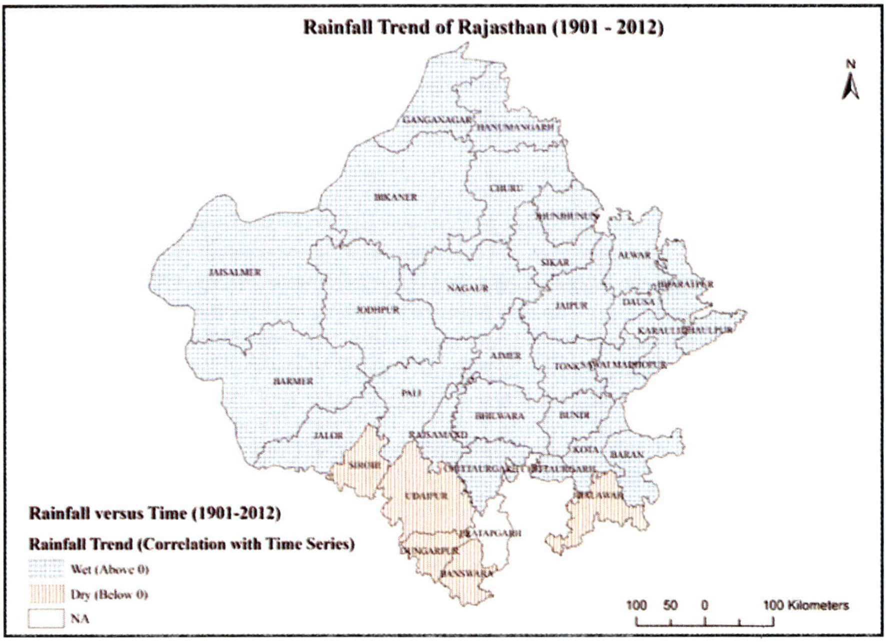

Fig. 2.14: Centennial Rainfall Trend in Rajasthan 1901-2012

Regression result of temperature and rainfall series shows that due to too much noise in the data, R^2 values are very weak (noise is clearly visible in temperature and rainfall *versus* time scatter plots), however, temperature-time series association is positive and significant for 17 districts having F value above 4 (Table 2.8). For rainfall series, F value is weak in all cases, hence no significant trend was identified.

Temperature and rainfall patterns of Rajasthan at district-level, were examined and the trends were modelled by the Autoregressive integrated moving average (ARIMA) model. Centennial temperature pattern of Ajmer shows no significant change (departure from the fit line) in the variability and the forecast line predicts no warming or cooling trend in near future (Fig. 2.15). Rainfall pattern depicts two distinct spells – high variability with less rainfall (1901-1920), and moderate variability with average rainfall (after 1920), however, no increasing or decreasing trend was reported (Fig. 2.16).

In case of Alwar no significant change in temperature was recorded although trend of fit line shows fluctuations in variability (Fig. 2.17). Perusal of Rainfall pattern shows phases of different variability but over all trend recorded no significant change (Fig. 2.18).

Temperature pattern of Banswara shows a continuous rising trend after 1950 and the model predicted a significant warming trend (Fig. 2.19). On the

Table 2.8: Regression result for temperature and rainfall patterns in Rajasthan

Sl. No.	*District*	*Correlation between temperature and time series (r)*	*Regression result for temperature series*		*Correlation between rainfall and time series (r)*	*Regressions result for rainfall series*	
			R^2	*F*		R^2	*F*
1.	Ajmer	0.099	0.010	0.987	0.090	0.008	0.823
2.	Alwar	0.001	0.000	0.000	0.190	0.036	3.749
3.	Banswara	0.466	0.217	27.709**	–0.074	0.006	0.554
4.	Baran	0.304	0.092	10.164**	0.023	0.001	0.051
5.	Barmer	0.399	0.159	18.888**	0.031	0.001	0.094
6.	Bharatpur	0.042	0.002	0.180	0.146	0.021	2.164
7.	Bhilwara	0.243	0.059	6.261*	0.062	0.004	0.384
8.	Bikaner	0.486	0.236	34.284**	0.042	0.002	0.183
9.	Bundi	0.200	0.040	4.13*	0.059	0.003	0.350
10.	Chittorgarh	0.334	0.112	12.568**	0.001	0.000	0.000
11.	Churu	0.112	0.013	1.276	0.071	0.005	0.504
12.	Dausa	–0.054	0.003	0.293	0.126	0.016	1.612
13.	Dholpur	0.135	0.018	1.848	0.107	0.012	1.168
14.	Dungarpur	0.430	0.185	22.732**	–0.109	0.012	1.212
15.	Ganganagar	0.190	0.036	3.727	0.055	0.003	0.298
16.	Hanumangarh	0.152	0.023	2.366	0.113	0.013	1.298
17.	Jaipur	0.017	0.000	0.033	0.161	0.026	2.851
18.	Jaisalmer	0.586	0.343	57.903**	0.022	0.000	0.052
19.	Jalore	0.356	0.127	14.487**	0.004	0.000	0.002
20.	Jhalawar	0.352	0.124	14.153**	–0.010	0.000	0.010
21.	Jhunjhunu	0.009	0.000	0.009	0.151	0.023	2.347
22.	Jodhpur	0.481	0.231	33.392**	0.059	0.003	0.369
23.	Karauli	0.098	0.010	0.974	0.108	0.012	1.182
24.	Kota	0.255	0.065	6.979**	0.037	0.001	0.137
25.	Nagaur	0.110	0.012	1.235	0.104	0.011	1.094
26.	Pali	0.289	0.084	9.114**	0.031	0.001	0.097
27.	Rajsamand	0.310	0.096	10.658**	0.003	0.000	0.001
28.	Sawai Madhopur	0.080	0.006	0.651	0.102	0.010	1.047
29.	Sikar	–0.020	0.000	0.040	0.155	0.024	2.450
30.	Sirohi	0.351	0.123	14.014**	–0.044	0.002	0.197
31.	Tonk	0.040	0.002	0.159	0.089	0.008	0.806
32.	Udaipur	0.360	0.130	14.935**	–0.076	0.006	0.579

**: $p<0.01$, *: $p<0.05$
Source: Computed by Author.

other hand 1951 seems a divide in rainfall trend where rainfall pattern showed increasing trend till 1951 and after that a decreasing trend was recorded, however, over all trend showed no significant change in the rainfall pattern (Fig. 2.20).

In case of Baran district temperature pattern showed fluctuations in variability and a significant warming trend was observed (Fig. 2.21). Rainfall pattern showed an increasing trend till 1961, and after that a decreasing trend was observed. However, no significant trend at centennial time scale for rainfall pattern was noticed (Fig. 2.22).

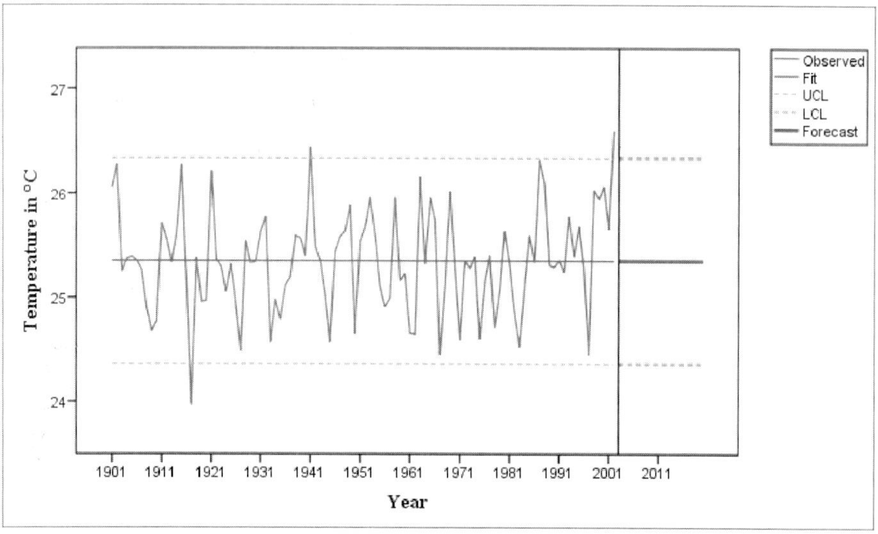

Fig. 2.15: Temperature and Rainfall Trend of 32 Districts of Rajasthan up to 2.78

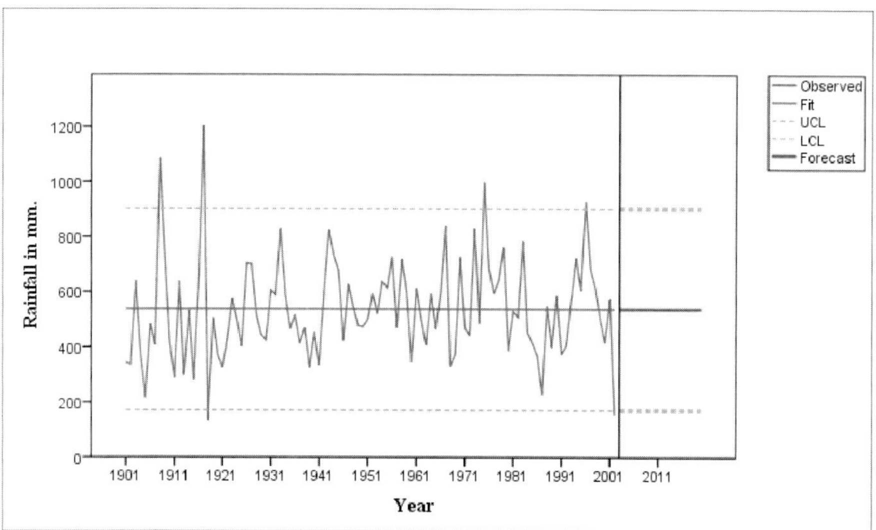

Fig. 2.16: Rainfall Trend in Ajmer

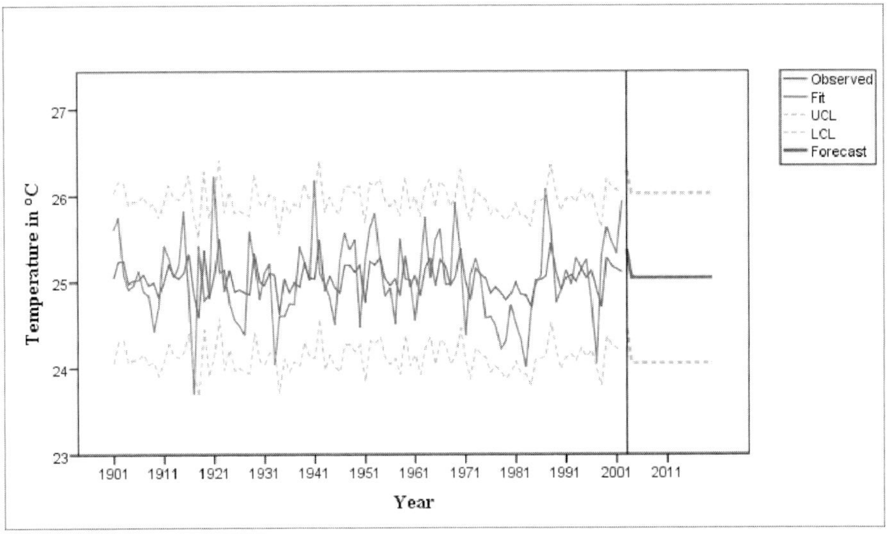

Fig. 2.17 Temperature Trend in Alwar

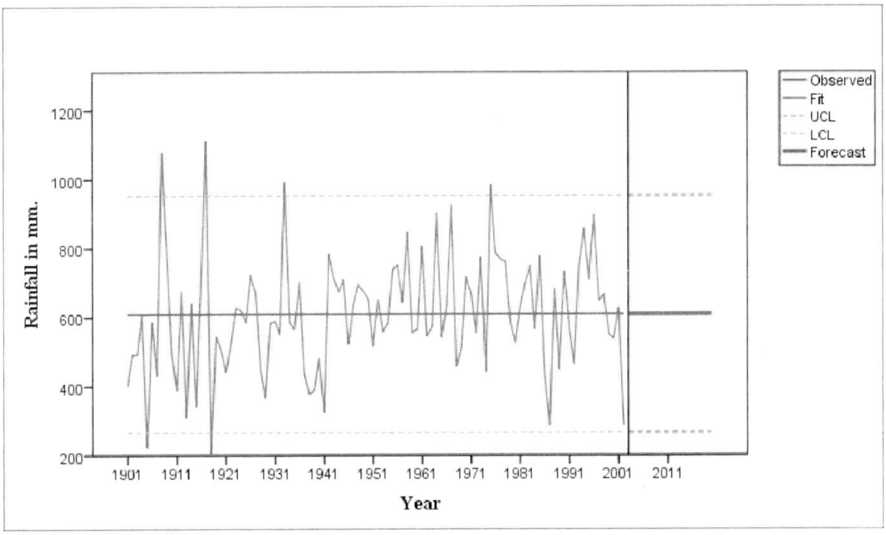

Fig. 2.18 Rainfall Trend in Alwar

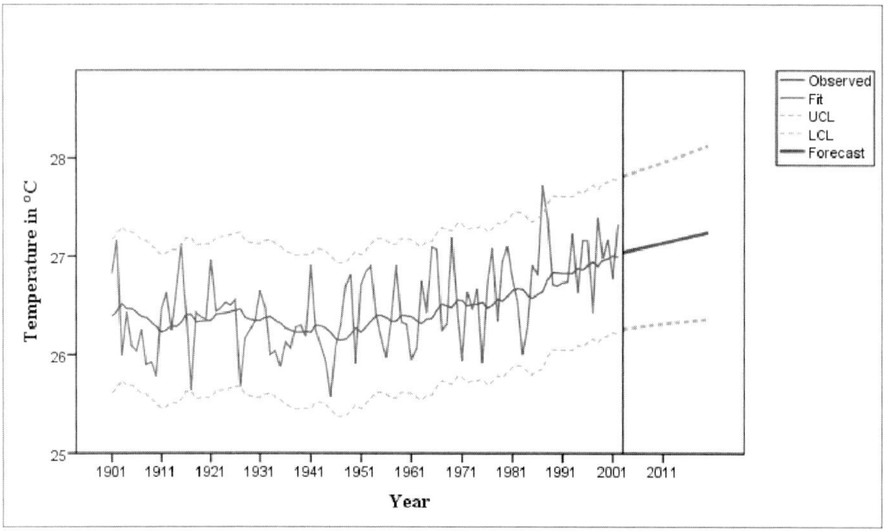

Fig. 2.19: Temperature Trend: Banswara

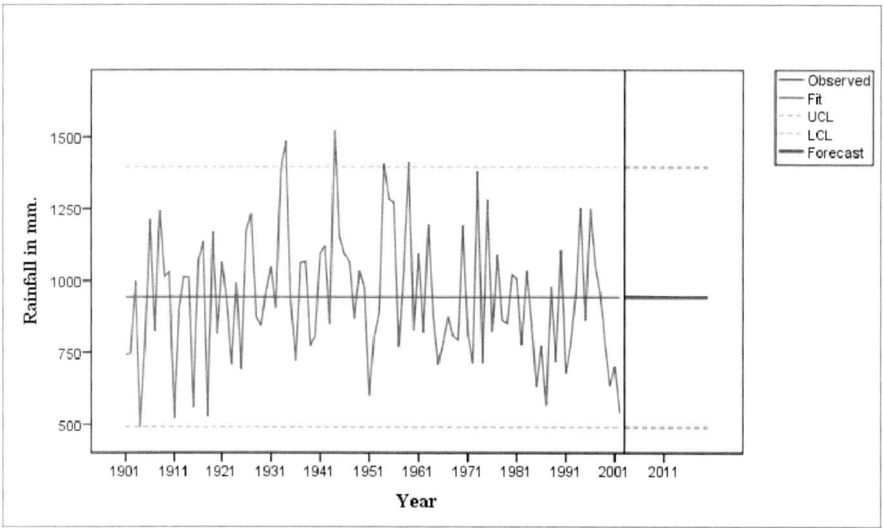

Fig. 2.20: Rainfall Trend: Banswara

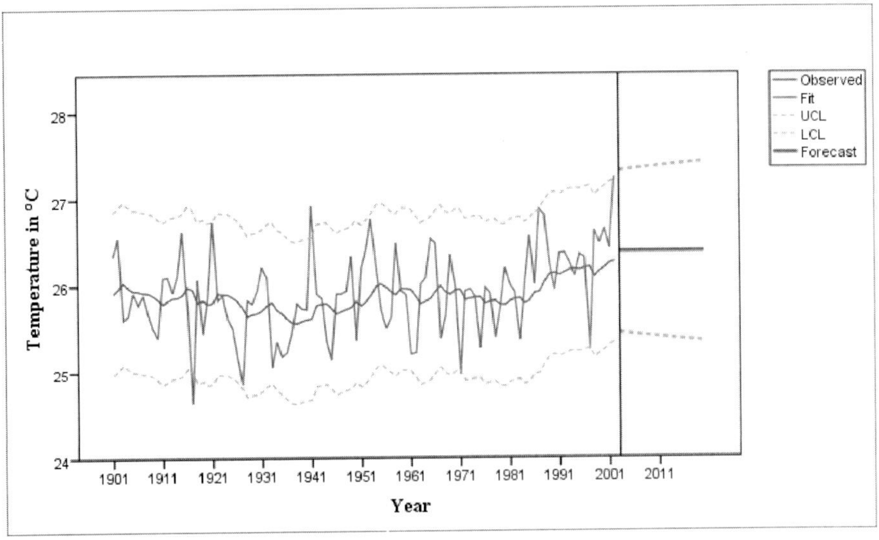

Fig. 2.21: Temperature Trend: Baran

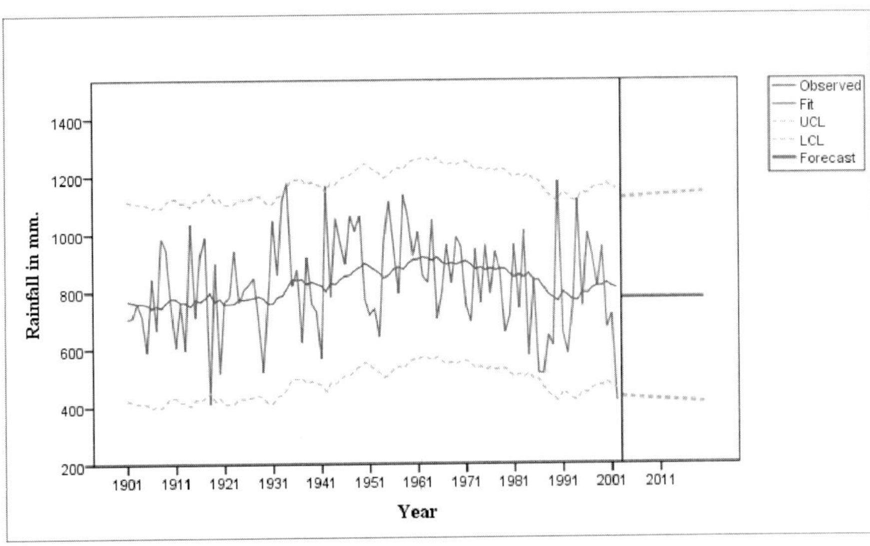

Fig. 2.22: Rainfall Trend: Baran

There are three distinct phases visible in the temperature pattern of Barmer, warming till 1950, cooling between 1950 to 1980 and a continuous warming trend after 1980. Based on the pattern, the forecast model predicted a significant warming trend for coming years (Fig. 2.23). Rainfall pattern showed almost equally spaced fluctuations and no significant trend was identified (Fig. 2.24).

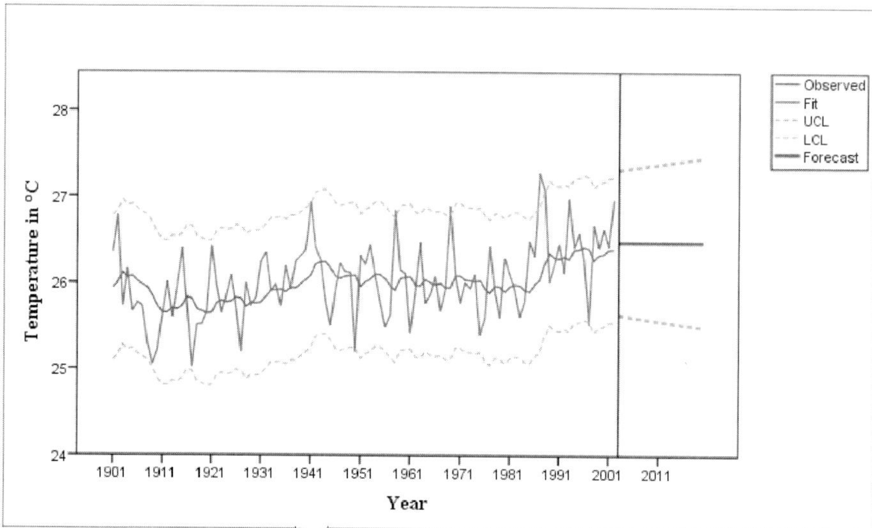

Fig. 2.23: Temperature Trend: Barmer

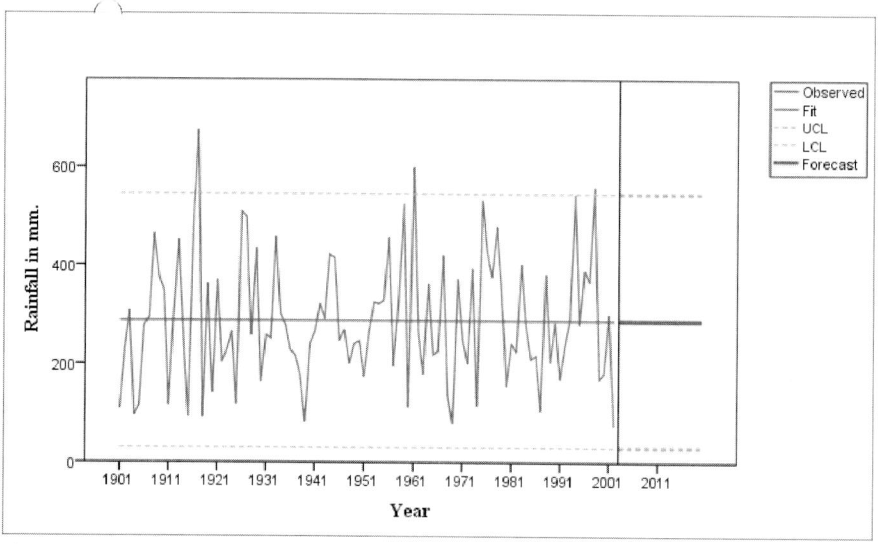

Fig. 2.24: Rainfall Trend: Barmer

Temperature pattern of Bharatpur did not show any significant change in the variability and no warming or cooling trend was gained at centennial level (Fig. 2.25). The year of 1951 seems the rainfall divide showing an increasing trend till 1951 and a declining trend after that, but over all there was no positive or negative trend forecasted (Fig. 2.26).

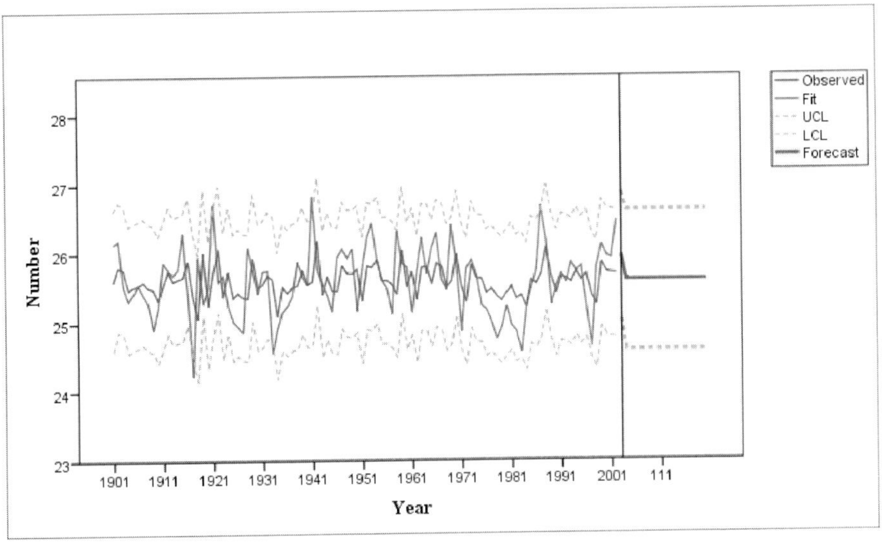

Fig. 2.25: Temperature Trend: Bharatpur

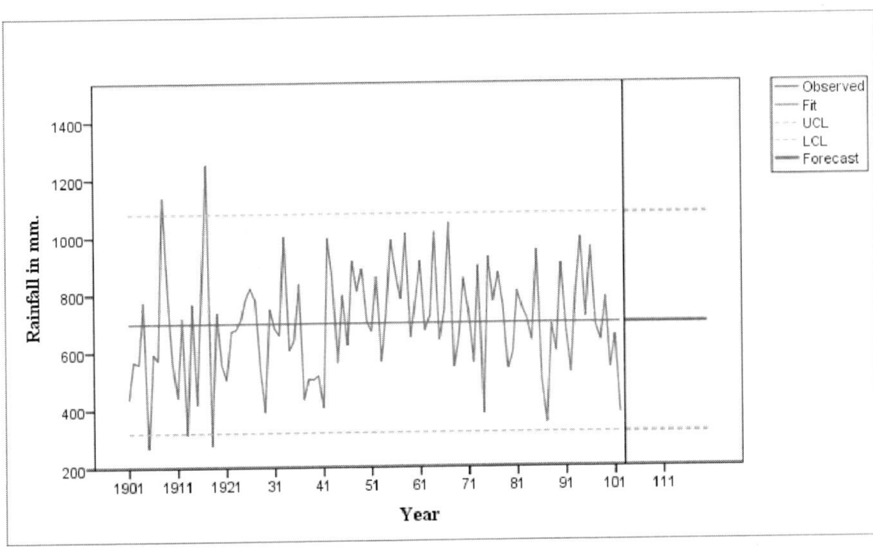

Fig. 2.26: Rainfall Trend: Bharatpur

District of Bhilwara showed a significant warming trend after 1980 and as the result a slight warming trend was predicted (Fig. 2.27). Rainfall pattern recorded no positive or negative trend however, periods of fluctuations and moderate rainfall were evident (Fig. 2.28).

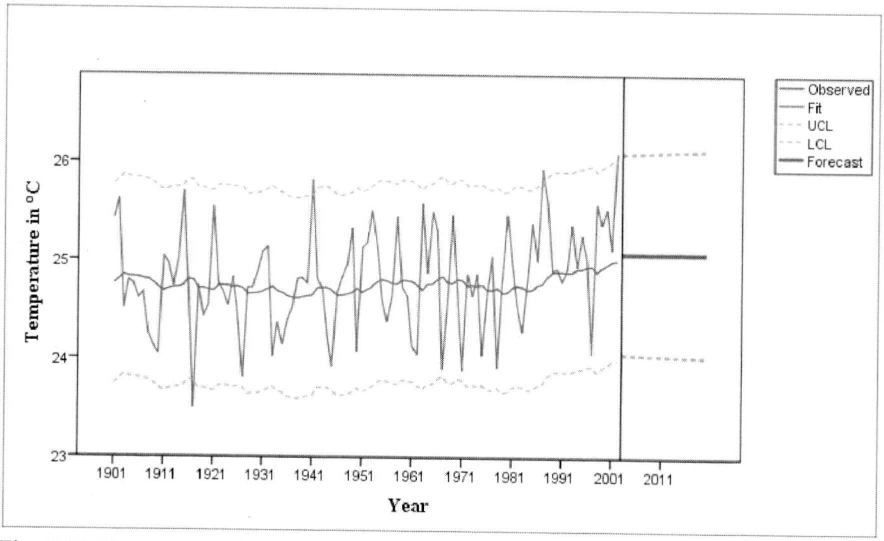

Fig. 2.27: Temperature Trend: Bhilwara

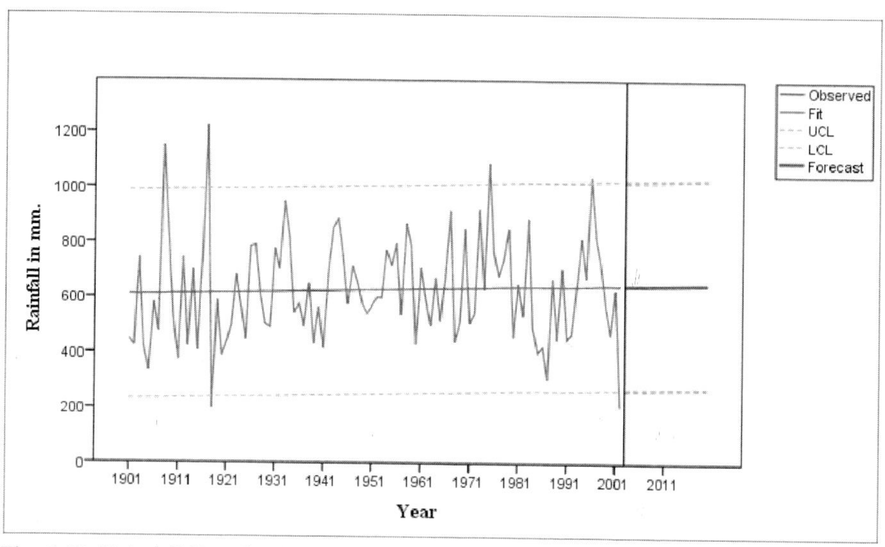

Fig. 2.28: Rainfall Trend: Bhilwara

Perusal of temperature pattern of Bikaner shows that up to 1941, temperature was in declining trend, between 1941-1981 it was almost stable and after 1981 it gained a rising trend. At centennial time scale a significant warming was identified between 95 per cent confidence limits (Fig. 2.29). The rainfall pattern showed no significant change in the variability pattern (Fig. 2.30).

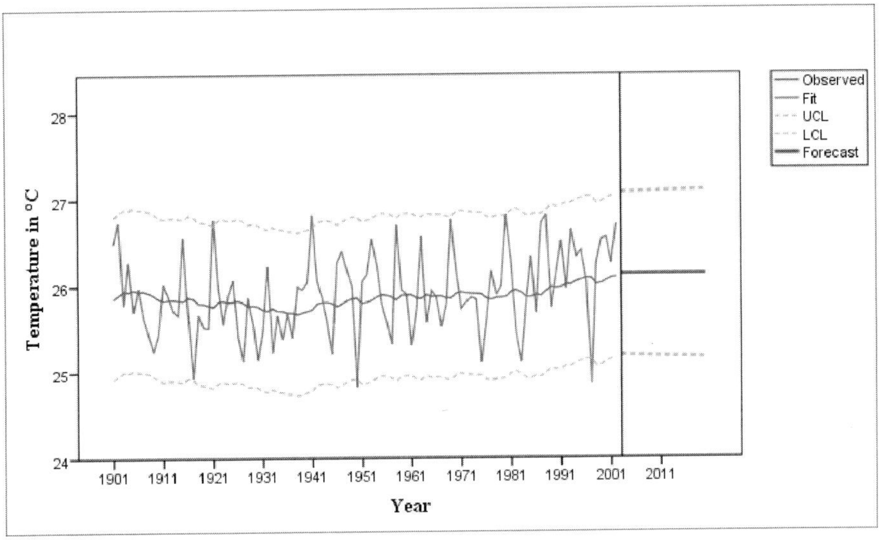

Fig. 2.29: Temperature Trend: Bikaner

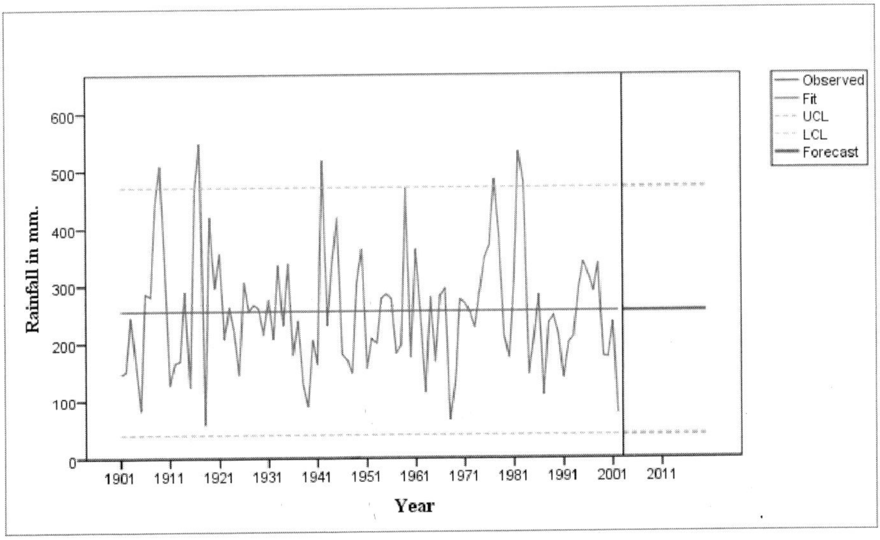

Fig. 2.30: Rainfall Trend: Bikaner

Temperature pattern fit line for Bundi district shows the variability in observed temperature and a moderate warming trend was forecasted (Fig. 2.31). Like other districts 1951 seems the rainfall divide for Bundi also, but no resultant wet or dry trend was observed at over all studied time scale (Fig. 2.32).

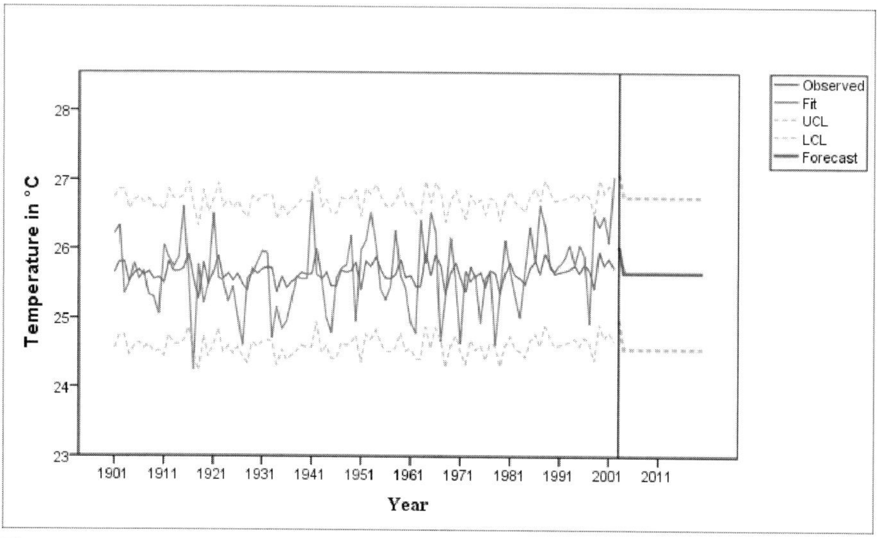

Fig. 2.31: Temperature Trend: Bundi

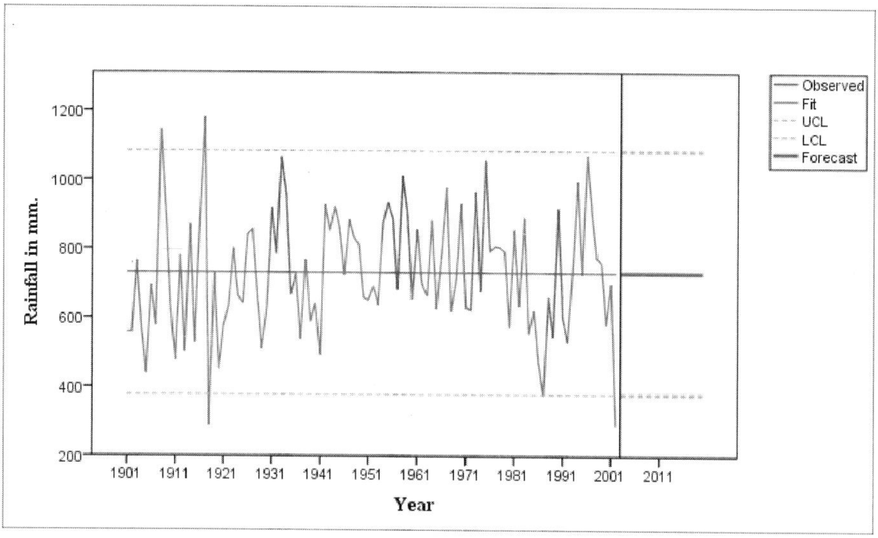

Fig. 2.32: Rainfall Trend: Bundi

Temperature pattern of Chittorgarh shows a rising trend after 1981 and a significant warming was recorded during the last century (Fig. 2.33). There was no significant trend noticed in rainfall pattern (Fig. 2.34)

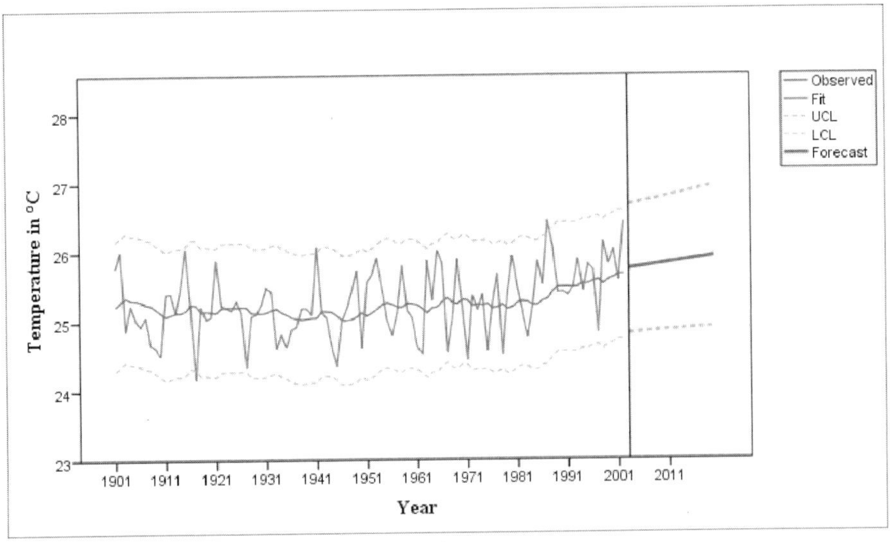

Fig. 2.33 : Temperature Trend: Chittorgarh

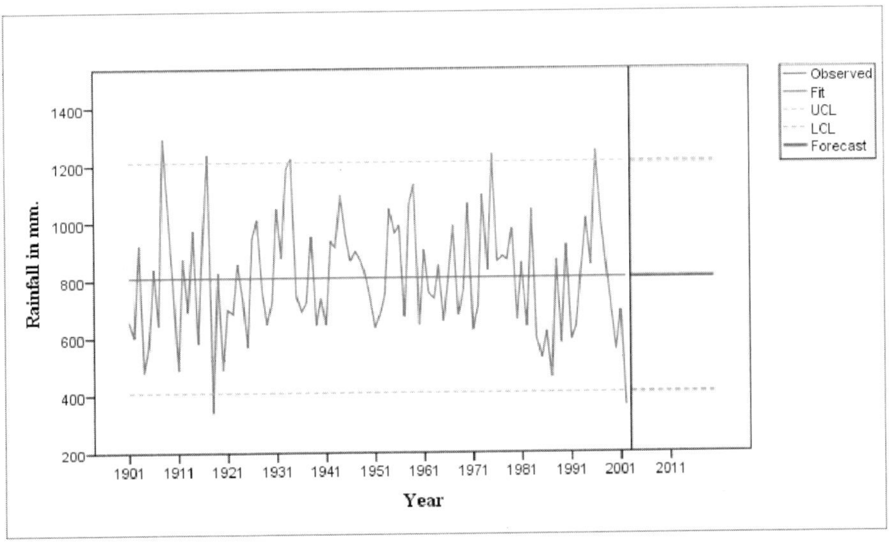

Fig. 2.34: Rainfall Trend: Chittorgarh

Churu showed no significant warming during the past century (Fig. 2.35). As far as rainfall is concerned, there was no dry or wet trend recorded (Fig. 2.36).

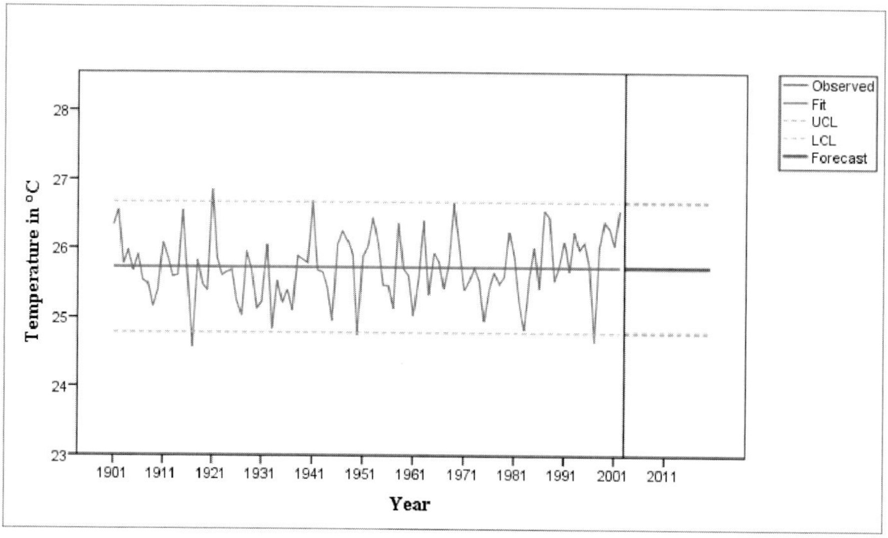

Fig. 2.35: Temperature Trend: Churu

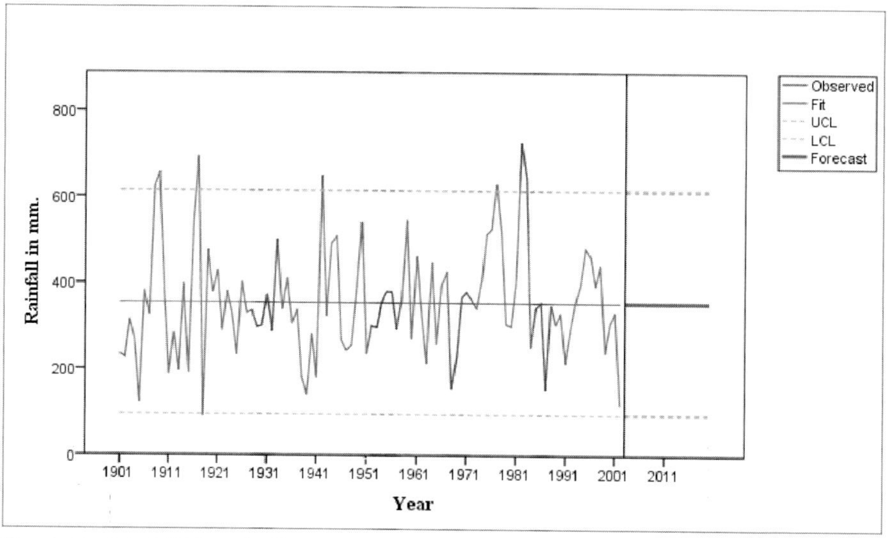

Fig. 2.36: Rainfall Trend: Chittorgarh

For Dausa, temperature variability was within confidence limits (95%) and no significant warming trend was noticed (Fig. 2.37). After 1921, rainfall variability showed almost a normal trend with moderate fluctuations and no up-down trend could be gained (Fig. 2.38).

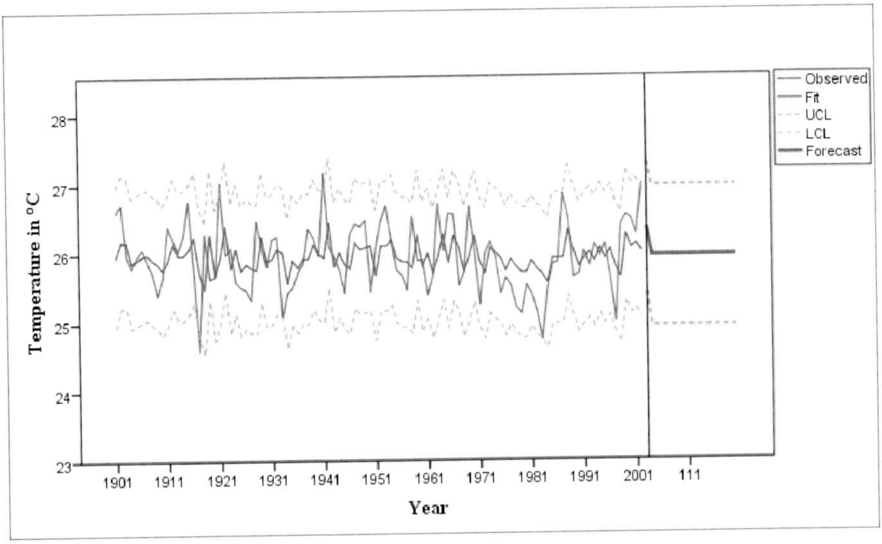

Fig. 2.37: Temperature Trend: Dausa

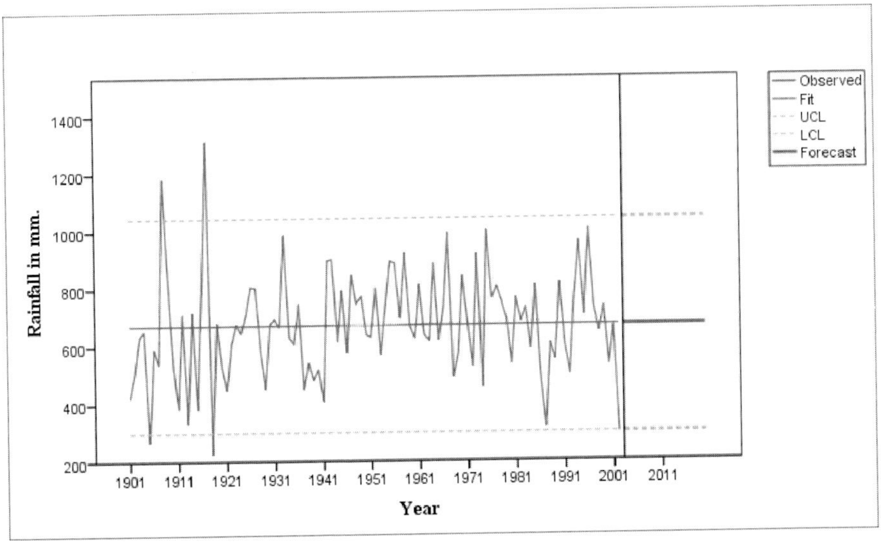

Fig. 2.38: Rainfall Trend: Dausa

Temperature pattern of Dholpur showed no significant warming trend and the fluctuations are within the confidence limit (Fig. 2.39). Although rainfall pattern recorded no positive or negative trend, but, four distinct phases of rainfall—no significant trend (up to 1940), increasing trend (1941 to 1971), declining trend (1971 to 1991), and again almost a stable trend after 1991, are evident (Fig. 2.40).

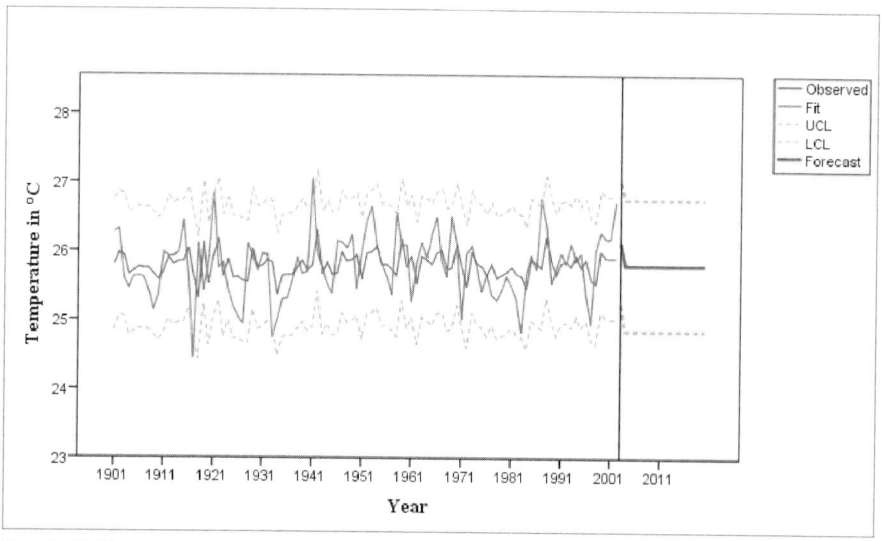

Fig. 2.39: Temperature Trend: Dhaulpur

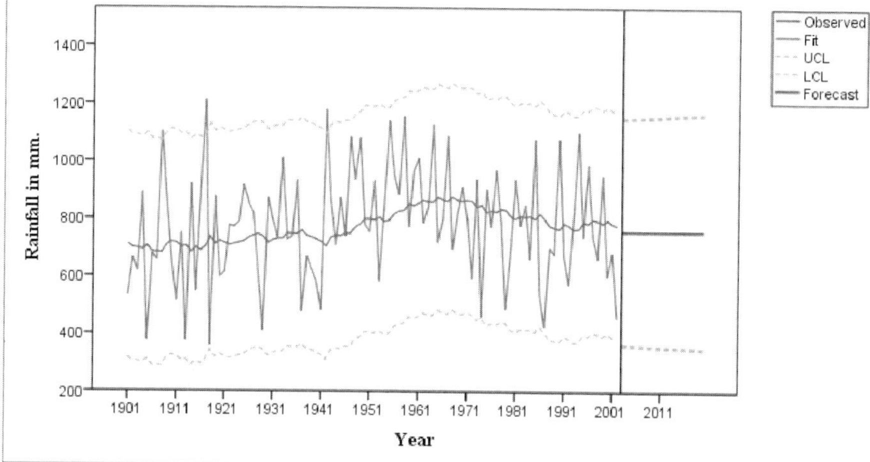

Fig. 2.40: Rainfall Trend: Dhaulpur

Temperature pattern of Dungarpur district has shown a clear warming trend during the past century and this trend is more evident after 1951 (Fig. 2.41), while rainfall pattern of the district recorded no significant trend during the period (Fig. 2.42).

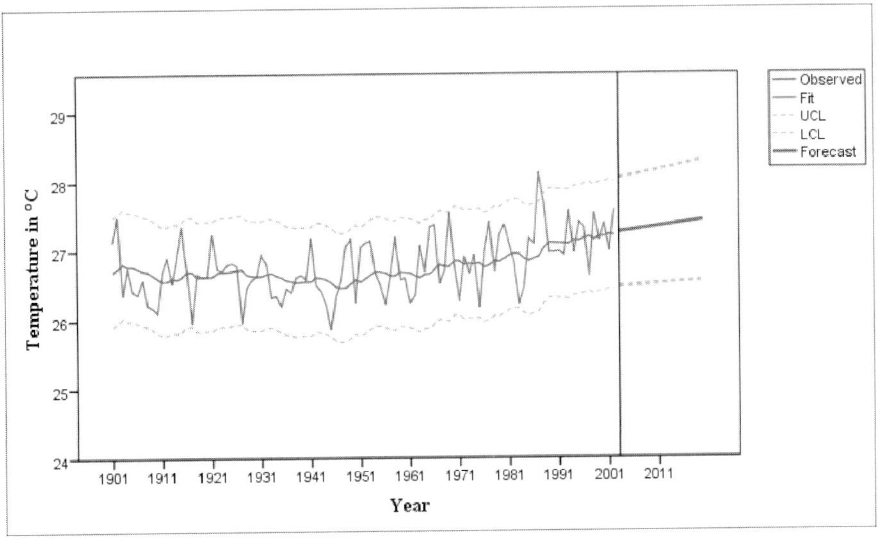

Fig. 2.41 Temperature Trend: Dungarpur

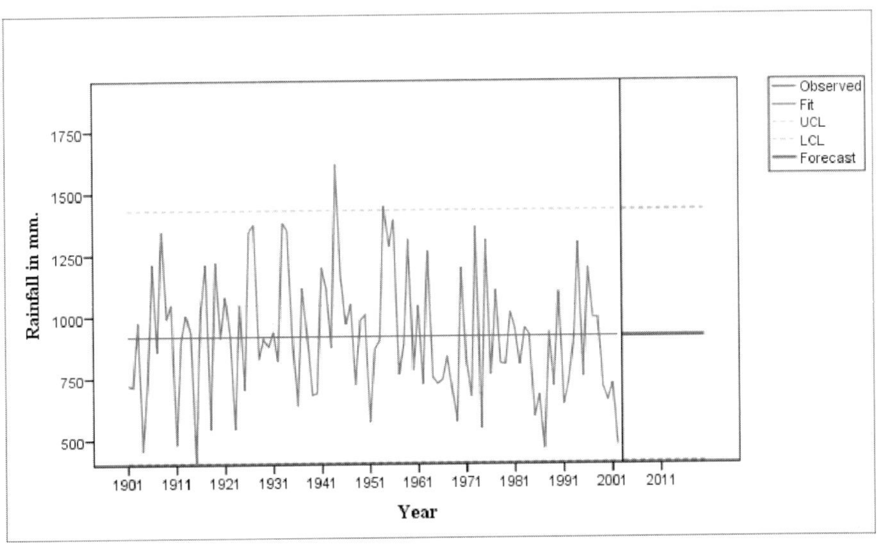

Fig. 2.42: Rainfall Trend: Dungarpur

Temperature record of Ganganagar shows no significant change in the variability (Fig. 2.43). The rainfall pattern too does not show any noticeable shift (Fig. 2.44).

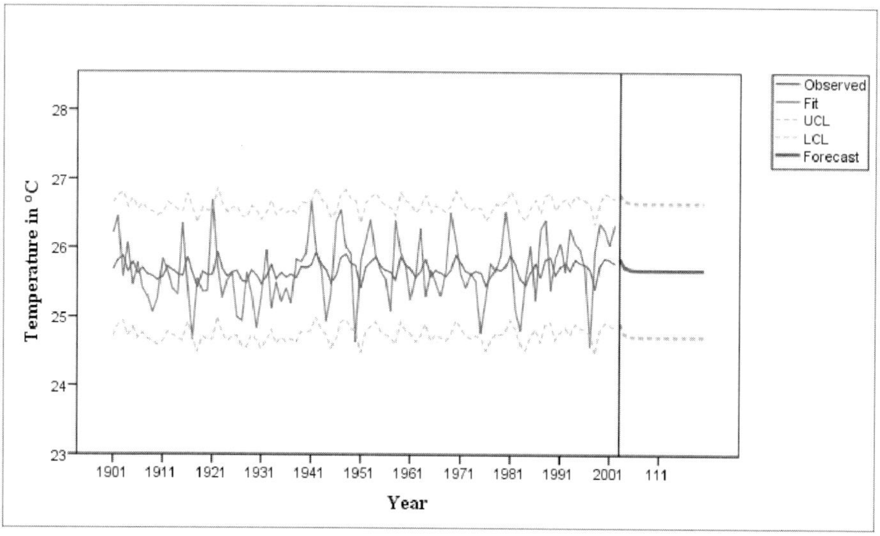

Fig. 2.43: Temperature Trend: Ganganagar

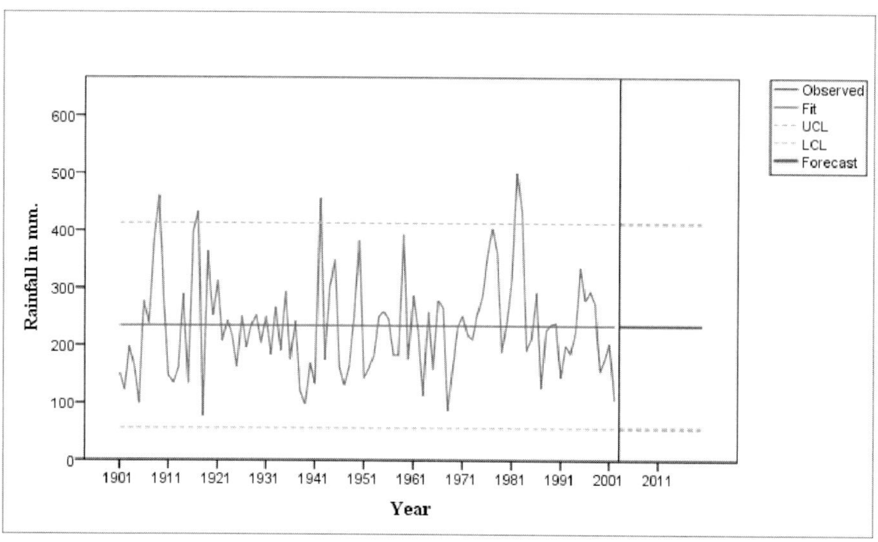

Fig. 2.44: Rainfall Trend: Ganganagar

Hanumangarh has shown no warming trend during the last century (Fig. 2.45). The rainfall pattern of the district too has not shown any up-down trend at centennial time scale, although fluctuations are evident (Fig. 2.46).

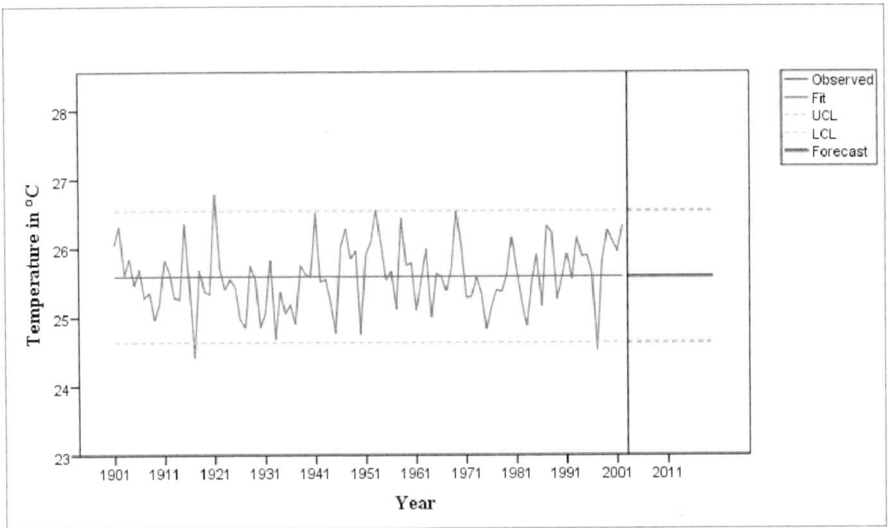

Fig. 2.45: Temperature Trend: Hanumangarh

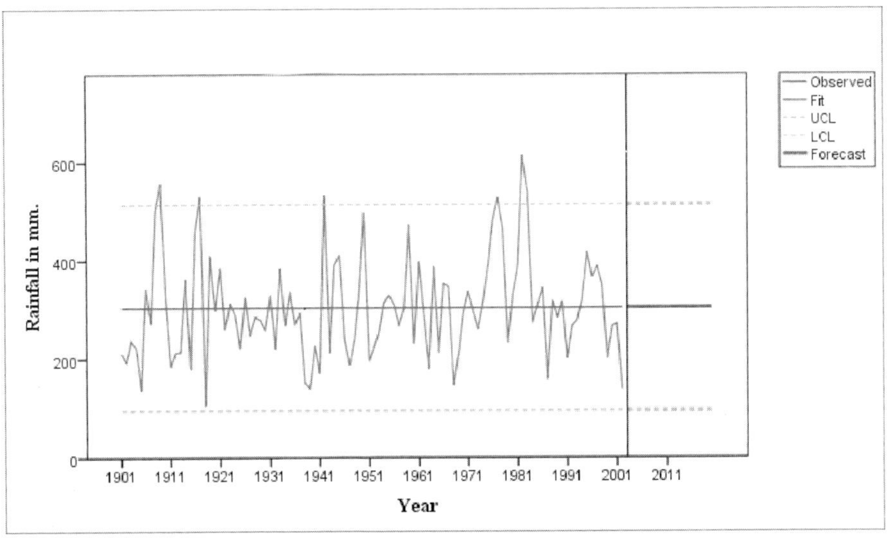

Fig. 2.46: Rainfall Trend: Hanumangarh

Temperature pattern of Jaipur shows no significant departure from the fit line and no warming or cooling trend was obtained with confidence (Fig. 2.47). After 1920, rainfall pattern shows moderate variability but no increasing or decreasing trend was predicted (Fig. 2.48)

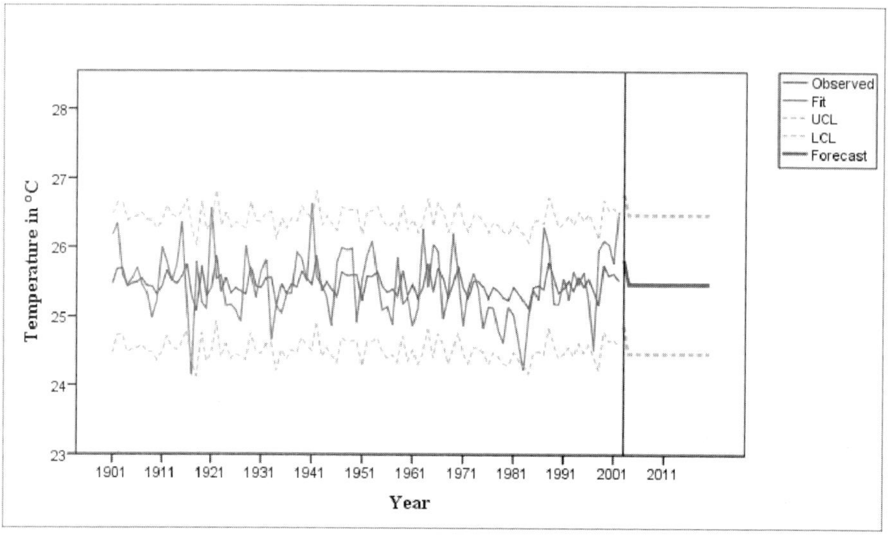

Fig. 2.47: Temperature Trend: Jaipur

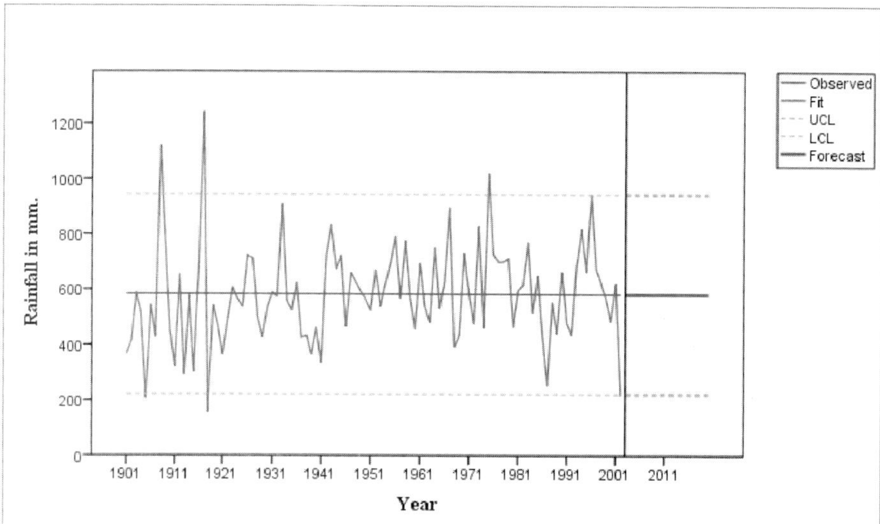

Fig. 2.48: Rainfall Trend: Jaipur

Temperature pattern of Jaisalmer shows a very significant warming trend at centennial scale. There are three distinct phases clearly visible—no change till 1931, warming trend between 1931-1951, cooling trend between 1951-1991, and again a steeper warming trend after 1991 (Fig. 2.49). Rainfall pattern indicated few years of heavy rainfall (observed rainfall line disobey upper confidence limit at some points), but at century level, there was no changing trend recorded (Fig. 2.50).

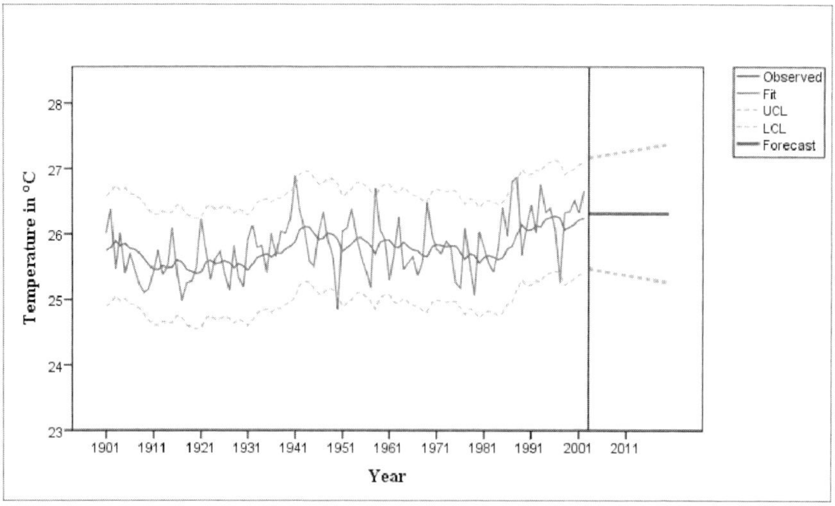

Fig. 2.49: Temperature Trend: Jaisalmer

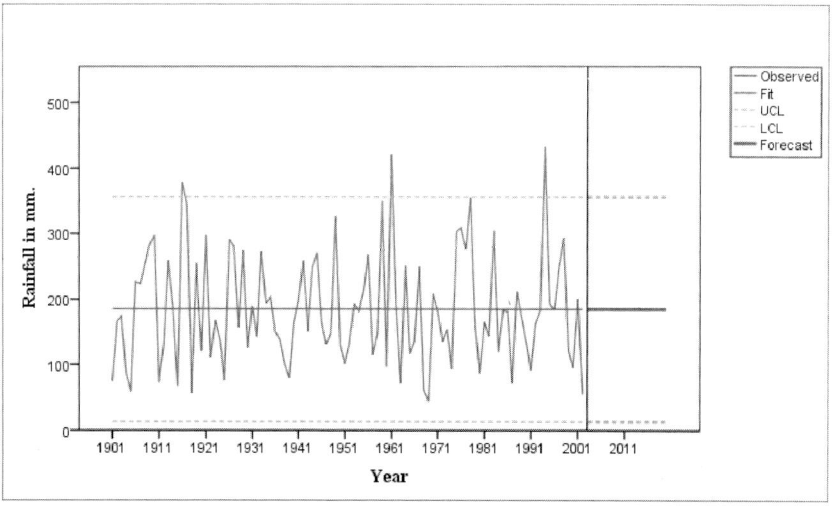

Fig. 2.50: Rainfall Trend: Jaisalmer

Past century temperature pattern of Jalore district shows significant warming trend (Fig. 2.51), although there was no noticeable variability in the rainfall pattern recorded (Fig. 2.52).

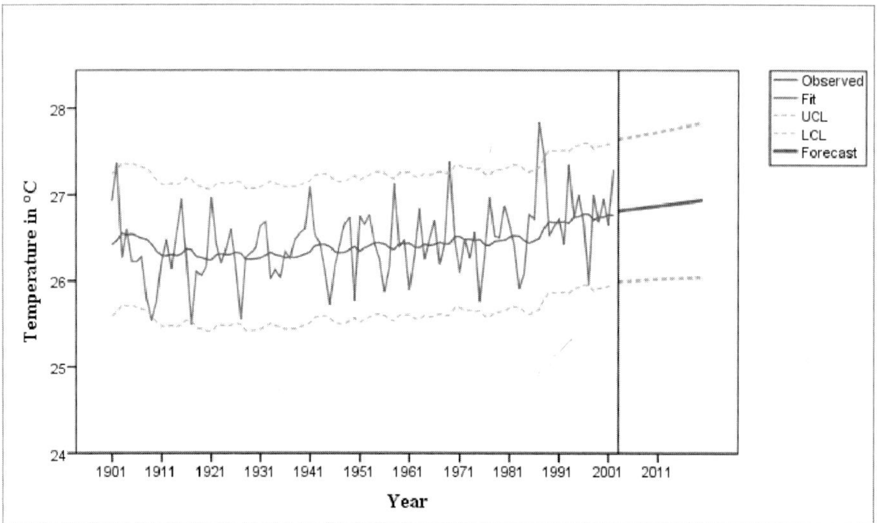

Fig. 2.51: Temperature Trend: Jalore

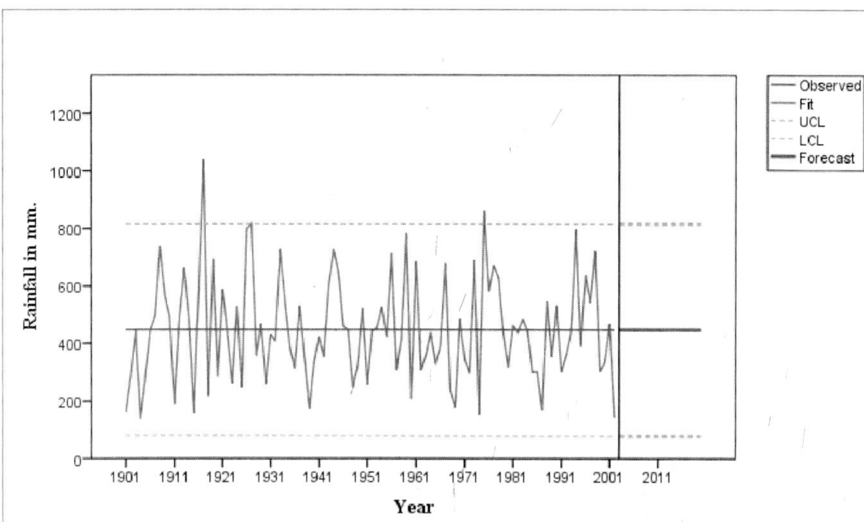

Fig. 2.52: Rainfall Trend: Jalore

Jhalawar showed almost a stable temperature trend till 1951 and a significant warming trend after that. At centennial scale the district has noticed significant temperature increase (Fig. 2.53). Fluctuations in rainfall is quite obvious in the pattern graph and an insignificant but negative trend was noticed in the rainfall-time correlation analysis, however, the model could not establish any significant increasing or declining trend (Fig. 2.54).

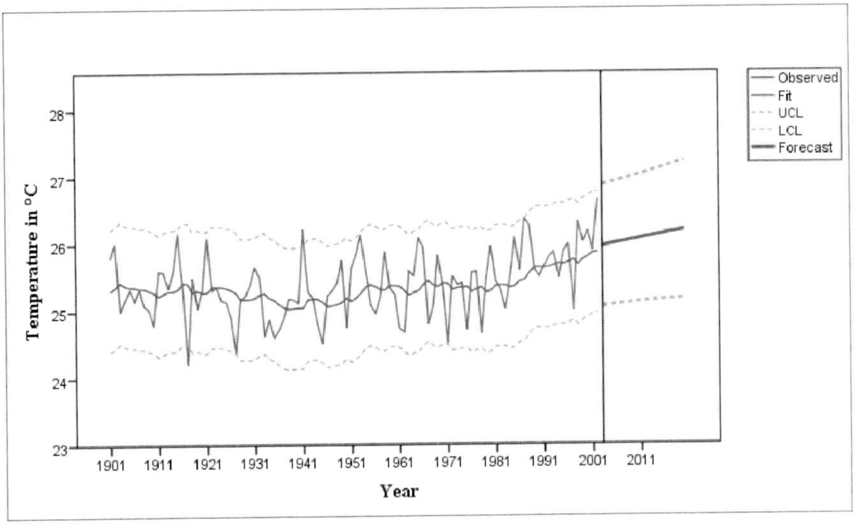

Fig. 2.53: Temperature Trend: Jhalawar

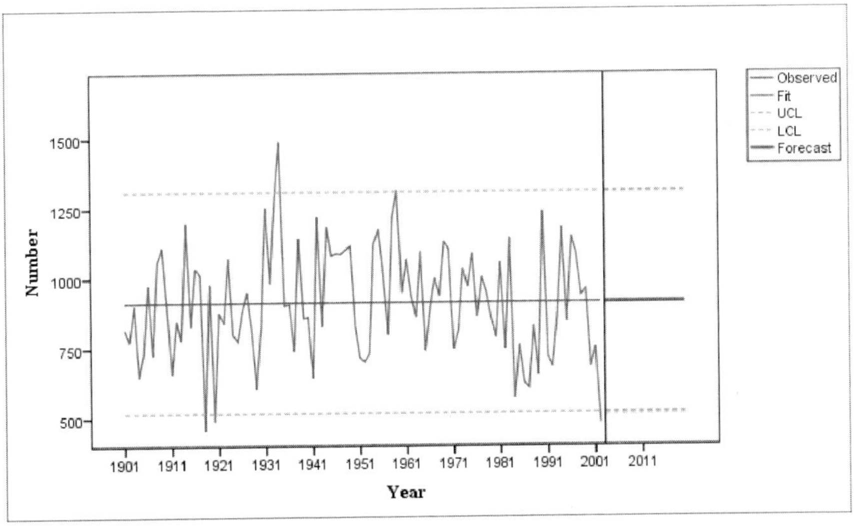

Fig. 2.54: Rainfall Trend: Jhalawar

Temperature pattern of Jhunjhunu showed no warming or cooling trend at century scale, however, some fluctuations beyond confidence limit were noticed (Fig. 2.55). The rainfall pattern showed varying variability with time but over all no significant trend was recorded (Fig. 2.56).

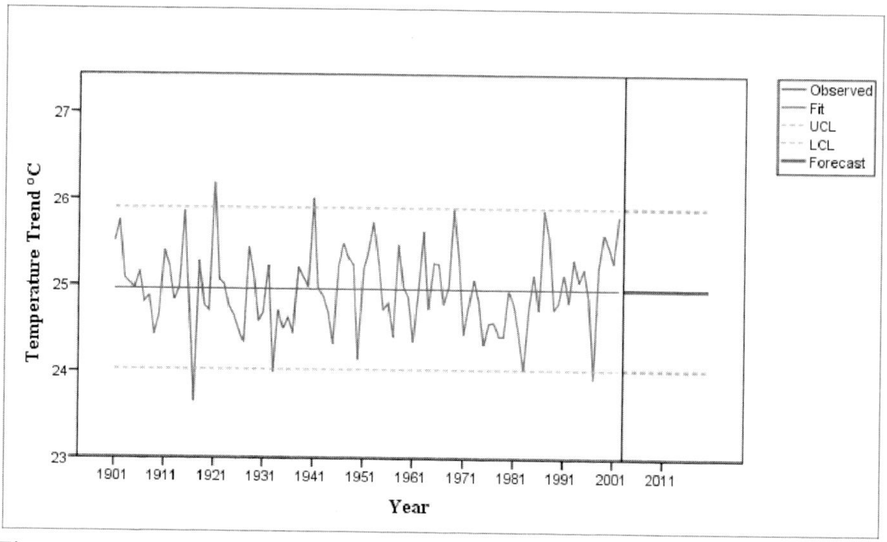

Fig. 2.55: Temperature Trend: Jhunjhunu

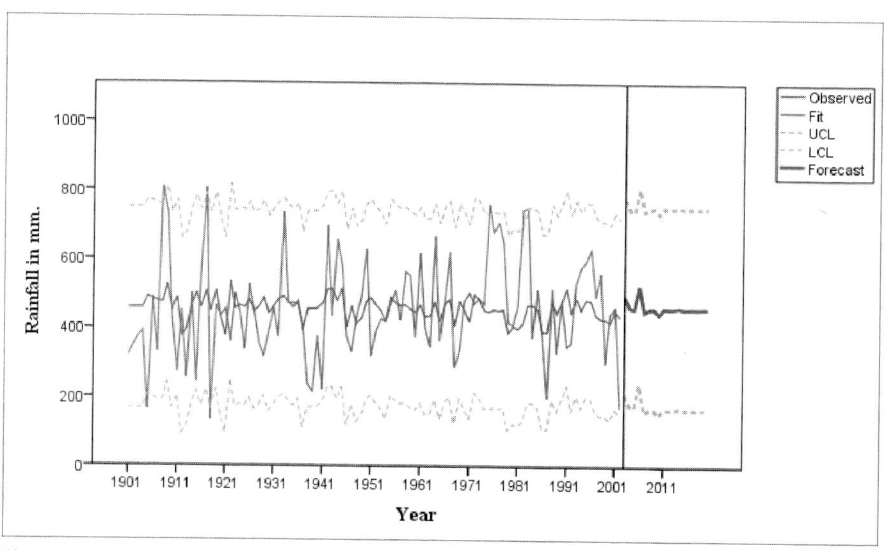

Fig. 2.56: Rainfall Trend: Jhunjhunu

Perusal of temperature record of Jodhpur portrays three different phases of temperature pattern—first between 1901-1941, second between 1941-1981, and third after 1981, and the model predicted a significant warming trend for coming years (Fig. 2.57). Rainfall record of the district did not identify any significant change in the pattern (Fig. 2.58).

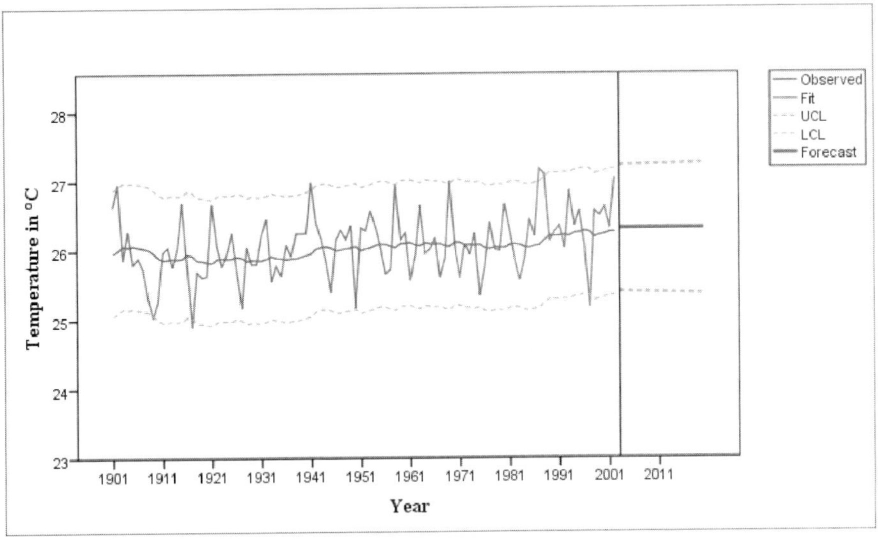

Fig. 2.57: Temperature Trend: Jodhpur

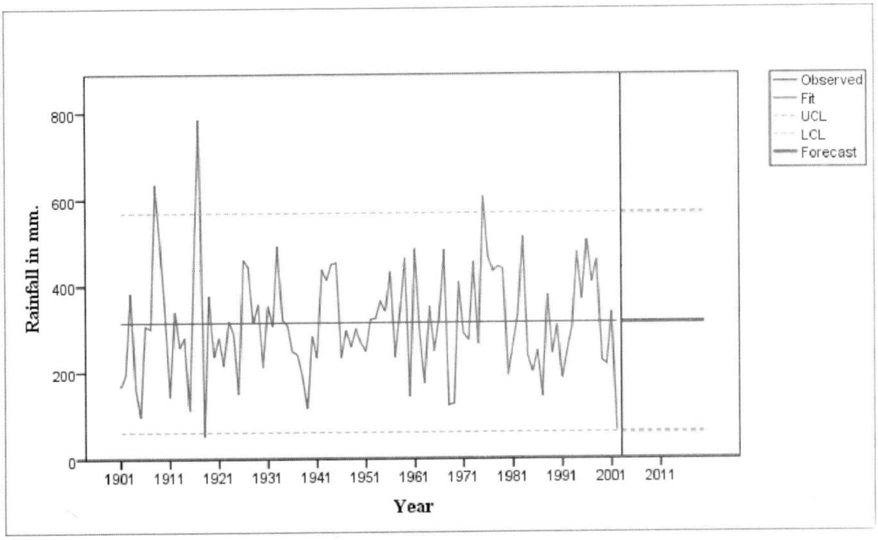

Fig. 2.58: Rainfall Trend: Jodhpur

Temperature record of Karauli did not report any significant warming or cooling trend and fluctuations were reported within the confidence limit (Fig. 2.59). Although 1941 seems a natural divide in rainfall pattern, modelling result did not report any increasing or decreasing trend for near future (Fig. 2.60).

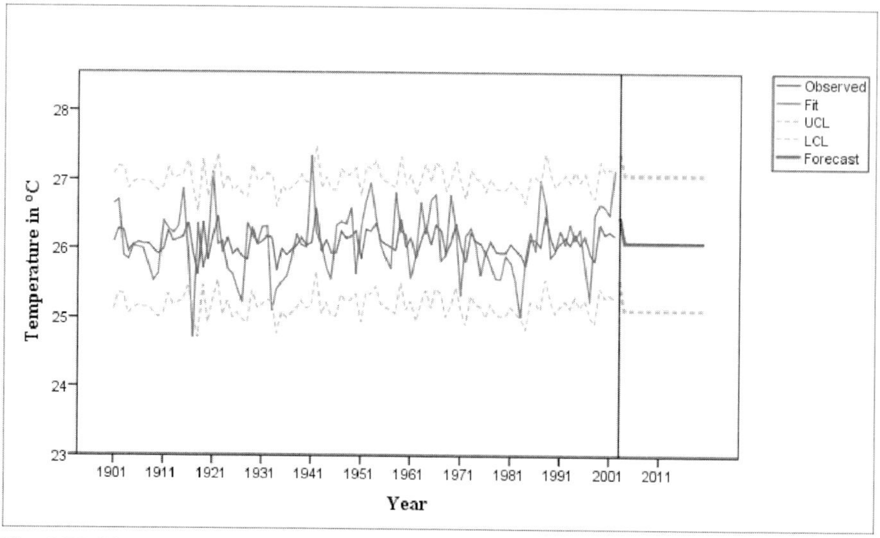

Fig. 2.59: Temperature Trend: Karauli

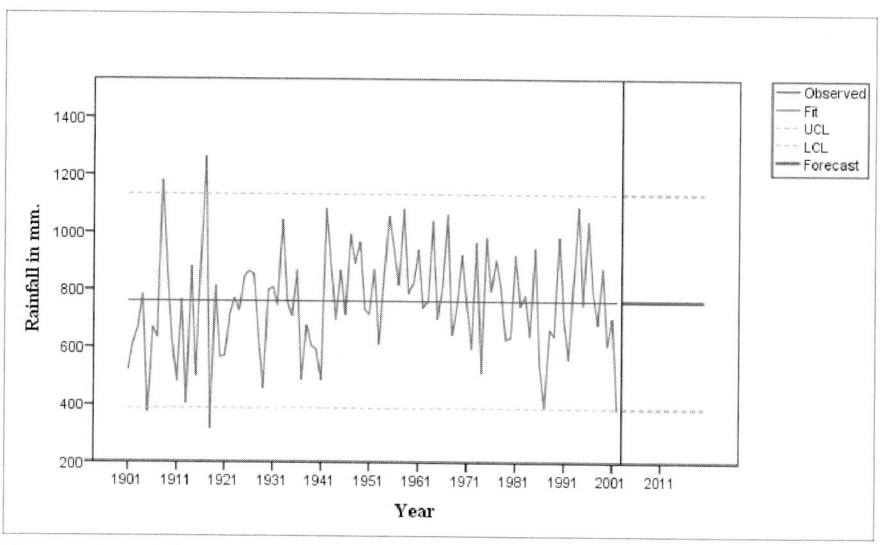

Fig. 2.60: Rainfall Trend: Karauli

Temperature pattern of Kota showed a significant warming after 1981, and noticeable rise in temperature was predicted (Fig. 2.61). The model reported no increasing or declining trend for rainfall on centennial scale (Fig. 2.62).

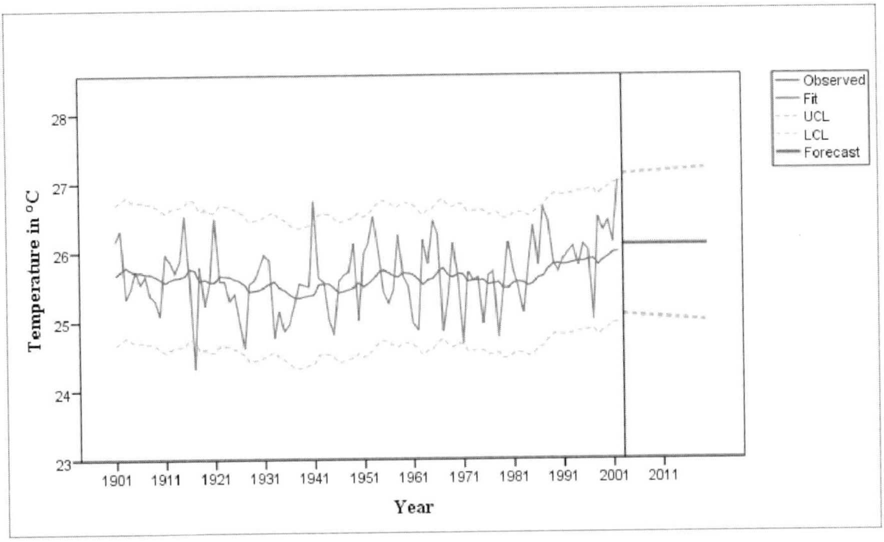

Fig. 2.61: Temperature Trend: Kota

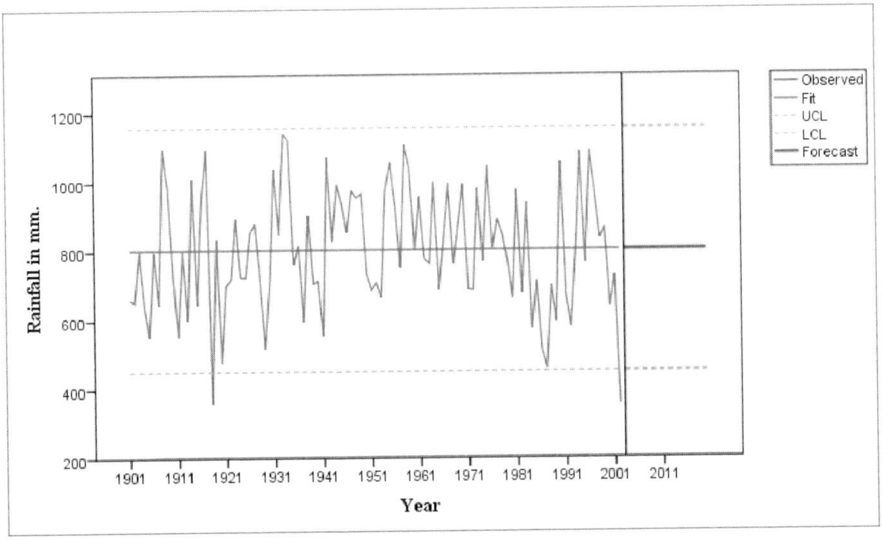

Fig. 2.62: Rainfall Trend: Kota

No significant change in the variability pattern of temperature of Nagaur was recorded and the model could not establish warming or cooling trend with confidence (Fig. 2.63). Rainfall record although showed some fluctuations beyond the confidence limit but over all no increasing or declining trend was noticed (Fig. 2.64).

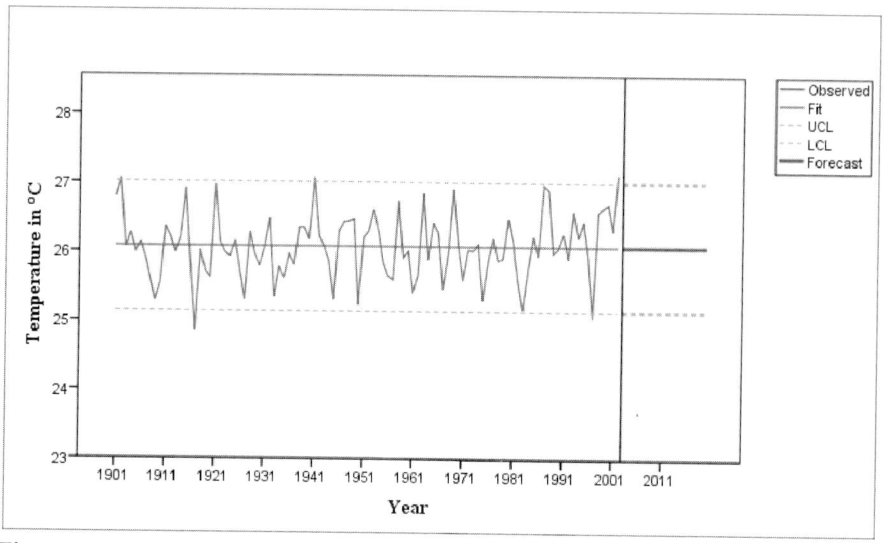

Fig. 2.63: Temperature Trend: Nagaur

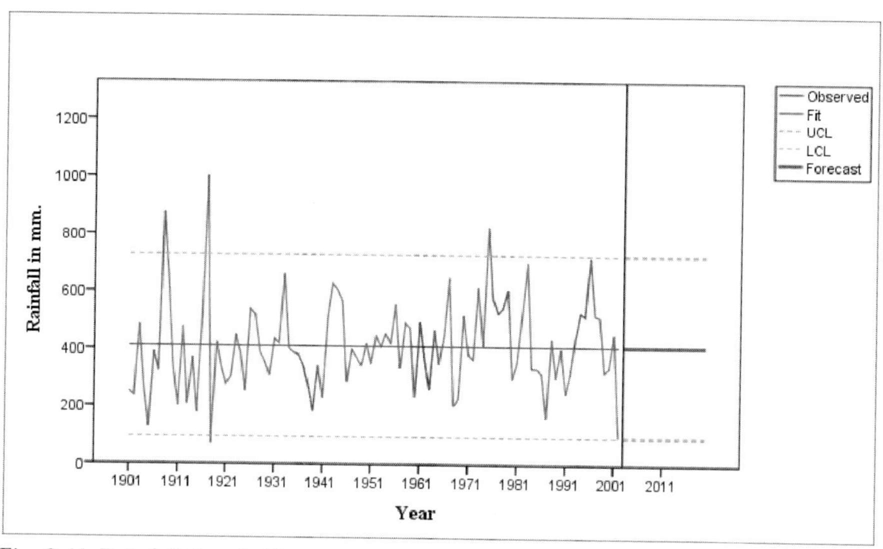

Fig. 2.64: Rainfall Trend: Nagaur

Pali has shown a continuous rise in the temperature and a significant warming trend was identified (Fig. 2.65). Rainfall pattern did not record any significant change yet some heavy rainfall years were observed (Fig. 2.66).

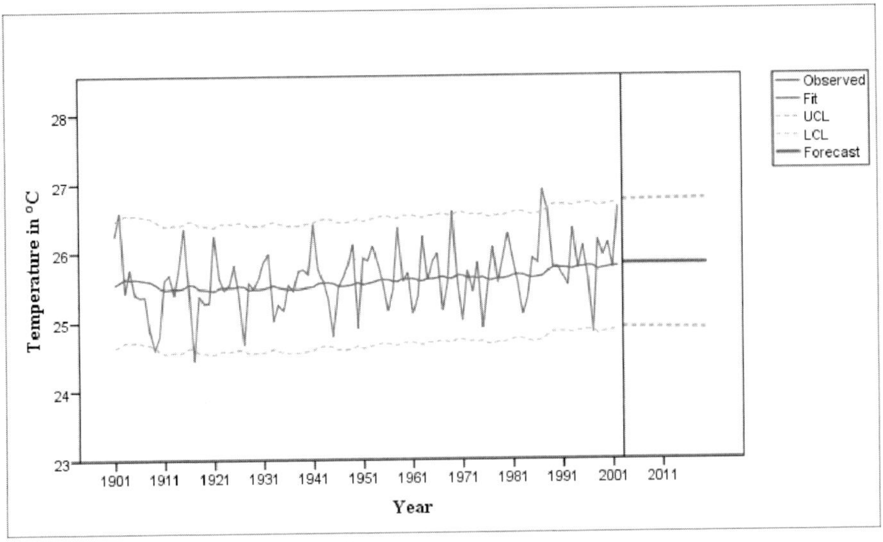

Fig. 2.65: Temperature Trend: Pali

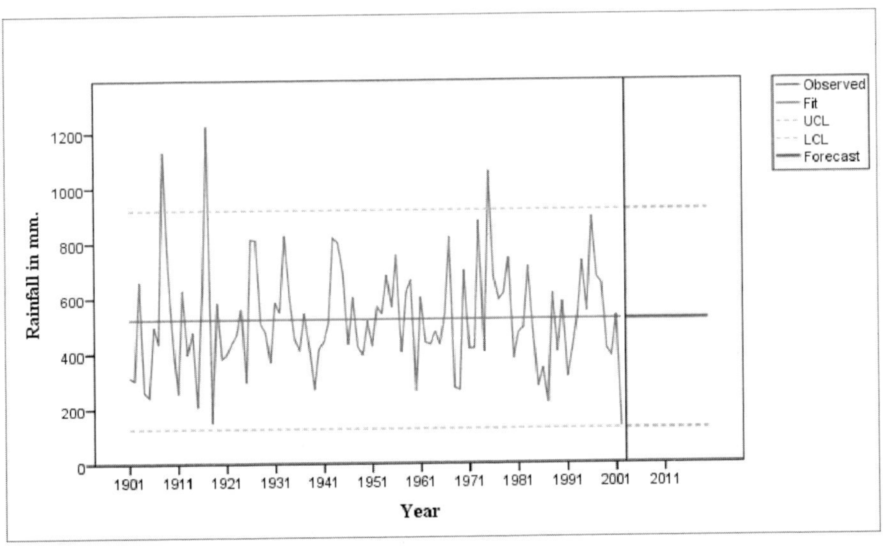

Fig. 2.66: Rainfall Trend: Pali

The district of Rajsamand showed a warming trend (Fig. 2.67), yet no significant change in the rainfall was observed (Fig. 2.68).

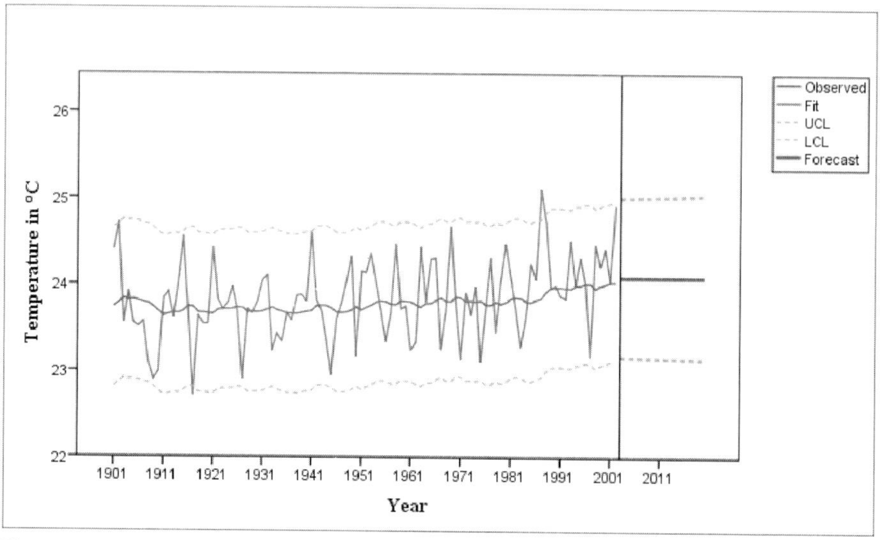

Fig. 2.67: Temperature Trend: Rajsamand

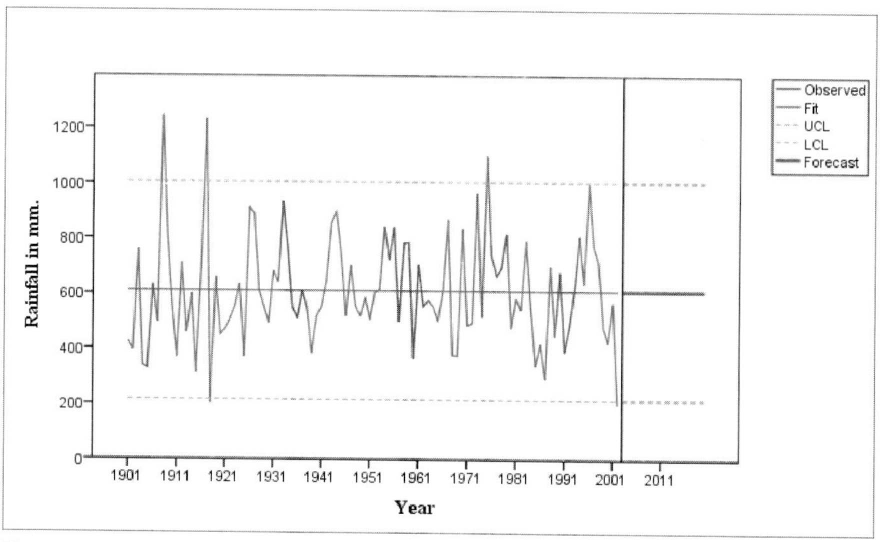

Fig. 2.68: Rainfall Trend: Rajsamand

Zigzag Temperature pattern fit line of Sawai Madhopur district shows fluctuations, however, at centennial level no significant departure from normal trend was recorded (Fig. 2.69). The rainfall pattern also indicate no significant change in the trend however, seasonality in pattern is easily visible (Fig. 2.70).

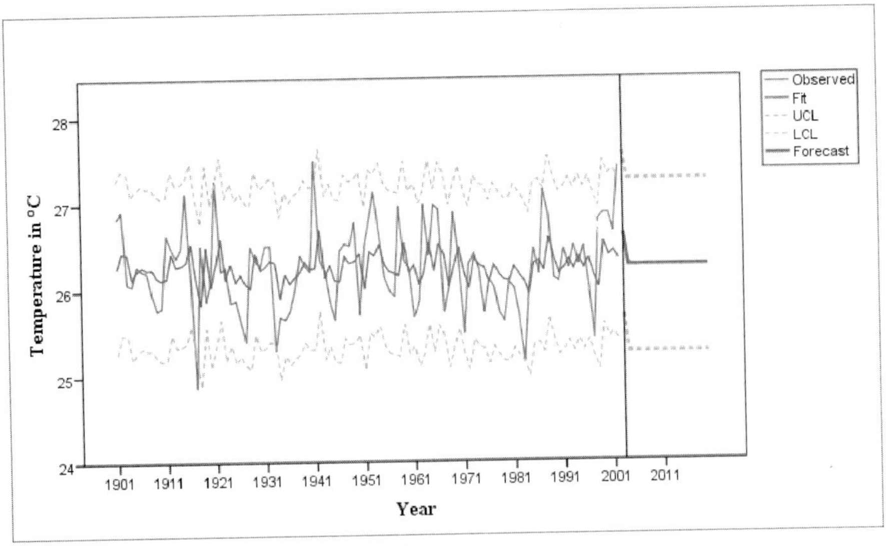

Fig. 2.69: Temperature Trend: Sawai Madhopur

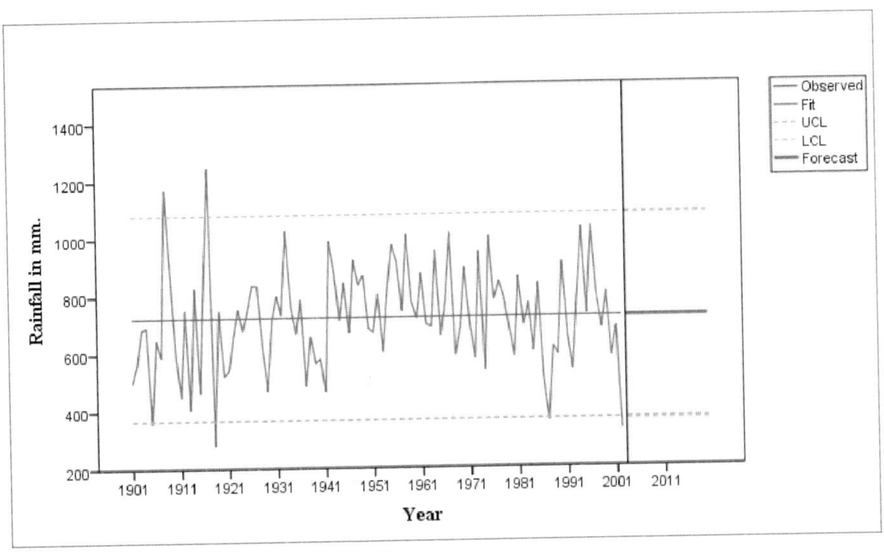

Fig. 2.70: Rainfall Trend: Sawai Madhopur

Sikar has not recorded any measurable departure from the centennial averaged temperature and no changing trend could be identified, however, a very weak but negative correlation between temperature and time series was obtained (Fig. 2.71). Rainfall pattern showed no noticeable trend although few years of high rainfall were reported (Fig. 2.72).

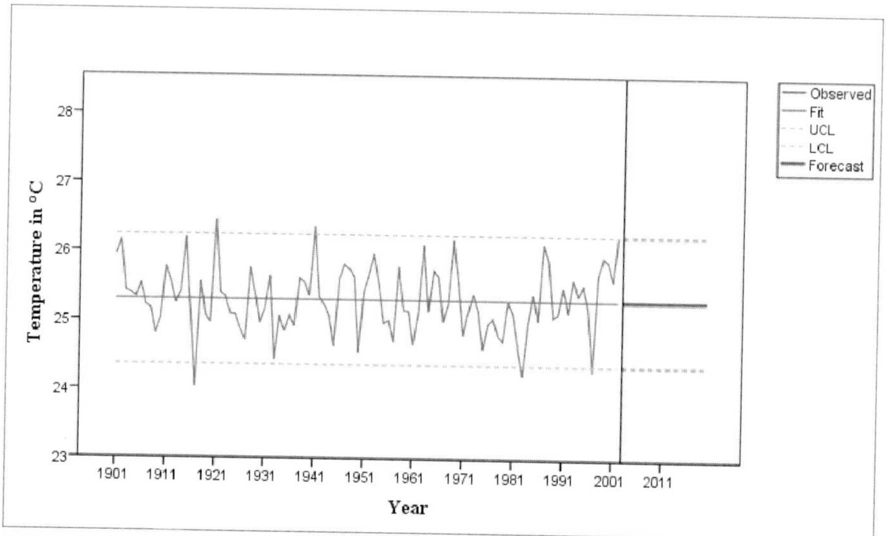

Fig. 2.71: Temperature Trend: Sikar

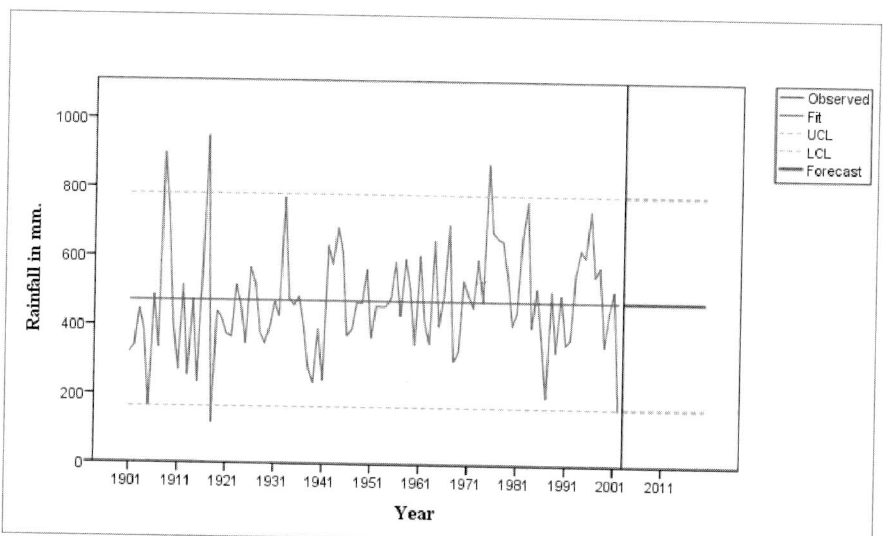

Fig. 2.72: Rainfall Trend: Sikar

Temperature pattern of Sirohi indicates a rising trend after 1951, and a significant warming trend was forecasted by the model (Fig. 2.73). Rainfall pattern of the district showed equally spaced variations, however, rainfall-time correlation produced insignificant but negative value showing an unnoticeable decline in the rainfall (Fig. 2.74).

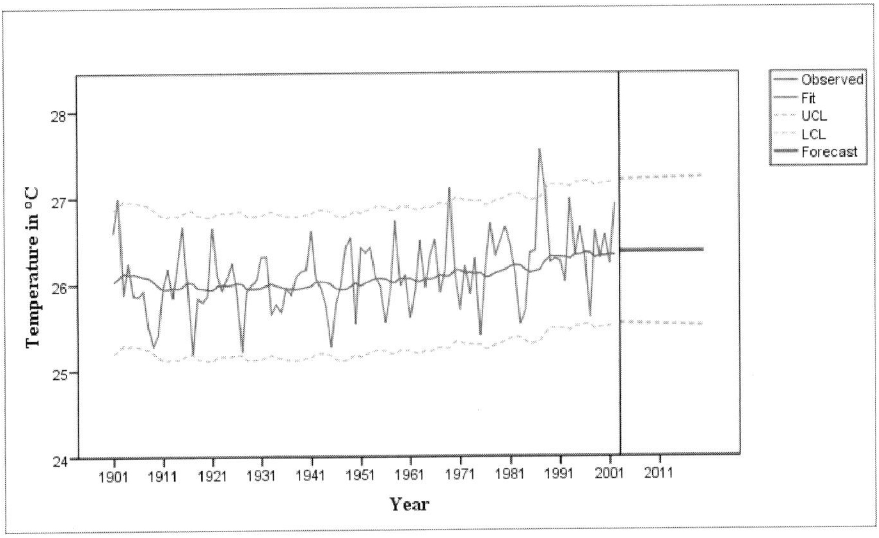

Fig. 2.73: Temperature Trend: Sirohi

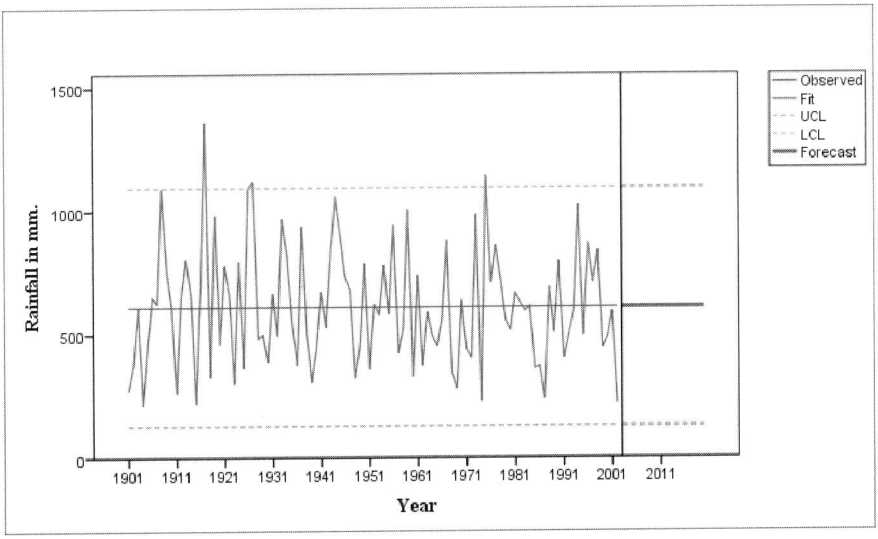

Fig. 2.74: Rainfall Trend: Sirohi

Temperature pattern of Tonk displays fluctuations in the confidence level in determining the real trend and the model forecasted no significant change from the normal trend (Fig. 2.75). Rainfall pattern showed no major shift in the variability and no trend was obtained (Fig. 2.76).

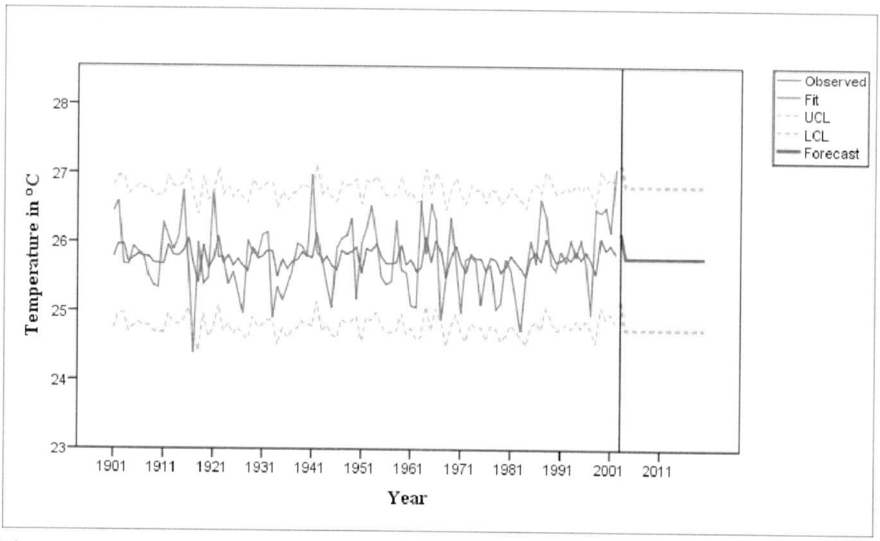

Fig. 2.75: Temperature Trend: Tonk

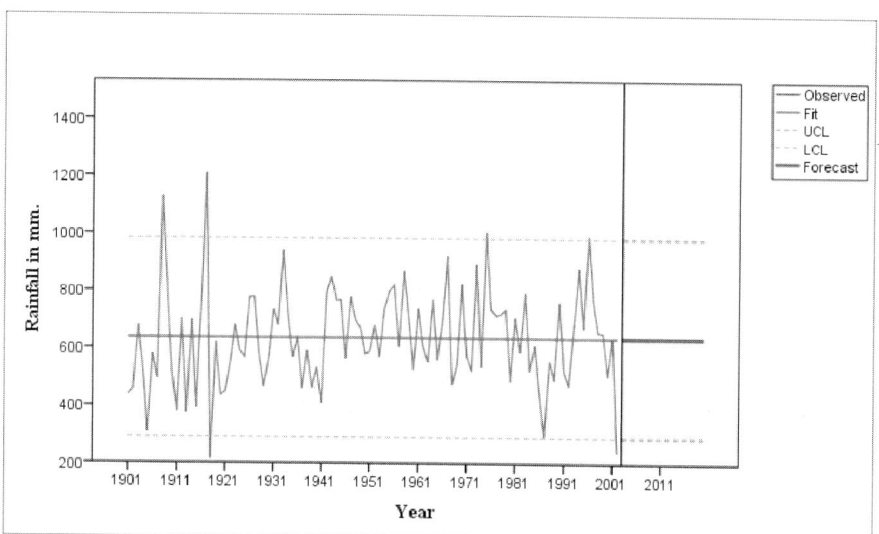

Fig. 2.76: Rainfall Trend: Tonk

Temperature record of Udaipur shows a continuous rising trend after 1951 and a significant warming trend was recorded during the last century (Fig. 2.77). Observed rainfall pattern follow the confidence limits and no significant shifting trend from the mean value was recorded, although a weak negative rainfall-time correlation was obtained in the analysis (Fig. 2.78).

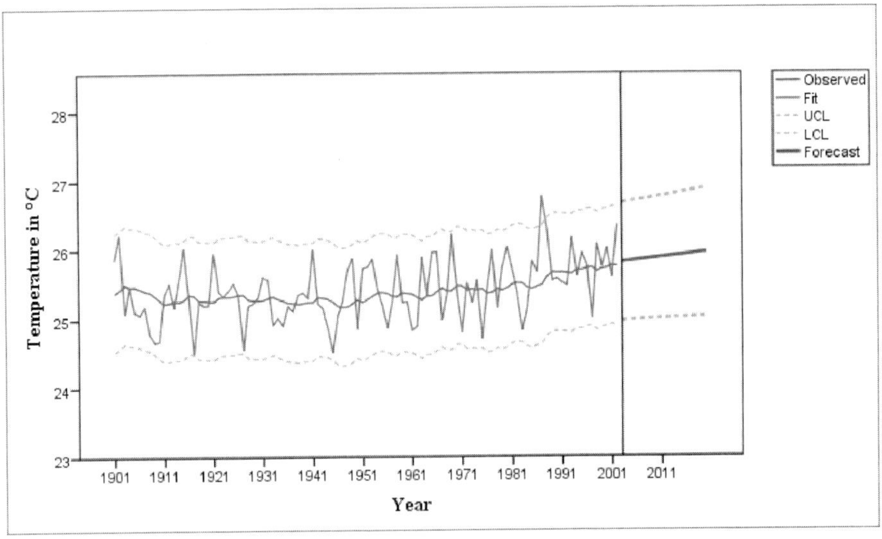

Fig. 2.77: Temperature Trend: Udaipur

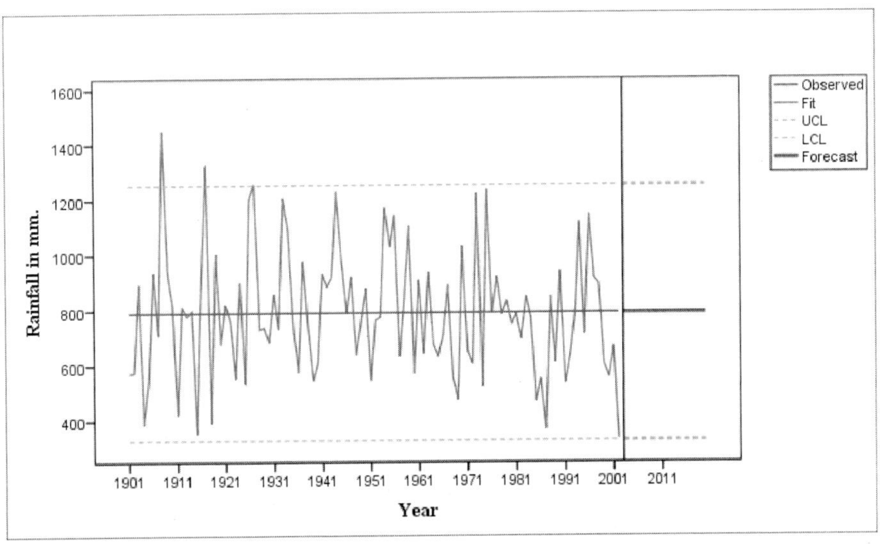

Fig. 2.78: Rainfall Trend: Udaipur

Causes of Climate Change in Rajasthan

Climate change is a dynamic phenomenon. Climate has been changing ever since the earth was born due to natural causes and anthropogenic factors. The natural factors have been mainly responsible for long-term climate changes. However, the short-term climate changes are generally attributed to anthropogenic factors such as energy generation, through transport, industrialization, urbanization, agricultural and increase in livestock population, land-use changes, forestry and population explosion.

The population of Rajasthan has been increasing at a fast rate than national average. The population in 1901 was only 10.29 million which maintained the same with slight decline in 1921. After 1921 phenomenal growth of population has been recorded. The population in 1951 was 15.79 million which increased to 34.26 million in 1981 and 68.62 million in 2011. The decadal growth rate of population recorded during the previous decade at 28.41 per cent was higher than the national average. However, the growth rate declined to 21.44 per cent between 2001-2011 in Rajasthan but still higher than the national level of 17.64 per cent. The growth rate of population between 2001-2011 has been recorded highest in Barmer (32.55%) followed by Jaisalmer (32.22%), Jodhpur (27.69%), Jaipur (26.91%), Banswara (26.58%), Jalore (26.31%), Dungarpur (25.39%), Bikaner (24.48%), Kota (24.34%), Udaipur (23.63%), Pratapgarh (22.84%), Dholpur (22.84%), Alwar (22.75%) and Sirohi (21.86%). Except Jaipur, Kota, Alwar, Dholpur and Dausa districts most of the western and southern districts have recorded growth rate of population higher than the state and national, meaning by that there is pressure of population of resources such as forests and land resources. The decadal growth rate of urban population in Rajasthan during 2001-2011 was recorded as 29.26 per cent but recorded lower than national level of 30.80 per cent. The future demographic projections suggest a further increase of more than 100 per cent by 2051 in Rajasthan, suggesting increase pressure on resources of Rajasthan.

The social, environmental and economic aspects of population growth are influenced not only by an excess of people but also by the geographical distribution of population in rural areas, cities and towns. The influx of people to urban areas in search of jobs has out stripped the capacity of urban centres of Rajasthan to provide basic services.

Greenhouse Gas Emission in Rajasthan

Total green house gas emission from the anthropogenic activities in Rajasthan in 1990 amounted to 18.6 million tons of CO_2, 827.9 thousand tons of CH_4 and 6.6 thousand tons of N_2O. In terms of CO_2 equivalent emissions amount to 38.0 million tons (Garg, Shukla). According to similar estimates aggregate emission from anthropogenic activities in Rajasthan in 1995 amounted to 27.0

million tons of the CO_2, 1044 thousand tons of CH_4, 10.5 thousand tons of N_2O. In terms of CO_2 equivalent, emissions in 1995 amount to 52.2 million tons. Rajasthan ranked 10th with 3.7 per cent of India's emission in 1990 and ranked 9th with 4.2 per cent of India's emission in 1995. On the sectoral basis, emission were estimated from the energy sector, agriculture sector, industrial processes, waste disposal activities and from land-use, land-use change and forestry sector. Key districts with emission in 1990 include Ganganagar, Jaipur, Ajmer, Jodhpur, Nagaur, Udaipur, Chittorgarh and Kota (Total Emission in these districts was greater than 1.5 million tons CO_2 equivalent). Table 2.9 summarizes district-wise emission for these two years. These estimates were made in the context of the first national communication to the UNFCCC. The only estimates or updated information on GHG emission is for the year 2007 as part of the INCCA publication India's green house gas emission 2007. These estimates suggest that there is about 3 per cent increase in 2007 as compared to

Table 2.9: Emissions inventory for districts for 1990 and 1995 in Rajasthan

Sl. No.	*District*	*CO_2*	*CO_2*	*CH_4*	*CH_4*	*N_2O*	*N_2O*	*NO_x*	*NO_x*	*SO_2*	*SO_2*	*CO_2 (Eq)*	*CO_2 (Eq)*
	Year	1990	1995	1990	1995	1990	1995	1990	1995	1990	1995	1990	1995
	Unit	*(MT)*	*(MT)*	*(Tons)*	*(Tons)*	*(Tons)*	*(Tons)*	*(Tons)*	*(Tons)*	*(Tons)*	*(Tons)*	*(MT)*	*(MT)*
1.	Ganganagar	0.81	0.65	46.9	63.4	0.63	1.36	6.3	7.1	7	4.1	2.03	2.43
2.	Bikaner	0.33	0.38	27.1	35.1	0.09	0.09	2.5	3.6	2	2.3	0.93	1.16
3.	Churu	0.26	0.22	26.5	37.5	0.13	0.12	2.2	2.5	1.7	1.5	0.86	1.06
4.	Jhunjhunu	0.31	0.28	21.2	27.9	0.15	0.19	2.4	2.8	16.5	23.3	0.81	0.93
5.	Alwar	0.41	0.5	34.5	43.7	0.34	0.54	3.9	5	3.1	4.5	1.26	1.61
6.	Bharatpur	0.29	0.27	26.7	32.9	0.26	0.46	2.5	3	2	1.8	0.93	1.12
7.	Dholpur	0.07	0.1	12.8	15.2	0.15	0.18	0.8	1.2	0.6	0.7	0.39	0.48
8.	Sawai Madhopur	0.31	0.29	37.1	47.1	0.41	0.64	3.2	3.7	2.2	2	1.26	1.5
9.	Jaipur	1.59	1.97	72.9	92.4	0.37	0.69	11.9	17.2	9.4	11.6	3.28	4.17
10.	Sikar	0.3	0.29	29	36.1	0.16	0.26	2.7	3.1	2	1.8	0.96	1.14
11.	Ajmer	1.26	1.63	30	40.3	0.15	0.23	4.8	7.1	5.5	7	1.94	2.57
12.	Tonk	0.16	0.17	23	26.7	0.13	0.25	1.5	1.9	1.1	1.1	0.68	0.82
13.	Jaisalmer	0.08	0.1	10.7	16.9	0.1	0.03	0.8	1.1	0.5	0.7	0.34	0.48
14.	Jodhpur	0.81	0.98	32.9	44.9	0.24	0.32	6.3	8.4	5.5	6.9	1.57	2.04
15.	Nagaur	1.01	1.17	38.1	47.3	0.21	0.39	4.4	5.2	4.4	4.8	1.9	2.31
16.	Pali	0.32	0.44	29.4	36	0.23	0.38	2.7	4.4	2.3	3	1.01	1.33
17.	Barmer	0.2	0.23	23	35.5	0.13	0.11	2.1	2.9	1.4	1.6	0.72	1.03
18.	Jalore	0.22	0.33	19.8	24.3	0.15	0.28	2.3	3.8	1.5	2.2	0.68	0.94
19.	Sirohi	0.66	0.87	10.4	12.4	0.09	0.18	2.1	3.6	2.5	3.5	0.9	1.19
20.	Bhilwara	0.29	0.34	39.6	47.3	0.28	0.39	2.7	3.8	6.9	7.2	1.22	1.47
21.	Udaipur	1.03	1.12	59.7	70	0.3	0.43	6.3	9.1	22.5	23.7	2.42	2.75
22.	Chittorgarh	3.24	4.44	38.5	43.2	0.54	0.58	6.4	9.3	11.3	15.1	4.22	5.55
23.	Dungarpur	0.1	0.09	22	24.3	0.12	0.13	1.2	1.4	0.7	0.6	0.6	0.65
24.	Banswara	0.14	0.13	28.6	30.9	0.34	0.4	1.8	2	1.1	1	0.85	0.91
25.	Bundi	0.41	0.5	20.7	23.5	0.28	0.44	1.8	2.3	1.7	2	0.94	1.14
26.	Kota	3.83	9.34	42.6	62.8	0.48	0.85	1.3	23.9	24.7	44	4.88	10.95
27.	Jhalawar	0.14	0.14	23.6	26.8	0.18	0.35	1.3	1.8	1	0.9	0.69	0.83

Source: Garg, Shukla, *Emission Inventory Book* 1-2.

1994. However, the study does not give State-wise estimates are not available. Some of the recent trends in activities that contribute to the green house emission are energy sector, electricity generation, transport, residential/ commercial and agricultural/fisheries, industry, agriculture and livestock, land-use change and forestry and urban waste.

The emission inventory for districts for 1990 and 1995 clearly indicate that the CO_2 has been showing an increasing trend in Bikaner, Alwar, Jaipur, Ajmer, Jodhpur, Pali, Jalore, Sirohi, Bhilwara, Udaipur, Chittorgarh and Kota. This increase may be attributed to increasing urbanization during 1990-1995 (Table 2.9).

3
Status of Water Resources

Climate change poses uncertainties to the supply and management of water resources. Climate change can impact surface water resources directly through changes in the major long-term climate variables such as temperature, rainfall and evapo-transpiration. The relationship between the changing climate variables and groundwater is more complicated. Greater variability in rainfall could result in frequent and prolonged periods of high or low groundwater levels, and saline intrusion in aquifers (Singh and Kumar, 2010). Thus, groundwater resources are related to climate change through the direct interaction with surface water resources, such as lakes and rivers, and indirectly through the recharge process. The direct effect of climate change on groundwater resources depends upon the change in the volume and distribution of groundwater recharge.

Rajasthan is one of the most water deprived States of India with respect to rainfall and per capita water availability. The study has shown that the State faces acute water scarcity challenges as rainfall patterns are impacted by climate change. Increasing agricultural and industrial demand is placing additional pressure on the dwindling resources. Being the largest State in India covering an area of 34.32 million hectares (10.4% of India's geographical area) but sharing only 1.16 per cent of its water resources.

Rajasthan is among the regions with greatest climate sensitivity and lowest adaptive capability. In Rajasthan 220603.74 sq. km. potential area under groundwater is available, however, 11.94 Billion Cubic Metre (BCM) is total annual replenishable. The State has capacity to harness and store 10.82 BCM or around 70 per cent of available water. Dependent on inflows into the rivers, 17.88 BCM is allocated through inter-State agreements, although not dependable due to political compulsions of the upper riparian states (*Rajasthan Water Assessment — Potential For Private Sector Interventions*, Sweta M. Huda, 2013). The details of water resources are given in Table 3.1.

Table 3.1: Water Resources of Rajasthan

Category	*Availability in BCM (as per percentage of economically usable water)* 1	*Usage in BCM (as per percentage of economically usable water)* 2	*Percentage used (2) (1)*
Internal Surface Water	21.71	—	5252
(a) Economically usable	16.05 (35.6)	11.29 (31.6)	70
(b) Economically non-usable	5.66	—	—
Ground Water	11.36 (25.2)	11.77 (39.3)	104
Inter-state/External	17.88 (29.2)	12.66 (35.4)	71
Total State Water Resources	**50.96**	—	**70***
Total Economically Usable Water Resources of State	**45.09 (100)**	**35.72 (100)**	**79**

* Total Water usage as percentage of total water resources.

Source: Planning Department, Government of Rajasthan and Water Sector Performance (Reddy).

Status of Ground Water Resources

In Rajasthan 90 per cent of drinking water and 73 per cent of water for irrigation is met out by groundwater. During last 60 years phenomenal development of groundwater resources has been made. As a result, the State has partially achieved self-sufficiency in foodgrain despite recurring drought conditions (nearly 50 droughts occurred). However, due to high growth rate of population, urbanization, industrialization and improvement in life style, unwise use of water in canal command areas, water resources are now under serious stress with declining per capita availability of water, deteriorating quality and lowering trend of groundwater levels.

Underground exploitation of water has shot up from 35 per cent in 1984, 166 in 2009 and 172 in 2011 and is likely to reach 189.1, 221.4 and 259.1 per cent in 2015, 2020 and 2025 respectively. In 16 districts of the 33 districts groundwater is over exploited (more than 100%) and the rates of exploitation are as high as 165 per cent in Jhunjhunu and 153 per cent in Jodhpur.

Table 3.2 gives a clear picture of temporal changes in groundwater blocks of Rajasthan. Out of 243 development blocks in the State only 10 per cent in 25 blocks are in safe category, and about 90 per cent are either over or critical (Table 3.2). During 1984 to 2011 there is a drastic increase in over exploited blocks from 12 to 172 respectively. Contrary to it, there has been unprecedented decline in blocks of safe category from 203 in 1984 to 25 in 2011.

The study of Tables 3.3 and 3.3(a) reveal that out of all the 33 districts of Rajasthan — Banswara, Dungarpur, Ganganagar, Hanumangarh are still under safe category while 12 block of Jaipur, all the blocks of Jalore, Alwar, Bhilwara,

Table 3.2: Temporal Change in Ground Water Blocks

Year	White (Safe) (<70%)	Gray (Semi-critical) (70 to 90%)	Dark (Critical) (90 to 100%)	Over exploited (> 100%)
1984	203	10	11	12
1988	122	42	18	44
1990	148	31	13	44
1992	149	19	15	53
1995	127	35	14	60
1998	135	34	26	41
2001	49	21	80	86
2004	32	14	50	140
2009	31	25	13	166
2011	25	20	24	172

Source: CGWB Report 2011.

Table 3.3: Categories of Blocks as on 31.03.2011

Sl. No.	District	No. of Blocks	Category of Blocks			
			Safe	Semi-Critical	Critical	Over Exploited
1.	Ajmer	8	0	0	0	8
2.	Alwar	14	0	0	0	14
3.	Banswara	8	7	1	0	0
4.	Baran	7	1	1	0	5
5.	Barmer	8	1	0	2	5
6.	Bharatpur	9	0	0	4	5
7.	Bhilwara	11	0	0	0	11
8.	Bikaner*	6	1	1	0	3
9.	Bundi	4	0	1	0	3
10.	Chittorgarh	11	0	0	0	11
11.	Churu*	6	1	2	0	2
12.	Dausa	5	0	0	0	5
13.	Dholpur	4	0	0	1	3
14.	Dungarpur	5	1	4	0	0
15.	Ganganagar	8	8	0	0	0
16.	Hanumangarh	3	3	0	0	0
17.	Jaipur	13	0	1	0	12
18.	Jaisalmer	3	0	0	1	2
19.	Jalore	8	0	0	0	8
20.	Jhalawar	6	0	0	1	5
21.	Jhunjhunu	8	0	0	0	8
22.	Jodhpur	10	1	1	0	8
23.	Karauli	5	0	0	2	3
24.	Kota	5	0	3	0	2
25.	Nagaur	11	0	0	2	9
26.	Pali	10	0	1	2	7
27.	Pratapgarh	5	1	0	1	3
28.	Rajsamand	7	0	0	0	7
29.	Sawai Madhopur	5	0	0	0	5
30.	Sikar	8	0	1	0	7
31.	Sirohi	5	0	0	2	3
32.	Tonk	6	0	2	1	3
33.	Udaipur	11	0	1	5	5
	Total	**243**	**25**	**20**	**24**	**172**

* Tara Nagar block of Churu district and Khajuwala block of Bikaner district have not been assessed due to saline groundwater.

Source: CGWB 2011.

Table 3.3 : A Comparison of Categories of Blocks As Computed on 1.03.2011 and 31.03.2009

Sl. No.	District	No. of Blocks		SAFE		Semi-Critical		Critical		Over Exploited	
		31.3.11	31.3.09	31.3.11	31.3.09	31.3.11	31.3.09	31.3.11	31.3.09	31.3.11	31.3.09
1.	Ajmer	8	8	0	0	0	0	0	0	8	8
2.	Alwar	14	14	0	0	0	0	0	0	14	14
3.	Banswara	8	8	7	7	1	1	0	0	0	0
4.	Baran	7	7	1	3	1	1	0	1	5	2
5.	Barmer	8	8	1	1	0	0	2	2	5	5
6.	Bharatpur	9	9	0	0	0	1	4	4	5	4
7.	Bhilwara	11	11	0	0	0	0	0	0	11	11
8.	Bikaner*	6	5	1	2	1	0	0	0	3	3
9.	Bundi	4	4	0	0	1	0	0	1	3	3
10.	Chittorgarh	11	11	0	0	0	0	0	0	11	11
11.	Churu*	6	6	1	3	2	0	0	0	2	2
12.	Dausa	5	5	0	0	0	0	0	0	5	5
13.	Dholpur	4	4	0	0	0	0	1	0	3	4
14.	Dungarpur	5	5	1	1	4	4	0	0	0	0
15.	Ganganagar	8	7	8	7	0	0	0	0	0	0
16.	Hanumangarh	3	3	3	3	0	0	0	0	0	0
17.	Jaipur	13	13	0	0	1	0	0	1	12	12
18.	Jaisalmer	3	3	0	1	0	0	1	0	2	2
19.	Jalore	8	7	0	0	0	0	0	0	8	7
20.	Jhalawar	6	6	0	0	0	0	1	1	5	5
21.	Jhunjhunu	8	8	0	0	0	0	0	0	8	8
22.	Jodhpur	10	9	1	1	1	0	0	1	8	7
23.	Karauli	5	5	0	0	0	0	2	1	3	4
24.	Kota	5	5	0	0	3	3	0	0	2	2
25.	Nagaur	11	11	0	0	0	1	2	1	9	9
26.	Pali	10	10	0	0	1	0	2	3	7	7
27.	Pratapgarh	5	5	1	1	0	0	1	1	3	3
28.	Rajsamand	7	7	0	0	0	0	0	0	7	7
29.	Sawai madhopur	5	5	0	0	0	0	0	1	5	4
30.	Sikar	8	8	0	1	1	0	0	0	7	7
31.	Sirohi	5	5	0	0	0	3	2	0	3	2
32.	Tonk	6	6	0	0	2	0	1	3	3	3
33.	Udaipur	11	11	0	0	1	2	5	4	5	5
	Total	**243**	**239**	**25**	**31**	**20**	**16**	**24**	**25**	**172**	**166**

* Tara Nagar Block of Churu District and Khajuwala Block of Bikaner District have not been assessed due to saline groundwater.

Source: CGWB 2011

Chittorgarh, Dausa, Jhunjhunu, Rajsamand, Sawai Madhopur have reached to the over exploited category (Figs. 3.1 and 3.2).

Changing groundwater from 1998 to 2011 is indicated in Table 3.4 It is evident from Table 3.4 that assessed replenishable groundwater resource in Rajasthan was 12.602 BCM which reduced to 11.86 BCM in 2009 and 11.94 BCM in 2011. The net groundwater available has increased from 10.38 BCM in 1998 to 10.79 BCM in 2009 and 10.82 BCM in 2011 due to good rainfall during this period. Total draft in 1998 was 8.708 BCM which increased to 14.52 BCM in 2009 and 14.84 BCM in 2011 increasing the stage of development to 137.07 per cent.

Table: 3.4: Changes in Ground Water Potentials

Sl. No		*1998*	*2004*	*2009*	*2011*
1	Assessed Replenishable Ground Water Resource (Provisional)	12.602 bcm	11.56 bcm	11.86 bcm	11.94 bcm
2	Net Ground Water Availability		10.38 bcm	10.79 bcm	10.82 bcm
3	Total Assessed Draft (Pumpage)	8.708 bcm	12.99 bcm	14.52 bcm	14.84 bcm
	Drinking, Domestic and Industrial use	.983 bcm (983 mcm)	1.39 bcm		17.09 bcm
	Irrigation	7.725 bcm	11.60 bcm		13.13 bcm
4	Projected Demand for Domestic and Industrial Use up to 2025		2.72 bcm		1.88 bcm
5	Stage of Development	69%	125%	135%	137.07%
6	Ground Water availability for future Irrigation		3.94 bcm		9.12 bcm

Source: CGWD Report-1998 to 2011.

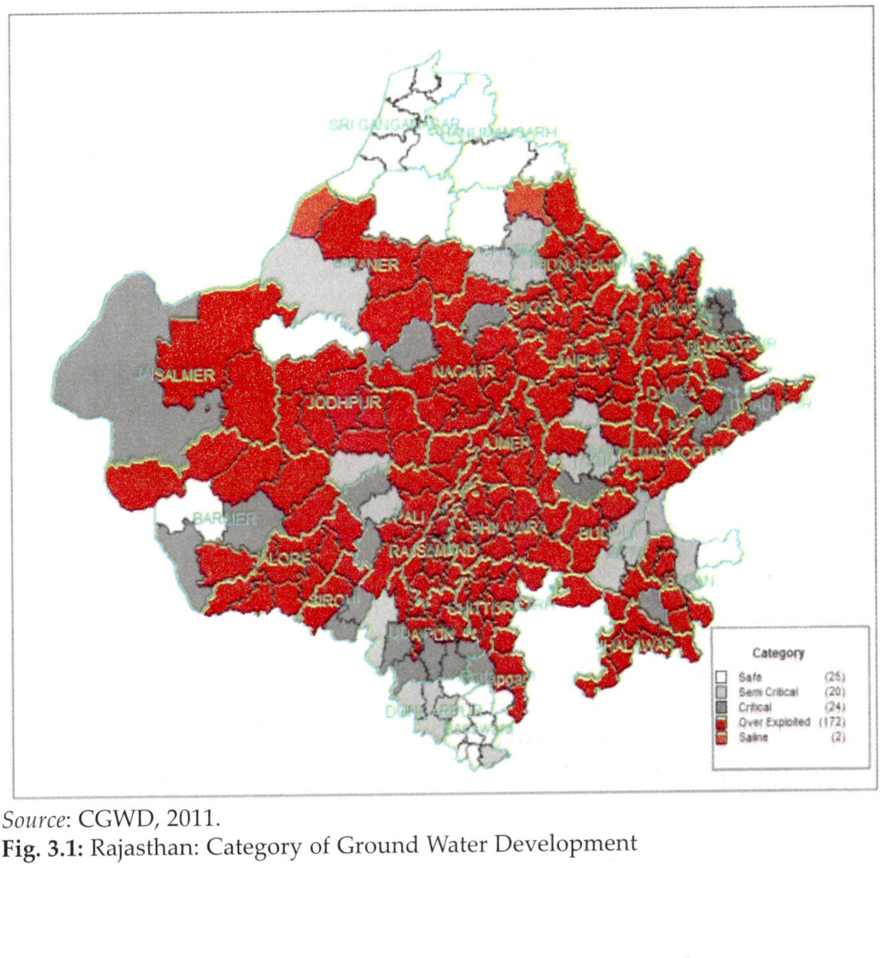

Source: CGWD, 2011.

Fig. 3.1: Rajasthan: Category of Ground Water Development

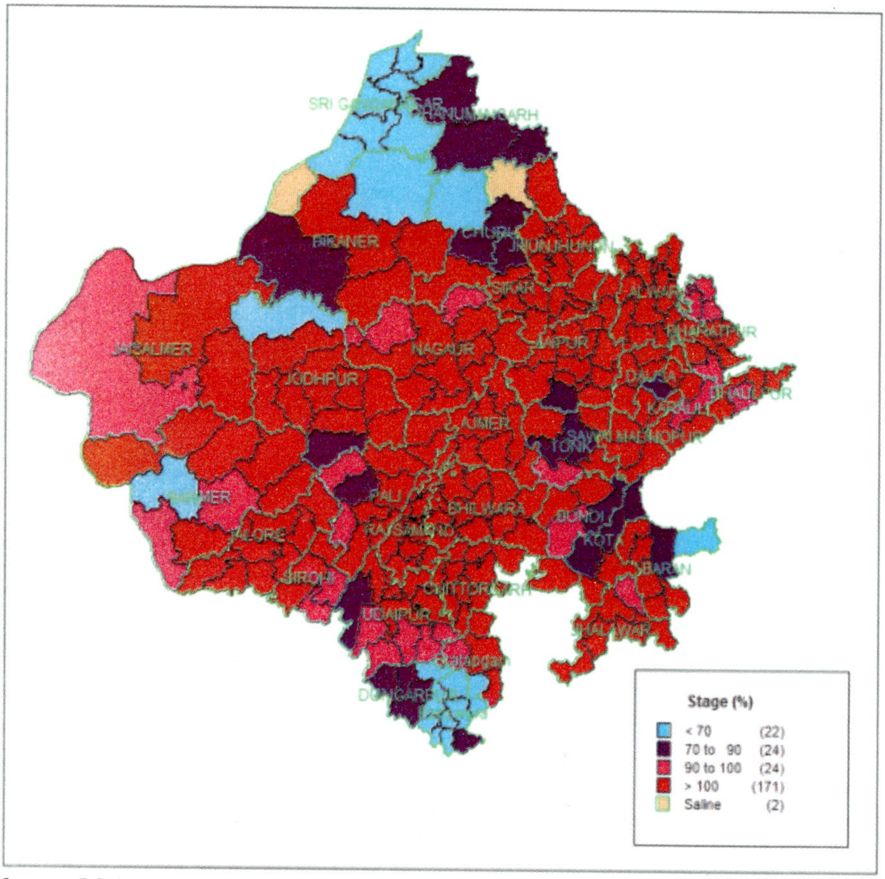

Source: CGWD, 2011.

Fig. 3.2: Rajasthan: Stage of Ground Water Development

The district-wise groundwater availability, utilization and stage of development is given in Table 3.5.

Status of Surface Water Resources and River Basins

In Rajasthan surface water resources are scare and confined to the south and south-east. The central and western parts are devoid of any drainage system and have very limited surface water, which too is not replenished due to low and erratic rainfall. High temperature and low humidity in these areas further add to rainwater loss due to evapo-transpiration. The study of drainage pattern in relation to climate change indicate the Luni along with west flowing rivers occupy about 60 per cent of the area in Rajasthan face acute water scarcity. River basins of Mahi and Sabarmati are likely to experience constant water shortage.

Table 3.5: Ground Water Resources of Rajasthan State as on 31st March, 2011

Sl. No.	Assessment Unit/District	Area	Potential Area	Total Annual Replenish-able Water Recharge	Natural Discharge During Non-Monsoon Season	Net Annual Ground Water Availability	Existing Gross Ground Water	Existing Gross Ground Water Draft for domestic and industrial water supply	Existing Gross Water Draft for All Uses	Net Ground water availability for future irrigation develop-ment	Ground-water Allocation for Domestic and Industrial Development	State of G.W. Development	Category
		Sq. Km.	Sq. Km.	MCM	MCM	MCM	MCM	MCM	MCM	MCM	MCM	%	
1.	Ajmer	8481.00	7466.76	355.11	33.05	322.06	415.47	46.91	462.38	00.00	46.91	143.57	OE
2.	Alwar	8720.46	6825.21	795.51	63.82	731.70	1215.52	95.66	1311.18	3.99	95.95	179.20	OE
3.	Banswara	4536.08	3979.96	257.62	26.96	230.66	92.31	19.61	111.92	166.87	28.53	48.52	Safe
4.	Baran	6955.31	6892.21	519.67	48.04	471.63	525.96	39.38	565.34	55.62	44.75	119.87	OE
5.	Barmer	28387.00	12734.65	272.53	25.24	247.29	239.01	67.66	306.66	21.17	68.47	124.01	OE
6.	Bharatpur	5044.10	3212.52	492.50	43.14	449.36	458.39	63.78	522.18	3.96	65.69	116.21	OE
7.	Bhilwara	10455.00	9354.85	464.32	44.76	419.55	501.98	37.96	541.78	0.00	39.80	129.13	OE
8.	Bikaner*	30381.75	13602.51	252.65	12.63	240.01	258.67	83.74	342.41	40.11	90.33	142.66	OE
9.	Bundi	5500.00	4240.18	422.65	59.70	362.95	336.69	27.11	363.80	53.28	34.72	100.23	OE
10.	Chittorgarh	7880.06	6095.00	352.76	34.17	318.60	432.27	13.79	446.06	00.00	13.79	140.01	OE
11.	Churu*	13792.95	5191.74	141.01	7.05	133.95	94.45	24.08	118.54	30.96	43.98	88.49	S.Critical
12.	Dausa	3420.17	3085.62	263.73	26.20	237.53	378.86	24.87	403.74	0.00	24.87	169.97	OE
13.	Dholpur	3009.05	2486.14	283.12	23.82	259.31	307.06	25.26	332.33	15.96	25.76	128.16	OE
14.	Dungarpur	3770.00	2634.13	141.03	12.91	128.12	83.96	8.82	92.78	0.00	44.16	72.41	S.Critical
15.	Ganganagar	11603.65	1545.64	407.98	40.80	367.19	156.17	5.45	161.63	200.68	10.33	44.02	Safe
16.	Hanumangarh	9579.54	1278.50	226.49	22.65	203.84	157.16	7.05	164.21	34.46	11.07	80.56	Safe
17.	Jaipur	11061	9994.67	712.38	66.85	645.53	1081.11	256.86	1337.97	9.79	257.33	207.27	OE
18.	Jaisalmer	38401.00	12090.04	67.92	6.33	61.59	93.56	28.93	122.49	18.74	29.61	198.89	OE

(Contd.)

(Contd.)

Sl. No.	Assessment Unit/District	Area	Potential Area	Total Annual Replenish-able Water Recharge	Natural Discharge During Non-Monsoon Season	Net Annual Ground Water Availability	Existing Gross Ground	Existing Gross Ground Water Draft for domestic and industrial water supply	Existing Gross Water Draft for All Uses	Net Ground water availability for future irrigation develop-ment	Ground-water Allocation for Domestic & Industrial Development	State of G.W. Development	Category
		Sq. Km.	Sq. Km.	MCM	MCM	MCM	MCM	MCM	MCM	MCM	MCM	%	
19.	Jalore	10640.00	8228.10	462.85	40.86	422.00	779.52	40.90	820.42	9.69	41.91	194.41	OE
20.	Jhalawar	6219.00	6106.16	441.71	29.48	412.23	476.01	15.83	491.83	4.79	19.40	119.31	OE
21.	Jhunjhunu	5928.00	5273.69	263.51	23.50	240.01	456.04	85.71	541.75	3.30	86.04	225.72	OE
22.	Jodhpur	22250.00	18867.92	429.52	42.39	387.13	727.00	103.97	783.33	47.61	113.03	202.34	OE
23	Karauli	5038.60	3902.42	373.14	36.17	336.97	411.77	50.64	462.41	12.39	49.84	137.23	OE
24.	Kota	5203.94	5123.17	566.89	53.59	513.30	420.97	48.91	469.88	77.76	62.51	91.54	Critical
25.	Nagaur	17718.25	16378.50	580.75	56.49	524.26	811.86	178.93	990.79	41.20	189.53	188.99	OE
26.	Pali	12357.00	7362.54	328.59	32.29	296.31	314.36	27.47	341.84	9.59	30.48	115.37	OE
27.	Pratapgarh	4359.80	2950.39	155.21	14.36	140.86	168.84	5.97	174.81	3.99	9.82	124.10	OE
28.	Rajsamand	4635.46	3540.09	118.72	11.87	106.85	102.70	15.25	117.95	0.63	15.25	110.39	OE
29.	Sawai Madhopur	5020.65	7263.46	321.48	30.37	291.11	370.52	58.42	428.94	2.09	83.75	147.35	OE
30.	Sikar	7880.85	4325.63	396.36	35.94	360.42	374.08	79.47	453.55	9.72	62.38	125.84	OE
31.	Sirohi	5136.00	4075.70	303.38	28.54	274.85	299.52	11.08	310.61	11.26	14.84	113.01	OE
32.	Tonk	7200.00	6524.72	483.53	44.59	438.95	357.45	76.35	433.80	18.41	91.04	98.83	Critical
33.	Udaipur	11760.60	7770.92	286.85	33.94	252.90	229.31	31.95	261.25	4.50	39.46	103.30	OE
	State Total	**342326.71**	**220603.74**	**11941.45**	**1112.49**	**10828.97**	**13133.1818**	**1709.8153**	**14842.9972**	**912.50**	**1885.32**	**137.07**	**OE**

* Tara Nagar Block of Churu District and Khajuwala Block of Bikaner District have not been assessed due to saline groundwater.

Source: C.G.W.D. 2011.

The basin-wise availability of surface water is indicated in Table 3.6. The surface water in all the basins is 25,931 million cubic metres (MCM) as shown in table 3.6. The economically usable water in all the basin, is 16,054 million cubic metre (MCM) having 50 per cent dependability.

Table 3.6: Economically Usable Water by River Basins (in MCM)

Basin-wise Availability of Surface Water (MCM)			
Basin	*Mean Availability*	*Economically usable water (at 50% dependability)*	*Percentage of usable water*
Shekhawati	221	105	47
Ruparail	210	180	85
Banganga	569	449	79
Gambhir	805	353	44
Parwati	226	138	61
Sabi	268	168	63
Banas	4,837	4,039	84
Chambal	11,541	5,203	45
Mahi	3,755	3,149	84
Sabarmati	960	800	83
Luni	1,224	452	37
West Banas	551	406	74
Sukli	190	112	59
Other Nalas	91	32	35
Outside Basin	483	468	97
Total	**25,931**	**16,054**	**62**

Source: Water Resource Planning Department, Government of Rajasthan, 1998, Tahel and WAPCOS.

The Chambal basin ranks first in terms of mean availability (11,541 MCM) of water. The Banas and Mahi rank second and third respectively. The percentage of usable water is 85 per cent in Ruparail other than outside basin. The Chambal basin ranks first in terms of economically usable water (5,203 MCM) (Table 3.6).

As indicated earlier in Table 3.2 that the block-wise groundwater situation is troubling. Similar to it, almost all river basins are over-exploited, except the West Banas and Sabarmati basins that are at critical/semi-critical stages (Table 3.7). The state of groundwater development in other nalas (258%) and Shekhawati (215%) are highest in comparison to other basins.

Ground Water Quality

Besides the scarcity of groundwater, Rajasthan is over dependent on

Table 3.7: Groundwater by River Basin (MCM)

	Basin-wise Availability of Groundwater (MCM)								
Basin	*Potential Zone (Sq. Km)*	*Net Annual Availability*	*Gross or Irrigation Use (1)*	*Gross for Domestic/ Industrial Use (2)*	*Gross for all uses (1)+(2)*	*Allocation for Domestic/Industrial Use (as on 2025)*	*Net Ground Water Available for Irrigation*	*State of Ground Water Development (%)*	*Category*
Shekhawati	9,495	504	972	112	1,084	243	(468)	215	Over Exploited
Ruparall	942	347	419	55	474	72	(144)	136	-do-
Banganga	6,593	606	814	90	904	144	(352)	149	-do-
Gambhir	3,616	349	409	47	456	73	(133)	131	-do-
Parwati	4,880	154	186	18	204	29	(79)	133	-do-
Sabi	400	426	644	44	688	64	(282)	161	-do-
Banas	41,090	2,291	2,815	389	3,204	795	(1,319)	140	-do-
Chambal	27,752	1,922	1,966	135	2,102	280	(325)	109	-do-
Mahi	12,062	528	439	37	476	110	(21)	90	-do-
Sabarmati	1,057	80	66	5	71	21	(4)	88	Semi-Critical
Luni	24,548	1,040	1,567	119	1,686	235	(762)	162	Over Exploited
West Banas	1,187	76	65	3	68	7	4	90	Critical
Sukli	867	48	53	1	54	2	(4)	113	Over Exploited
Other Nalas	1,750	100	250	9	259	16	(165)	258	Over Exploited
Outside Basin	80,330	2,063	2,317	463	2,781	689	(943)	135	Over Exploited

Source: Water Resources Planning Department

groundwater. The quality of groundwater has also progressively deteriorated. Tremendous use of groundwater has brought adverse changes the geochemistry of water. Natural contaminates such as fluoride, nitrate and total dissolve salt are increasing in groundwater making it unfit for drinking and posing risk to health. The major problems associated with groundwater quality are presence of fluoride, nitrate and salinity. The worst affected districts with 50 per cent or more concentration of fluoride are Tonk, Churu, Barmer, Pali, Sirohi, Jalore, Nagaur and Rajsamand (Fig. 3.3).

The fluoride in these districts is more than 1.5 ppm. The nitrate concentration more than 100 ppm is found in Churu, Nagaur and Jhunjhunu districts. Total Disolved Salt (TDS) more than 2000 ppm are generally found in Churu, Barmer and Bharatpur districts. Bhilwara, Jodhpur, Baran and Jaipur districts have more concentration of iron (1 ppm) in groundwater. Table 3.8 presents a comparison of the magnitude of the problem with the rest of the country and highlights the worst affected districts in Rajasthan.

There are three major causes of high concentration of fluoride, nitrate and TDS and deterioration of surface water quality are declining groundwater level due to fluctuating climate, sewage waste in urban areas and industrial affluent

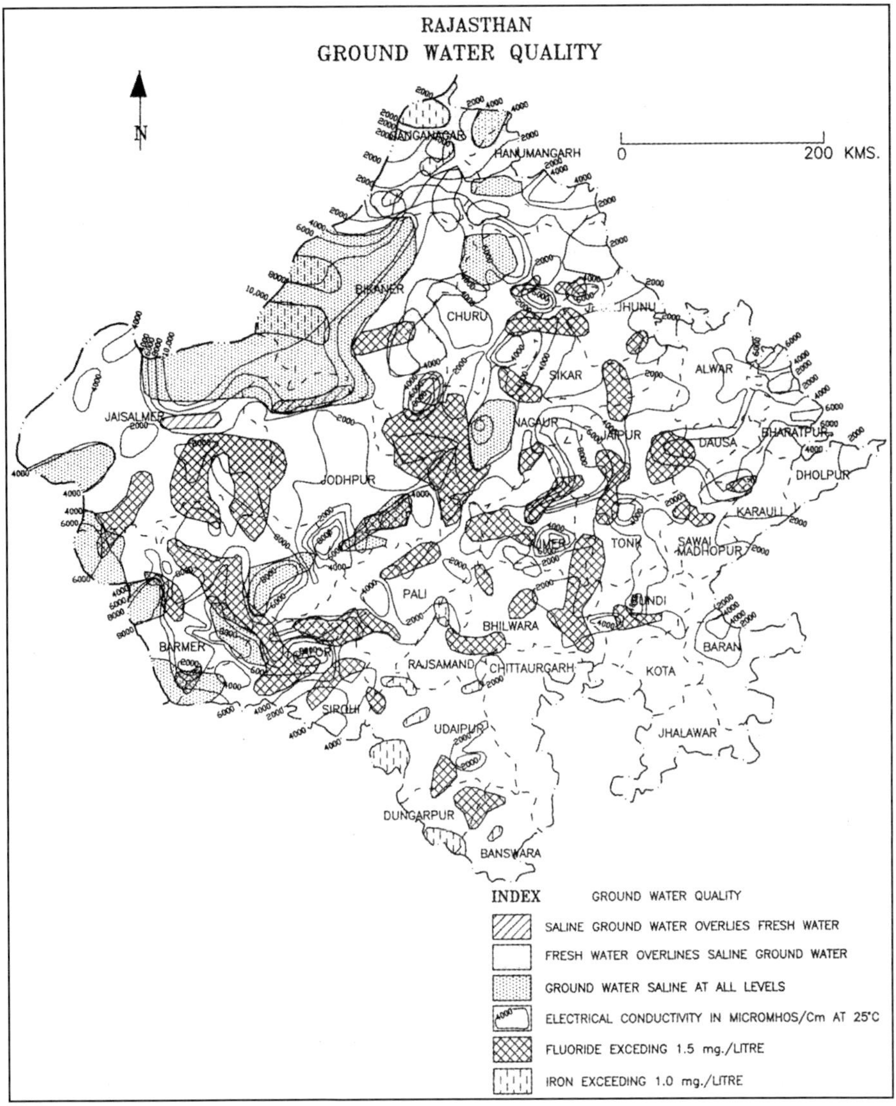

Source: Data Source, CGWD, 2011.

Fig. 3.3: Rajasthan: Ground Water Quality

in industrial estate of Rajasthan. None of the towns in Rajasthan, except Jaipur, have sewage system collection, treatment and disposal system.

Table 3.8: Groundwater Quality in India and Rajasthan

Particulars	*India*	*Rajasthan*	*Percentage of Country*	*Main Affected Districts in Rajasthan**
		Villages/ Habitations		
Only Fluoride	17,986	7,130	39.64	Jaipur, Tonk, Nagaur, Ajmer, Bhilwara, Sirohi and Pali
Only Salinity	22,985	18,924	82.42	Churu, Bharatpur, Barmer, Jhunjhunu, Nagaur and Ajmer
Only Nitrate	2,758	624	22.62	Jaipur, Nagaur, Barmer, Udaipur, Jodhpur, Churu, Alwar and Tonk
Only Iron	56,144	46	0.08	
Only Arsenic	4,314	5	0.12	
Total	**101,160**	**26,729**	**25.6**	

* From Environment Management and Action Plan of SWRPD Report, February 2013.
Source: Draft Annual Plan (2013-14), Planning Department, Government of Rajasthan.

Water Demand-Supply Gap

The share of water use in agriculture accounts for 83 per cent in Rajasthan. The share of water for drinking purpose and industrial/other use account only 11 per cent and 6 per cent respectively. Fuelled by increasing population and improved awareness of sanitation, non-agricultural demand for water is expected to go up to 8.07 billion cubic metres (BCM) in 2045 against 3.28 BCM in 1995 (Report on Water Assessment and Potential for Private Sector Intervention, IFC, World Bank, 2013). The Vyas Committee Report clearly shows the rising gap between demand and supply of water in the state. The current deficiency is 8 BCM and is expected to go up to 9 BCM by 2015 (Table 3.9).

The average per capita water availability in the State is said to be in the range of 640 to 780 cubic metres on the basis of projected population (as on July 2009). This is against the generally accepted requirement of 1000 cubic metres. The availability is expected to further drop below 450 cubic metres by 2050, which is below the international minimum of 500 cubic metres or absolute water scarcity.

In terms of supply, 88.5 per cent of villages and habitations in the State receive less than 40 litres per capita per day (ipcd) of water. Of these, 8 per cent receive less than 10 ipcd, 33 per cent between 10 to 20 ipcd, and 41 per cent between 20 to 30 ipcd. The supply-demand gap is also reflected in the proportion of households reporting purchase of water in rural Rajasthan, which is the highest in the country (Reddy and Mahendra, 2003).

Table 3.9: Rajasthan Water Demand/Supply Gap (in billion cubic metres)

Purpose/Year	*2005*			*2015*			*2045*		
	SW	*GW*	*Total*	*SW*	*GW*	*Total*	*SW*	*GW*	*Total*
Demand									
Domestic	0.5	2.1	2.6	1.0	2.2	3.2	2.5	2.2	4.7
Livestock	0.1	0.8	0.9	0.3	0.8	1.1	0.5	0.8	1.3
Irrigation	20.0	15.9	35.9	26.0	14.0	40.0	36.0	13.1	49.1
Others	0.3	0.4	0.7	0.4	0.4	0.8	1.0	1.0	2.0
Total	**20.9**	**19.2**	**40.1**	**27.7**	**17.4**	**45.1**	**40.0**	**17.1**	**57.1**
Availability									
Intra-state	5.8	7.5		8.0	7.5		16.9	7.5	
Inter-State	1.2			13.0			15.0		
Sub-Total			25.5			28.5			39.4
Recycled Water									
Domestic	0.8			1.1			1.1		
Irrigation		6.0			6.5			7.2	
Sub-Total			6.8			7.6			8.3
Total	**18.8**	**13.6**	**32.4**	**22.1**	**14.0**	**36.1**	**33.0**	**14.7**	**47.7**
Shortage			7.7			9.0			9.4

Source: Vyas Committee Report.

Industrial Demand

Rajasthan is undergoing rapid transformation in industrial sector. Two important projects in Rajasthan, namely Dedicated Freight Corridor (DFC, Railways) and Delhi-Mumbai Industrial Corridor (DMIC) could bring rapid industrial development to Rajasthan since 50 per cent of the state's area across 22 districts lies in the area of influence of these corridors. Hence, water availability from these projects, translating into further pressure on limited water resources. The data reveals that in 1995 the demand in industrial sector was 45.5 MCM per year, projected to grow to 138.44 MCM/year in 2045 (PHED Report 2010). The Rajasthan State Industrial Development Corporation Limited (RIICO) has developed 257 industrial estates including 6 growth centres and 10 industrial development centres in the state to provide infrastructure, which would require additional demand for industries.

Status of Drinking Water Supply

It has been discussed earlier that Rajasthan has acute drinking water problem in some areas. The groundwater table fluctuates rapidly due to unique geological structure. Reservoirs generally dry up and create water crisis during the years of drought. The western Rajasthan often face drought conditions due to scanty and erratic rainfall. In addition to harsh geographical conditions, the high growth rate of population (highest in the country at 21.44 per cent decadal growth) and its distribution over sparsely populated clusters have aggravated the problem of providing safe drinking water. About 70 per cent of habitations in the state face problems of getting potable drinking water (Sweta Mirdha Hooda, Report for IFC, 2013).

In Rajasthan at present 85 per cent of all rural water supply schemes are dependent on groundwater. During last 2-3 decades severe depletion in groundwater table due to over drafting for agriculture has led to deterioration of water quality making a large number of drinking water sources unsustainable. The poor condition of rural water assets and infrastructure is reflected in the extent of water loss due to leakage in rural water supply schemes, which is estimated at 40 to 50 per cent. There are about 1,22,250 rural habitations have problem of water quality. The details of rural drinking water coverage is given in Table 3.10.

Table 3.10: Village Rural Drinking Water Supply by Scheme

Type of Scheme	*No. of villages covered*	*Percentage*
Piped water supply scheme	2075	5.22
Pump and tank water supply scheme	3803	9.57
Regional water supply scheme	11,510	28.96
Hand pump scheme	19,214	48.33
Total sanitation schemes	1,711	4.30
Diggis [43] and others	1,438	3.62
Total	**39,751**	**100**

Source: Draft Annual Plan 2013-14. Planning Department, Government of Rajasthan.

The Government of Rajasthan has taken up a number of drinking water supply projects based on surface water resources. Under the Bharat Nirman Scheme, 77,052 habitations were identified as having drinking water issues as on April 2005. As of March 2012, of these 50,323 habitations have been identified as problematic habitations; many of these have multiple quality problems. As per the draft of Planning Commission, Government of Rajasthan, the surface water source-based projects have solved the drinking water problems in 14,314 of these habitations. The remaining habitations are to be covered on completion of the major water supply projects that are currently underway.

Status of Urban Drinking Water Supply

The supply of safe, hygienic drinking water to urban population is undoubtedly a top priority for the State Government. All 297 towns (2011) are covered by piped water supply from 124 schemes. Of these schemes, 14 are based on surface water, 54 on both surface and groundwater, and the remaining 156 are entirely dependent on groundwater. However, around 30 per cent of the peripheral population in urban areas still remains uncovered. According to the 2011 Census, only 82.6 per cent urban households in Rajasthan had tap supply, and 78.2 per cent households had the source of water located within their premises.

Rajasthan State Water Policy 2010

Rajasthan Government has brought out water policy after National Water Policy. The first policy document was released in 2003 including State specific issues. The current State Water Policy (SWP) draft issued in February 2009 goes further in outlining policy direction on the integrated basis of water resource (IWRM) principles to address mounting pressure on the state's rapidly deteriorating water resources. The draft was subsequently adopted by the Government as the SWP in February 2010.

For water resources management and planning purposes the order of priorities of water allocation is in order of human drinking water, livestock drinking water, other domestic, commercial and municipal water uses, agriculture, power generation, environment and ecological, industrial, non-consumptive uses including tourist uses and other in the order of highest priority to lowest priority respectively. In the new policy document, drinking water has been highest priority and agriculture is placed on 4th rank. In this document the State government has adopted a radical shift from predominantly engineering-based solution to local community based water management solutions. This largely involves a combination of 'bottom-up' decision-making and 'top-down' technical support within a much more holistic conceptual framework for water management.

The preceding discussion has brought out the fact the water resources of Rajasthan are deteriorating at the fast rate. The climate change poses uncertainties to the supply and management of water, the most precious resource of ecologically vulnerable state like Rajasthan. The critical situation of water resources in the light of above analysis is summarized below and how climate change will act as an additional stressor in future?

- Rajasthan is one of the most water deprived State with respect of rainfall and per capita water availability. The state faces acute water scarcity challenges as rainfall patterns are impacted by climate change.

- Surface water resources are limited and in a precarious situation. The rivers are rainfed and there is no perennial river in the state except the Chambal and Mahi.
- Per capita annual availability of water is about 0.000780 BCM against minimum requirement of 0.001 BCM. It is anticipated that the water availability would further fall in future.
- The deficit between demand and supply is 8 BCM at present and likely to increase to 9 BCM by 2015. Non-agricultural water demand is expected to increase implying that the state will slip from scarcity to an absolute scarcity zone.
- With the rising demand for water for both agriculture and non-agriculture purposes the share of water for agriculture is set to reduce to 70 per cent from 83 per cent by 2050. This will have serious implications for agriculture.
- The stage of groundwater exploitation development which was 35 per cent in 1984 has reached to 166 per cent in 2009 out of 239 blocks and 172 per cent in 2011 out of 243 blocks.
- The scanty and irregular rainfall trends from 1901 to 2012 have resulted in poor recharge to groundwater. Climate change projections in declining rainfall and increasing mean temperatures in future which will further limit the recharge of groundwater resources.
- Indiscriminate use of groundwater has brought out changes in quality (fluoride, nitrate and TDS) making groundwater unsafe for health. Climate change will further increase groundwater extraction due to less availability of surface water and rising demand which could further deteriorate groundwater quality and have a serious implication for human health.
- The efficiency of water use in the state is low and unallocated for water loss is very high. About 2/3 of the water used for irrigation is affected by inefficiency. Losses in surface water irrigation are much higher as compared to those in groundwater irrigation. Hence, Rajasthan needs more holistic and integrated management of its water resources. Collective action is necessary to bring together all sectors and stakeholders to work together towards integrated and participatry solutions for water resource planning and management in view of climate change.

4
Land Use, Land Use Change and Agricultural Productivity

Land-use and land cover changes are linked to climate change and other environmental factors. The land surface is an important component of the earth's climate as a source of moisture and energy for the atmosphere, and therefore, any perturbation in land surface properties can cause significant changes. On the other hand, changes in land-use may affect the rainfall and hydrological cycle through the partitioning of the incoming solar radiation into turbulent sensible and latent heat fluxes. Therefore, land-use changes and climate change are interlinked. On the one hand climate change affects terrestrial ecosystem and from other these ecosystems can also affect the regional climate by impacting net radiation and surface heat fluxes (Chase *et al.*, 2000; Foley *et al.*, 2003).

In view of above information on the rate and kind of change in the use of land resources it is essential for the proper planning, management and regulations of the use of such resources. Knowledge about existing land-use and trends of change is essential in order to address the issue of haphazard and uncontrolled growth. Land-use is a dynamic phenomenon and hence the study of changes in land-use are all the more important.

In Rajasthan several attempts have been made by scholars to analyze the land-use and land-use changes on the basis of secondary data both at tehsil and district-levels and different time frame. However, a recent study on influence of land-use/land cover (LULC) changes on atmospheric dynamics over the arid region of Rajasthan State has been attempted by Shailesh *et al.* (2013). The study is based on Landset MSS (1972-73) and IRS-P6 AWiFS (2006-07). The study reveals that significant changes are observed in LULC pattern over Rajasthan during 34 years time period, especially in the desert regions encompassing Indira Gandhi Canal Command area in western Rajasthan. The IRS data as per this study reveals that the scrubland in arid areas have been transformed into cropland which present a considerable increase over central western Rajasthan, due to irrigation facility provided by the IGC transforming the barren desert to green land. The chapter indicates that the large increase in cropland is shown in most of the districts of western Rajasthan (54% to 60%). However, the increase of cropland during 34 years in eastern Rajasthan is 14 per cent to 4 per cent. The large increase in cropland in western Rajasthan is attributed to irrigation by IGC, while in eastern Rajasthan changes in precipitation, and comparatively less irrigation facilities are main causes of small increase in cropland.

In the present study land-use changes in Rajasthan have been analyzed on the basis of secondary data of land-use at district-level from 1985-86, 1995-96, 2005-06 and 2009-10. The decadal changes in land-use have been thus analyzed. During this timeframe significant changes in the land-use have been observed especially in the Canal Command areas like the Indira Gandhi, Mahi and Chambal Command areas. Land-use changes in Rajasthan are closely linked with changing pattern of rainfall and temperature.

General Land Use, Land Use Changes

The study of land-use has been done under the following major land-use classes as per Revenue Department, Government of Rajasthan:

- Forests,
- Area not available for cultivation,
- Other uncultivated land exclusively fallow land,
- Fallow land,
- Net area sown ,
- Total cropped area, and
- Area sown more than once.

Forests

The general land-use changes in Rajasthan is evident from Table 4.1. It is evident from Table 4.1 that the forest area of Rajasthan during 2009-10 was 27.35 lakh hectares accounting 7.98 per cent of the total geographical area. The comparative study of Table 4.1 during 1985-86, 1995-96, 2005-06 and 2009-10 indicates increasing trend from 6.50 per cent in the year 1985-86, 7.18 per cent in 1995-96, 7.81 per cent in 2005-06 and 7.98 per cent in 2009-10 (Figs. 4.1 and 4.2). It is interesting to note that India's state of Forest Report, 2011 for Rajasthan presents a different picture. As per this report the forest area in Rajasthan accounts for 9.54 per cent of the total geographical area. In contrast to above 9.54 per cent of forests area, the satellite data of October-December 2008 accounts only 4.70 per cent area under forest cover in Rajasthan. Thus, there is a wide gap in data which needs a serious consideration.

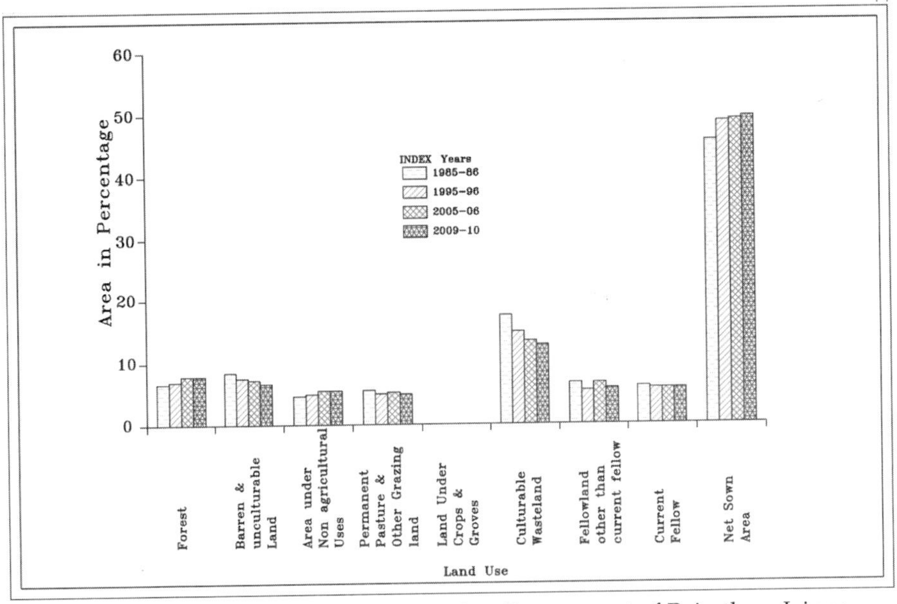

Source: Directorate of Economic and Statistics, Government of Rajasthan, Jaipur.
Fig. 4.1: Rajasthan: Decadal change in land-use

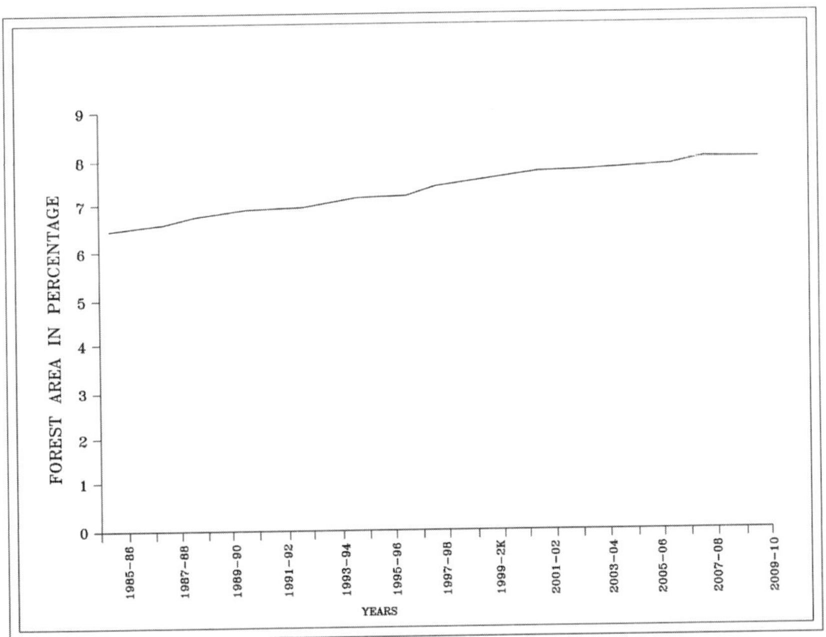

Source: Directorate of Economic and Statistics, Government of Rajasthan, Jaipur.
Fig. 4.2: Rajasthan: Temporal change in forest area

Table 4.1: Land Use Changes in Rajasthan

Land use	1985-86	1995-96	2005-06	2009-10
1. Forests	6.50	7.18	7.81	7.98
2. Area not available for cultivation	12.66	12.66	12.44	12.45
(a) Barren and unculturable land	8.22	7.76	7.12	6.69
(b) Land put to non-agricultural uses	4.44	4.90	5.32	5.76
3. Other uncultivated land excluding fallow land	22.95	20.05	18.44	18.06
(a) Permanent Pastures and other grazing land	5.37	5.10	4.98	4.95
(b) Land under Misc. tree crops and groves not included in net sown area	0.09	0.05	0.06	0.05
(c) Culturable wasteland	17.49	14.90	13.40	13.06
4. Fallow land (a+b)	12.39	11.71	12.18	11.97
(a) Fallow land other than current Fallow	6.50	5.75	6.61	5.97
(b) Current Fallow	5.89	5.95	5.57	6.00
5. Net sown area	45.45	48.41	49.13	49.53
6. Total Cropped Area	52.97	57.46	63.38	63.51
7. Area sown more than once	7.51	9.04	14.28	13.92

Sources: Directorate of Economics and Statistics-2009-10.
Agricultural Statistics of Rajasthan (2009-10).

The spatial distribution of forest area at district-level varies considerably (Fig. 4.3). Maximum forest area above 20 per cent is recorded in Sirohi (29.14%), Kota (26.84%), Bundi (23.03%) and Banswara (20.12%) in the year 1985-86. The south-eastern districts record maximum area under forest in 1985-86 (Fig. 4.3). All the districts of western, northern and north-eastern Rajasthan have the forest area under 1 per cent. The data of 1995-96 reveals that the area under forest has increased in Sirohi (29.33%), Udaipur (26.28%), Bundi (29.90%), Baran (30.01%) and Sawai Madhopur (24.02%) over 1985-86. The districts falling under the category of more than 20 per cent forest cover in the year 2005-06 were Banswara (22.31%), Baran (30.90%), Bundi (24.28%), Karauli (34.16%), Kota (23.61%), Sirohi (30.01%), and Udaipur (28.35%). The lowest per cent of area below 2 per cent was reported in Barmer (1.13%), Churu (0.47%), Jaisalmer (0.68%), Jodhpur (0.31%) and Nagaur (1.04%). The highest forest cover in 2009-10 was reported in the districts like Karauli (34.20%), Baran (30.99%), Sirohi (30.02%), Pratapgarh (29.38%), and Udaipur (28.60%) (Table 4.2). The lowest area was reported in Barmer (1.14%), Churu (0.84%), Hanumangarh (1.90%), Jaisalmer (1.17%), and Nagaur (1.05%). However, during 1985-86 to 2009-10, there has been an increase in the forest area even in western desert districts. It is pertinent to note that the drastic change in forest area in Alwar district from 2.12 per cent in 1985-86 to 10.15 per cent in 2009-10 has been recorded, which may be due to conservation of forest reverse in Sariska Tiger Reserve.

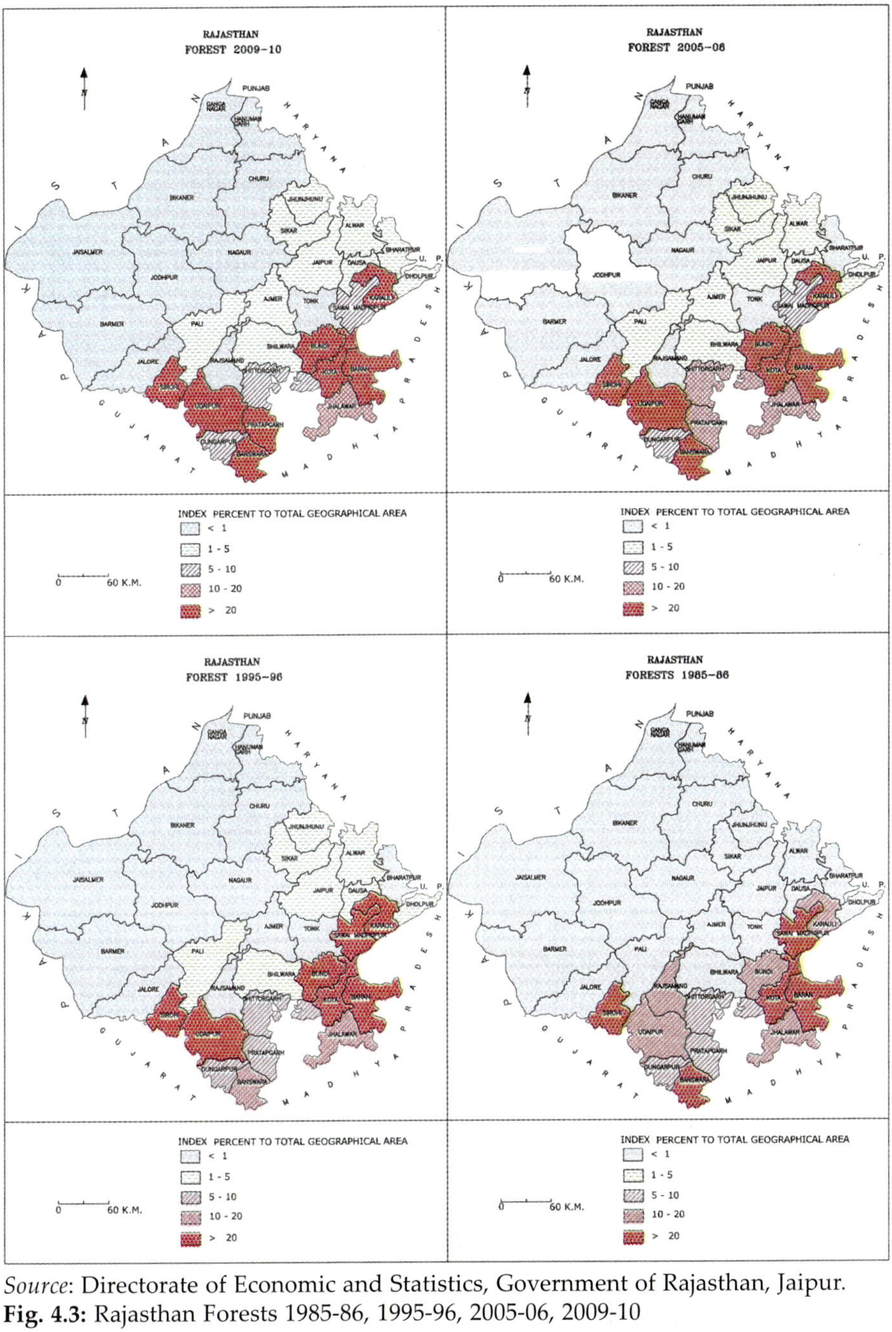

Source: Directorate of Economic and Statistics, Government of Rajasthan, Jaipur.

Fig. 4.3: Rajasthan Forests 1985-86, 1995-96, 2005-06, 2009-10

Table 4.2: Distribution of Forest Area in Rajasthan

Sl. No.	*Name of District*	*1985-86*	*1995-96*	*2005-06*	*2009-10*
1.	Ajmer	4.45	5.80	6.60	6.69
2.	Alwar	2.12	6.34	10.14	10.15
3.	Banswara	20.12	19.88	22.31	20.12
4.	Baran	—	30.01	30.90	30.99
5.	Barmer	0.72	0.84	1.13	1.14
6.	Bharatpur	5.13	5.09	5.98	6.64
7.	Bhilwara	5.60	6.07	7.10	7.14
8.	Bikaner	2.34	2.83	2.72	3.09
9.	Bundi	32.03	23.90	24.28	24.52
10.	Chittorgarh	13.56	17.19	18.87	16.00
11.	Churu	0.30	0.40	0.47	0.84
12.	Dausa	—	6.62	7.19	7.67
13.	Dholpur	3.94	6.93	8.99	9.03
14.	Dungarpur	14.75	15.42	15.94	16.13
15.	Hanumangarh	—	0.76	1.86	1.90
16.	Jaipur	4.71	6.49	7.33	7.44
17.	Jaisalmer	0.43	0.59	0.68	1.17
18.	Jalore	1.70	1.77	2.06	2.14
19.	Jhalawar	18.08	18.24	19.61	19.78
20.	Jhunjhunu	5.64	6.68	6.71	6.72
21.	Jodhpur	0.22	0.30	0.31	0.32
22.	Karauli	—	—	34.16	34.20
23.	Kota	26.84	22.56	23.61	24.07
24.	Nagaur	0.86	0.83	1.04	1.05
25.	Pali	5.96	6.41	6.65	7.02
26.	Pratapgarh	—	—	—	29.38
27.	Rajsamand	—	5.14	5.42	5.72
28.	Sawai Madhopur	22.78	24.02	15.84	16.10
29.	Sikar	4.35	7.61	7.89	7.89
30.	Sirohi	29.14	29.33	30.01	30.02
31.	Sri Ganganagar	2.66	5.11	5.53	5.53
32.	Tonk	3.25	3.37	3.79	3.05
33.	Udaipur	19.96	26.28	28.35	28.60

Sources: Directorate of Economics and Statistics-1985-86 to 2009-10; Agricultural Statistics of Rajasthan (2009-10).

Area Not Available for Cultivation

Under this category area under non-agricultural uses (building, roads, pathway, social forestry, railways, water wayes) barren and unculturable land (mountains, desert and hills) are included. The area under this category in Rajasthan was about 12.45 per cent in 2009-10 as against 12.66 per cent in 1985-86, 12.66 per cent in 1995-96 and 12.44 per cent in 2005-06. The barren and unculturable land accounts for 6.69 per cent during 2009-10 as against 8.22 per cent in 1985-86, 7.76 per cent in 1995-96 and 7.12 per cent in 2005-06, meaning that unculturable wasteland in Rajasthan is witnessing declining trend during four decades. In contrast to this the area under non-agricultural uses is showing.

Table 4.3: Distribution of Area Not Available for Cultivation (Rajasthan)

Sl. No.	*Name of District*	*1985-86*	*1995-96*	*2005-06*	*2009-10*
1.	Ajmer	22.93	16.90	16.48	16.54
2.	Alwar	22.06	18.19	16.33	16.61
3.	Banswara	18.16	17.38	13.65	13.87
4.	Baran	—	9.50	9.16	9.32
5.	Barmer	6.96	7.25	7.05	7.03
6.	Bharatpur	11.49	11.50	10.61	10.62
7.	Bhilwara	21.89	20.52	20.19	20.19
8.	Bikaner	5.08	10.06	9.63	9.86
9.	Bundi	17.17	15.58	15.43	15.59
10.	Chittorgarh	18.96	14.96	13.38	15.22
11.	Churu	4.31	4.36	4.64	4.75
12.	Dausa	—	11.53	11.07	10.71
13.	Dholpur	29.75	26.74	24.86	24.83
14.	Dungarpur	25.08	26.40	24.78	24.47
15.	Hanumangarh	—	6.32	5.82	5.84
16.	Jaipur	13.48	11.96	11.86	12.23
17.	Jaisalmer	11.06	11.78	12.92	13.21
18.	Jalore	11.31	11.90	11.62	11.61
19.	Jhalawar	10.16	10.32	9.68	9.79
20.	Jhunjhunu	6.02	5.96	6.31	6.43
21.	Jodhpur	9.87	9.82	9.95	9.98
22.	Karauli	—	—	14.22	14.24
23.	Kota	11.92	13.18	12.73	12.80
24.	Nagaur	7.97	8.23	8.23	8.24
25.	Pali	16.34	16.64	16.39	16.04
26.	Pratapgarh	—	—	—	9.49
27.	Rajsamand	—	28.33	28.12	27.93
28.	Sawai Madhopur	15.47	15.39	13.72	13.53
29.	Sikar	8.87	6.21	6.72	30.56
30.	Sirohi	20.22	19.86	19.29	19.30
31.	Sri Ganganagar	6.30	6.32	6.46	6.49
32.	Tonk	9.79	9.78	10.56	10.26
33.	Udaipur	35.22	35.68	33.76	33.94

Sources: Directorate of Economics and Statistics-2009-10;
Agricultural Statistics of Rajasthan (2009-10).

an increasing trend from 4.44 per cent in 1985-86, 4.90 per cent in 1995-96, 5.12 per cent in 2005-06, and 5.76 per cent in 2009-10 (Table 4.1).

The spatial distribution of area not available for cultivation indicates that only two districts of southern Aravallis and Dholpur district of eastern Rajasthan reported percentage more than 25 per cent from 1985-86 to 1995-96. In the year 2005-06 and 2009-10 there has been a decline of area under this category in Dholpur district and increase in Sikar district (Fig. 4.4). The districts which fall in the highest category over 25 per cent during 1985-86 were Udaipur (35.22%), Dungarpur (25.08%) and Dholpur (29.75%). These districts show declining trend in 1995-96, 2005-06 and 2009-10 (Table 4.3). Table 4.3 further

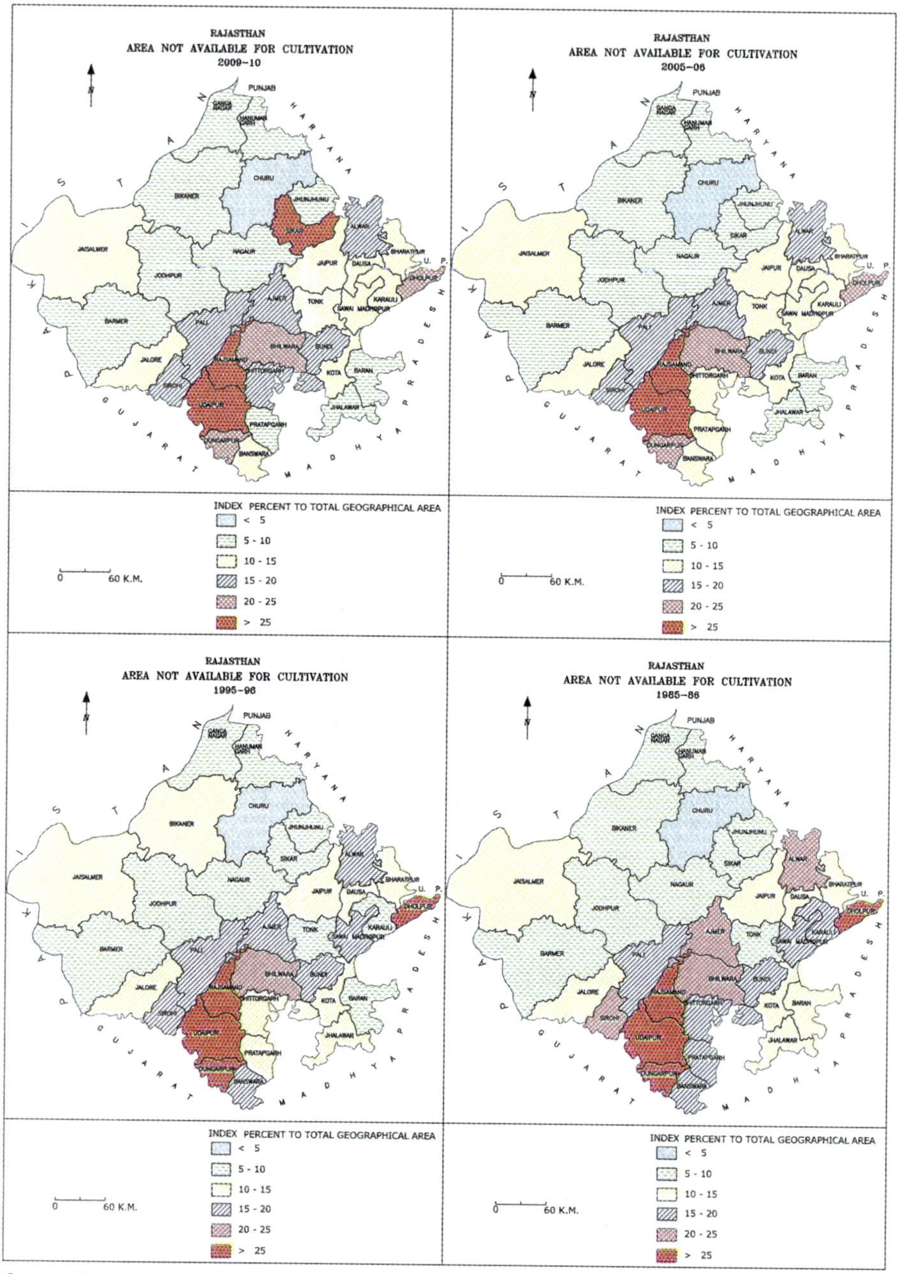

Source: Directorate of Economic and Statistics, Government of Rajasthan, Jaipur.

Fig. 4.4: Rajasthan: Area not Available of Culitivation1985-86, 1995-96, 2005-06, 2009-10.

indicates that in most of the districts there has been a declining trend except in Churu, Jaisalmer, Jhunjhunu, Karauli, Nagaur, Sikar, Sri Ganganagar in which increasing trend has been witnessed.

Other Uncultivated Land Excluding Fellow Land

This category covers permanent pasture and other grazing lands etc. and not included in net sown area and culturable wasteland. An extent of 6189617 hectares of land comes under this category or 18.06 per cent area as against 22.95 per cent in 1985-86, 22.05 per cent in 1995-96 and 18.44 per cent in 2005-06 indicating a declining trend (Table 4.1).

The spatial distribution of this land-use category shows that during 1985-86, 1995-96 the districts of Jaisalmer, Bikaner, Hanumangarh (1995-96) and Pali were under the higher percentage of area in western Rajasthan. However, the districts of Rajsamand, Bhilwara and Chittorgarh were under higher percentage of area in eastern Rajasthan during 1985-86 and 1995-96 (Fig. 4.5). The area during 2005-06 and 2009-10 has been declined in most of districts. The district Jaisalmer is having highest area under this category but it has been shown a declining trend from 78.71 per cent in 1985-86, 75.34 per cent in 1995-96, 67.80 per cent in 2005-06, and 66.00 per cent in 2009-10 (Table 4.4). The change is attributed to pressure of population on land resources and extension of agricultural activities on marginal lands and common property resources like pastures and grazing lands.

Fallow Land

This category of land-use includes both current fallow and fallow land other than current fallow. The fallow land in Rajasthan is showing a declining trend from 12.39 per cent in 1985-86, 11.71 per cent in 1995-96, 12.18 per cent in 2005-06 and 11.97 per cent in 2009-10 except in 2005-06 (Table 4.1).

The spatial distribution of fallow land at district-level reveals that the districts having higher area under fallow lands pertains to western Rajasthan. In the year 1985-86 Jodhpur and Churu ranked first having higher percentage (more than 25%) of area under fallow land. In 1995-96 Jodhpur had area more than 25 per cent. In 2005-06 Sri Ganganagar had higher percentage along with Jodhpur. The area under this category declined in 2009-10 in all the districts. Most of the districts fall under less than 20 per cent except Sri Ganganagar which has more than 20 per cent (Fig. 4.6).

The comparative Table 4.5 reveals that in most of the districts the area under fallow land was below 51.98 per cent in Churu and 33.18 per cent in Jodhpur in the year 1985-86. During 1995-96 there has been a declining trend in all the districts being highest only at Barmer (22.95%). During 2005-06 and 2009-10, there is a declining trend in most of the districts except Ajmer, A

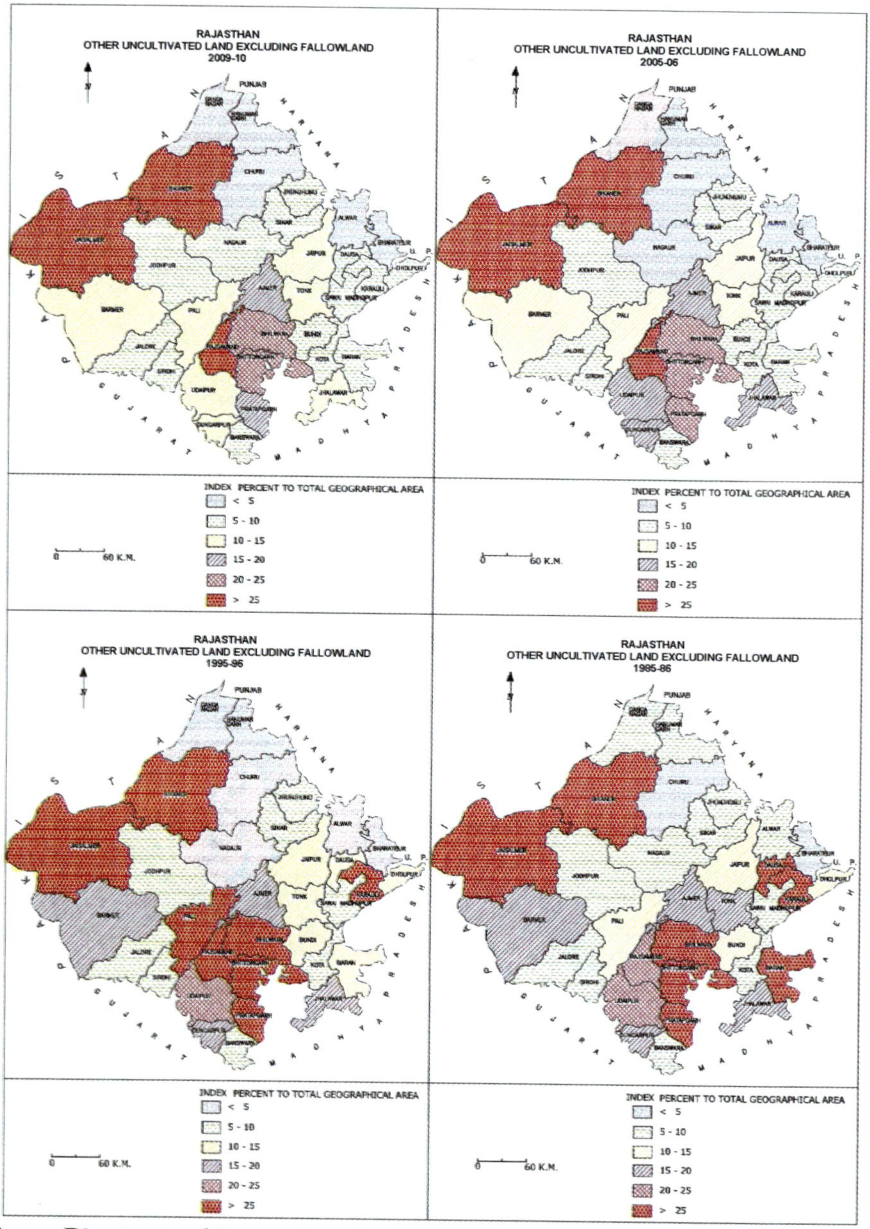

Source: Directorate of Economic and Statistics, Government of Rajasthan, Jaipur.

Fig. 4.5: Rajasthan: Other Uncultivated Land Excluding Fallowland 1985-86, 1995-96, 2005-06, 2009-10.

Table 4.4: Other uncultivated land excluding Fallow Land (Rajasthan)

Sl. No.	*Name of District*	*1985-86*	*1995-96*	*2005-06*	*2009-10*
1.	Ajmer	19.97	18.26	17.64	18.10
2.	Alwar	5.33	4.64	4.13	3.95
3.	Banswara	9.23	7.83	8.65	7.99
4.	Baran	—	10.06	8.05	7.17
5.	Barmer	17.76	15.17	14.63	14.14
6.	Bharatpur	3.24	2.41	2.34	2.13
7.	Bhilwara	31.82	27.83	24.62	24.27
8.	Bikaner	48.57	33.49	27.25	27.35
9.	Bundi	10.17	10.29	9.35	9.18
10.	Chittorgarh	28.67	25.69	22.32	23.69
11.	Churu	4.54	3.65	3.49	3.59
12.	Dausa	—	10.51	9.91	9.70
13.	Dholpur	11.93	10.88	9.70	9.16
14.	Dungarpur	18.10	15.94	15.33	14.79
15.	Hanumangarh	—	1.57	0.9	0.78
16.	Jaipur	13.40	10.23	10.12	10.29
17.	Jaisalmer	78.71	75.34	67.80	66.00
18.	Jalore	7.50	6.83	7.34	7.78
19.	Jhalawar	18.92	18.14	15.96	14.89
20.	Jhunjhunu	8.66	7.88	7.84	7.69
21.	Jodhpur	8.95	6.47	7.18	6.01
22.	Karauli	—	—	8.82	8.57
23.	Kota	9.22	7.75	6.97	7.48
24.	Nagaur	5.74	4.90	4.94	5.05
25.	Pali	11.33	43.44	11.09	10.93
26.	Pratapgarh	—	—	—	15.13
27.	Rajsamand	—	38.72	38.43	38.76
28.	Sawai Madhopur	5.55	8.30	7.49	7.40
29.	Sikar	7.95	6.86	6.44	6.53
30.	Sirohi	8.95	8.12	8.48	8.51
31.	Sri Ganganagar	6.79	4.87	3.82	4.20
32.	Tonk	16.23	13.81	12.00	11.92
33.	Udaipur	21.27	20.51	15.06	14.68

Sources: Directorate of Economics and Statistics-2009-10;
Agricultural Statistics of Rajasthan (2009-10).

Chittorgarh, Hanumangarh, Jaisalmer, Jhunjhunu, Rajsamand, Sawai Madhopur, Sirohi, Sri Ganganagar, Tonk, and Udaipur where increasing trend is evident.

Net Area Sown

The net area sown represents total area sown with crops and orchards accounting area sown more than once in the same year. The net area sown during 2009-10 was 169.74 lakh hectares or 49.53 per cent. Table 4.1 reveals that the net area sown in Rajasthan is showing an increasing trend from 45.45

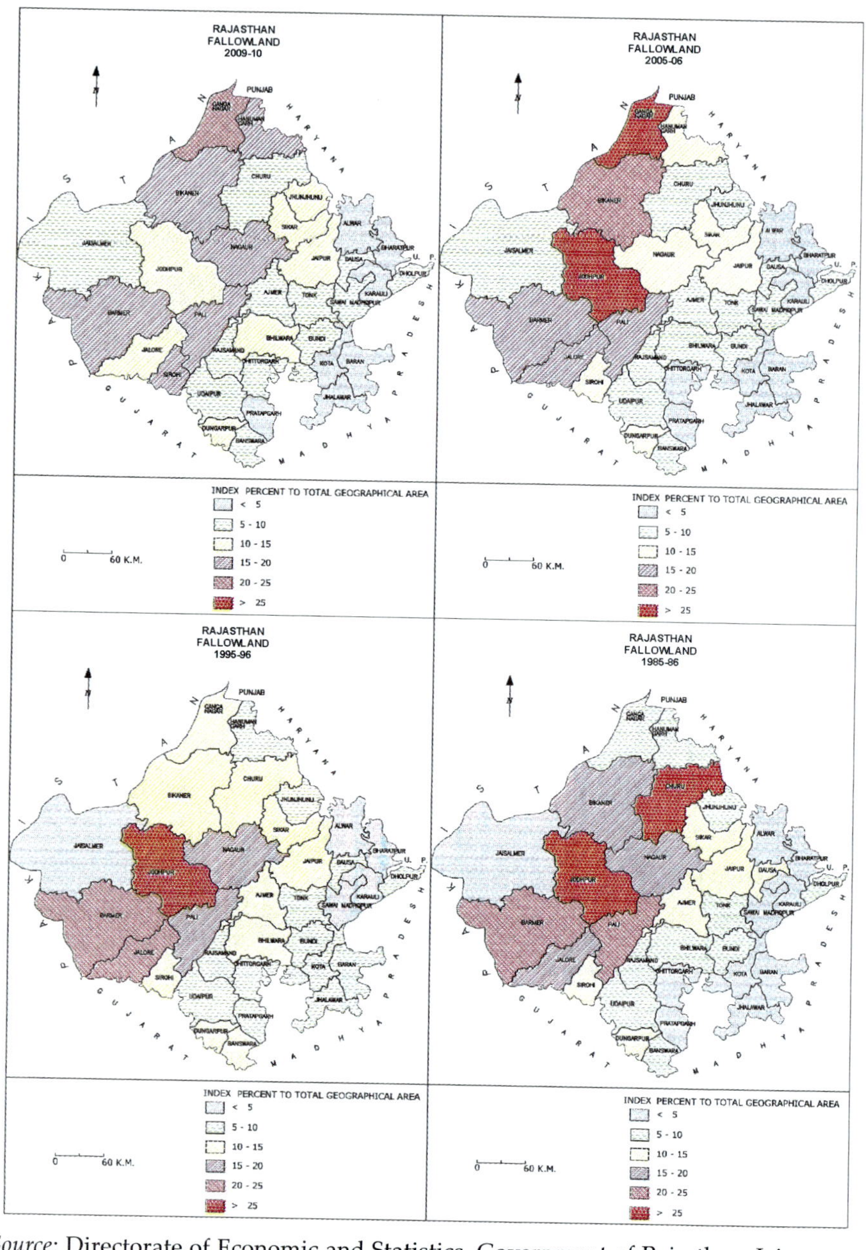

Source: Directorate of Economic and Statistics, Government of Rajasthan, Jaipur.

Fig. 4.6: Rajasthan: Distribution of Fallowland 1985-86, 1995-96, 2005-06, 2009-10.

Table 4.5: Distribution of Fallow Land (Rajasthan)

Sl. No.	*Name of District*	*1985-86*	*1995-96*	*2005-06*	*2009-10*
1.	Ajmer	14.54	12.91	8.15	9.14
2.	Alwar	4.84	4.53	4.45	4.84
3.	Banswara	8.64	10.16	8.90	8.57
4.	Baran	—	6.65	4.17	4.04
5.	Barmer	20.14	22.95	20.00	17.27
6.	Bharatpur	4.51	3.85	2.85	2.91
7.	Bhilwara	9.33	12.39	9.35	11.15
8.	Bikaner	15.71	10.76	20.75	15.10
9.	Bundi	7.27	7.09	6.27	7.55
10.	Chittorgarh	3.91	5.00	3.64	5.42
11.	Churu	51.98	10.51	7.09	9.75
12.	Dausa	—	7.59	5.92	5.51
13.	Dholpur	7.12	7.67	6.42	5.96
14.	Dungarpur	10.86	10.97	12.16	10.01
15.	Hanumangarh	—	5.09	12.53	15.86
16.	Jaipur	11.79	11.98	14.40	10.94
17.	Jaisalmer	4.90	2.92	5.14	5.37
18.	Jalore	19.00	21.35	16.90	11.45
19.	Jhalawar	3.77	5.03	4.58	3.57
20.	Jhunjhunu	6.12	5.26	6.66	10.84
21.	Jodhpur	33.18	27.30	26.28	13.36
22.	Karauli	—	—	3.49	2.96
23.	Kota	4.31	5.97	3.91	3.59
24.	Nagaur	19.14	18.16	14.69	15.61
25.	Pali	20.11	16.47	16.59	17.28
26.	Pratapgarh	—	—	—	4.06
27.	Rajsamand	—	6.69	6.35	7.89
28.	Sawai Madhopur	4.02	4.76	5.31	5.31
29.	Sikar	12.68	12.40	10.31	12.62
30.	Sirohi	10.92	14.18	12.45	15.56
31.	Sri Ganganagar	8.13	14.03	25.13	22.01
32.	Tonk	7.29	7.16	7.44	9.76
33.	Udaipur	5.43	6.64	5.46	5.76

Sources: Directorate of Economics and Statistics-2009-10;
Agricultural Statistics of Rajasthan (2009-10).

per cent in 1985-86, 48.41 per cent in 1995-96, 49.13 per cent in 2005-06 and 49.53 per cent in 2009-10.

The spatial distribution of net sown area at district-level indicates that the main concentration of area having higher percentage under this category is found in north-western districts and north-eastern districts except Jalore in 1985-86, 2005-06 and 2009-10 (Fig. 4.7).

Table 4.6 indicates that in most of the districts of Rajasthan the net sown area is showing increasing trend except in Ajmer, Alwar, Bhilwara, Chittorgarh, Hanumangarh, Sirohi and Sikar in which a declining trend is evident from 2005-06. This is indeed a matter of concern that Hanumangarh district fall under Indira Gandhi Canal Command area is showing such a trend. Other studies have reported increase in cropland based on satellite data of 2008 (Shailesh *et al.*, 2013).

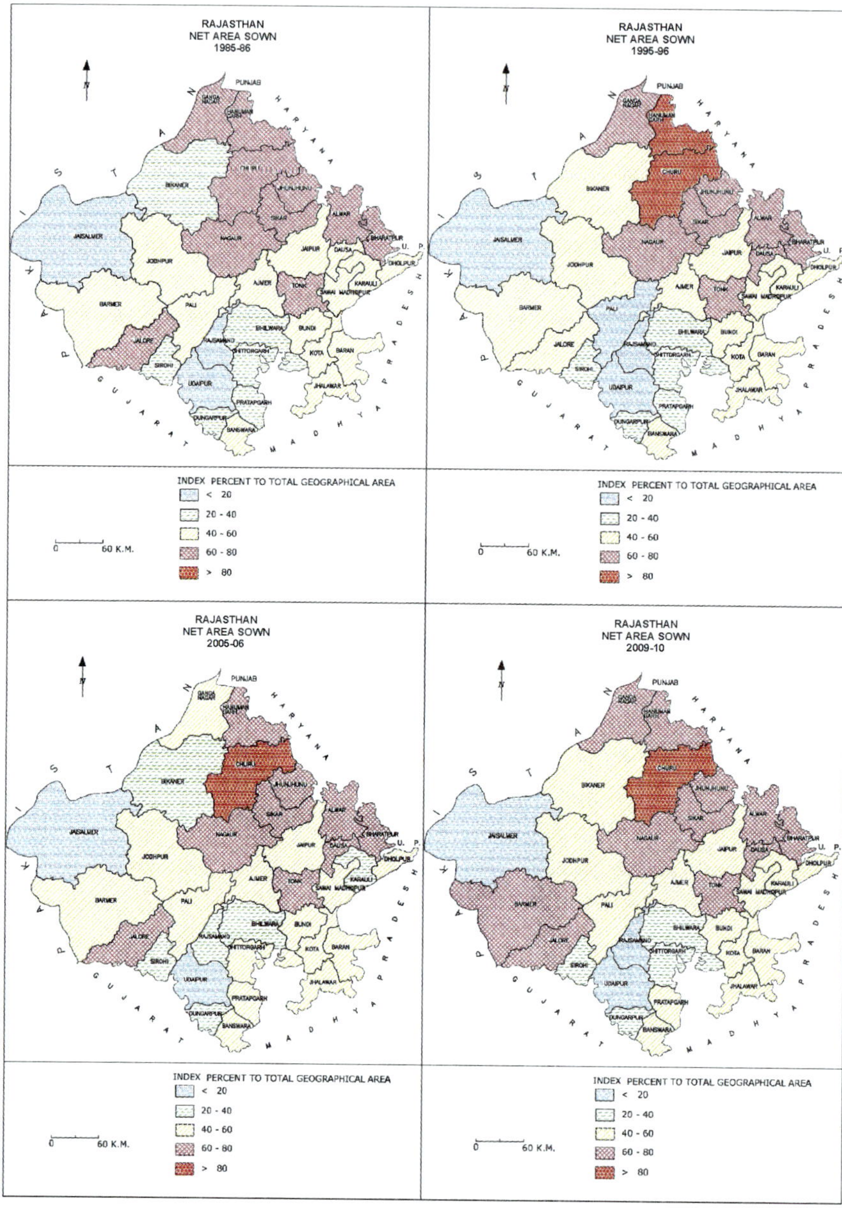

Source: Directorate of Economic and Statistics, Government of Rajasthan, Jaipur.

Fig. 4.7: Rajasthan: Distribution of Net Area Sown 1985-86, 1995-96, 2005-06, 2009-10.

Table 4.6: Distribution of Net Area Sown (Rajasthan)

Sl.No.	*Name of District*	*1985-86*	*1995-96*	*2005-06*	*2009-10*
1.	Ajmer	43.60	46.10	51.11	49.51
2.	Alwar	65.63	66.20	64.92	64.43
3.	Banswara	43.32	44.74	46.47	49.43
4.	Baran	—	43.76	47.71	48.45
5.	Barmer	54.39	53.78	57.17	60.40
6.	Bharatpur	75.60	77.13	78.20	78.15
7.	Bhilwara	31.32	33.18	38.72	37.22
8.	Bikaner	28.27	42.84	39.64	44.58
9.	Bundi	42.33	43.07	44.64	43.14
10.	Chittorgarh	34.88	37.14	41.77	39.65
11.	Churu	75.63	81.07	84.29	81.41
12.	Dausa	—	63.74	65.89	66.40
13.	Dholpur	47.23	47.76	50.01	51.04
14.	Dungarpur	31.18	31.25	31.67	34.57
15.	Hanumangarh	—	86.23	78.81	75.61
16.	Jaipur	56.60	59.33	56.27	59.08
17.	Jaisalmer	4.87	9.35	13.45	14.24
18.	Jalore	60.45	58.14	62.06	61.66
19.	Jhalawar	49.05	48.26	50.14	51.97
20.	Jhunjhunu	73.53	62.71	72.46	68.32
21.	Jodhpur	47.75	56.10	56.25	55.68
22.	Karauli	—	—	38.94	40.01
23.	Kota	47.69	50.52	52.76	41.68
24.	Nagaur	66.25	67.86	71.08	70.04
25.	Pali	46.23	4.94	49.26	48.71
26.	Pratapgarh	—	—	—	41.91
27.	Rajsamand	—	2.11	21.66	19.68
28.	Sawai Madhopur	48.36	47.51	57.62	58.34
29.	Sikar	66.12	66.90	68.62	66.06
30.	Sirohi	30.75	28.83	29.75	26.59
31.	Sri Ganganagar	76.10	69.04	59.03	61.75
32.	Tonk	63.42	65.86	66.19	64.19
33.	Udaipur	18.10	15.76	17.35	17.00

Sources: Directorate of Economics and Statistics-2009-10;
Agricultural Statistics of Rajasthan (2009-10).

Total Cropped Area

Gross area shown presents total cropped area cultivated under all food and non-food crops, including the area sown more than once during the agricultural year. The total cropped area during 2009-10 was 217.45 lakh hectares against 227.71 lakh hectares in 2008-09 recording a decrease of 4.51 per cent due to unfavourable weather conditions in 2009-10. The total cropped area in the state is showing an increasing trend from 52.97 per cent in 1985-86, 57.46 per cent in 1995-96, 63.38 per cent in 2005-06 and 63.51 per cent in 2009-10 (Table 4.1).

The spatial distribution of total cropped area in Rajasthan at district-level reveals that the districts falling in northern and north-eastern part have higher percentage above 80 per cent under this category (Fig. 4.8). The increase in south eastern districts from 1985-86 to 2009-10 may also be observed and increase in area of desert districts is also evident during 1985-86 to 2009-10 (Fig. 4.8).

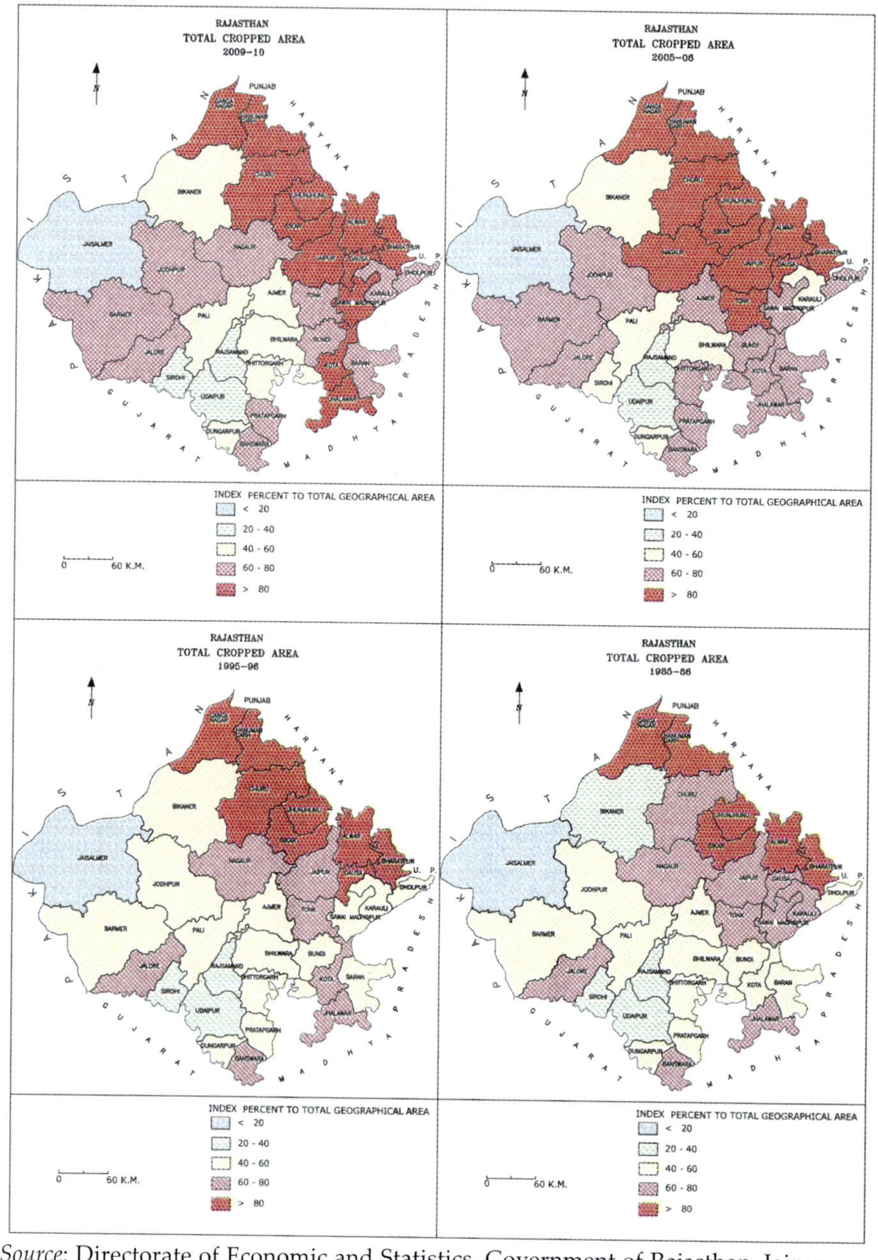

Source: Directorate of Economic and Statistics, Government of Rajasthan, Jaipur.

Fig. 4.8: Total Cropped Area of Rajasthan 1985-86, 1995-96, 2005-06, 2009-10.

The district-wise analysis of total cropped area reveals that it has shown an increasing trend during 1985-86 to 2009-10 in most of the districts except Rajsamand, Jaipur, Nagaur, Pali, Sirohi, Tonk and Udaipur (Table 4.7).

Table 4.7: Distribution of Total Cropped Area (Rajasthan)

Sl. No.	*Name of District*	*1985-86*	*1995-96*	*2005-06*	*2009-10*
1.	Ajmer	50.52	54.49	61.73	51.75
2.	Alwar	97.28	90.11	104.02	110.36
3.	Banswara	60.35	64.56	66.86	71.55
4.	Baran	—	52.41	71.59	78.19
5.	Barmer	54.99	56.03	60.01	64.59
6.	Bharatpur	100.66	97.88	112.83	119.01
7.	Bhilwara	41.87	45.61	54.83	43.38
8.	Bikaner	29.00	45.48	43.85	49.38
9.	Bundi	55.61	59.60	67.20	65.94
10.	Chittorgarh	51.66	52.58	65.27	56.27
11.	Churu	77.82	82.77	102.67	92.58
12.	Dausa	—	87.87	102.47	111.76
13.	Dholpur	59.23	59.04	69.57	76.19
14.	Dungarpur	42.59	40.66	45.27	48.35
15.	Hanumangarh	—	107.38	109.62	104.46
16.	Jaipur	73.33	75.08	89.50	84.50
17.	Jaisalmer	4.87	9.51	15.05	16.31
18.	Jalore	67.90	72.94	74.49	78.02
19.	Jhalawar	65.27	70.96	73.57	91.58
20.	Jhunjhunu	95.15	97.76	116.00	103.77
21.	Jodhpur	48.70	58.13	61.05	62.13
22.	Karauli	—	—	59.38	69.10
23.	Kota	57.97	66.08	79.81	84.18
24.	Nagaur	68.53	73.66	84.06	79.99
25.	Pali	49.06	56.76	53.31	51.08
26.	Pratapgarh	—	—	—	65.06
27.	Rajsamand	—	30.48	27.83	21.91
28.	Sawai Madhopur	62.11	58.06	71.69	85.42
29.	Sikar	80.33	81.12	98.69	92.35
30.	Sirohi	36.33	36.11	40.70	33.05
31.	Sri Ganganagar	90.20	95.50	80.90	86.57
32.	Tonk	71.03	73.59	85.20	79.63
33.	Udaipur	25.77	21.69	25.33	20.65

Sources: Directorate of Economics and Statistics-2009-10; Agricultural Statistics of Rajasthan (2009-10).

Area Sown More than Once

It is an area on which crops are cultivated more than once during agricultural year. During 2009-10, it was 47.70 lakh hectares or 13.92 per cent of the total reporting area.

The spatial distribution at the district-level indicates that Jodhpur district has shown highest growth in area sown more than once. In 2009-10 it attained

56.73 per cent increase as against 0.94 per cent in 1985-86, 2.03 in 1995-96 and 4.79 in 2005-06 (Fig.4.9).

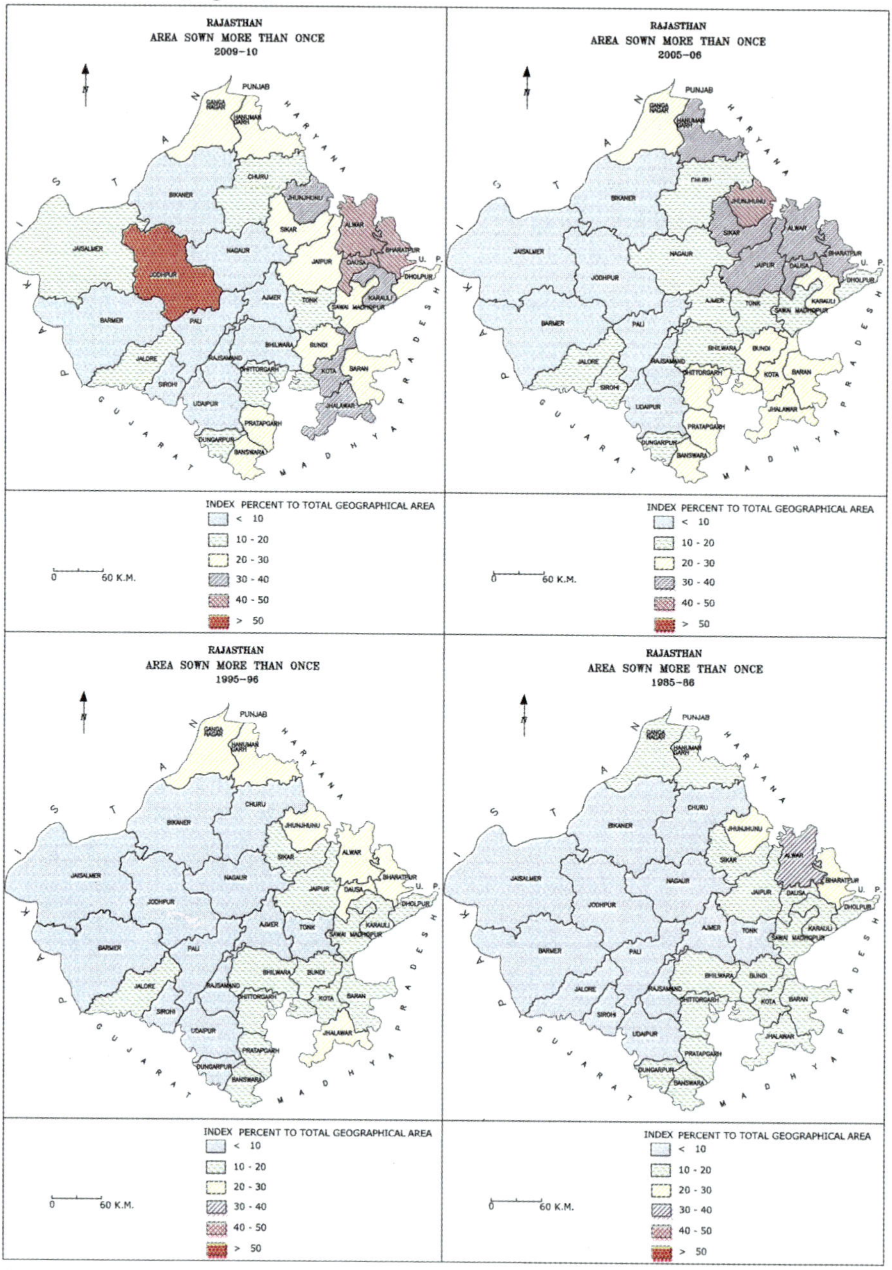

Source: Directorate of Economic and Statistics, Government of Rajasthan, Jaipur.

Fig. 4.9: Rajasthan Area Sown More than Once 1985-86, 1995-96, 2005-06, 2009-10.

Table 4.8 further indicates that out of all the 33 districts of Rajasthan, nearly 14 districts are showing a declining trend in this category during 1985-86, 1995-96, 2005-06 and 2009-10. However, the increasing trend has been reported in 19 districts (Table 4.8).

Table 4.8: Distribution of Area sown for more than once (Rajasthan)

Sl.No.	*Name of District*	*1985-86*	*1995-96*	*2005-06*	*2009-10*
1.	Ajmer	6.92	8.38	10.61	2.24
2.	Alwar	31.65	23.83	39.09	45.93
3.	Banswara	17.42	19.81	20.39	22.11
4.	Baran	—	12.93	23.88	29.73
5.	Barmer	0.60	2.24	2.96	4.22
6.	Bharatpur	25.05	20.74	34.62	40.86
7.	Bhilwara	10.55	12.43	16.10	6.15
8.	Bikaner	0.79	2.64	4.20	4.79
9.	Bundi	13.28	16.52	22.56	25.42
10.	Chittorgarh	16.78	15.44	23.49	16.62
11.	Churu	2.19	1.68	18.37	11.17
12.	Dausa	—	24.12	36.57	45.36
13.	Dholpur	12.00	11.27	19.55	25.18
14.	Dungarpur	11.40	9.40	13.60	13.77
15.	Hanumangarh	—	21.14	30.81	28.84
16.	Jaipur	16.73	15.75	33.22	25.41
17.	Jaisalmer	.00026	0.16	1.60	12.07
18.	Jalore	7.44	14.80	12.43	16.30
19.	Jhalawar	16.22	22.69	23.42	39.61
20.	Jhunjhunu	21.62	24.46	43.54	35.44
21.	Jodhpur	0.94	2.03	4.79	56.73
22.	Karauli	—	—	20.44	33.23
23.	Kota	10.27	15.55	27.04	33.67
24.	Nagaur	2.22	5.79	12.97	1.65
25.	Pali	2.83	7.29	4.05	2.37
26.	Pratapgarh	—	—	—	23.14
27.	Rajsamand	—	9.37	5.64	2.22
28.	Sawai Madhopur	13.74	10.55	14.07	27.77
29.	Sikar	14.20	14.21	30.06	26.28
30.	Sirohi	5.57	7.62	10.94	6.45
31.	Sri Ganganagar	14.09	23.46	21.86	24.82
32.	Tonk	7.61	7.73	19.00	15.43
33.	Udaipur	7.66	5.93	7.97	3.64

Sources: Directorate of Economics and Statistics 2009-10.
Agricultural Statistics of Rajasthan (2009-10).

Area, Production and Productivity

About 65 per cent population (56.5 million) of the State is dependent on agriculture and allied activities for their livelihood. The three major canal irrigation projects (Indira Gandhi, Chambal and Mahi), offer great help for agricultural development of the State. Agriculture in Rajasthan in primarily rainfed covering India 13.27 per cent of available land. The diversity and variability in climatic conditions of the State creates potentiality to develop certain belts of horticulture crops. The arid part of the state which receives not more than annual rainfall of 300 mm thrives on agriculture that is done with irrigation systems and painstaking efforts of the poor cultivators of Rajasthan. As a major portion of the state is parched, the risk and instability in agricultural production and productivity are quite high.

The AERC Report on the State Rajasthan Agriculture-2011-12 (December 2012) has brought out the fact that the economy of Rajasthan is largely agrarian in nature with high level of fluctuation in agricultural production and productivity that has resulted in wide fluctuation in Gross State Domestic Product (GSDP) of the state over the years. Despite this, the state economy has exhibited a healthy growth path during the recent past. The State's GSDP at current prices has increased by more than double, i.e. from Rs. 142236.14 crore in 2005-06 to Rs. 368319.52 crore in 2011-12. This has made Rajasthan as one of the fastest growing states of India (Mrutyunjay Swain *et al.*, 2012).

It is beyond doubt that agriculture forms the source of livelihood of the major population of the state, it is largely dependent on rainfall. Since, the rainfall amount is very scanty, sporadic and highly erratic, the expansion of irrigation provisions and efficient water management are major challenging task for the policy-makers. The Draft State Agriculture Policy (Government of Rajasthan, 2012a) highlighted following major challenges for agricultural sector in Rajasthan:

1. Frequent droughts leading to decline in productivity and reduce performance and even death of animals;
2. Climate change and global warming;
3. Strengthening of comprehensive technology-based developmental approach to promote dryland arid agriculture;
4. Deteriorating soil health including imbalanced use of fertilizers, micro-nutrient deficiency, lack of organic matter content, inadequate soil microbial flora and fauna;
5. Low productivity, unfavourable prices and practically very little value addition, distress sales, rising cost of cultivation;
6. Lack of efforts for stabilization of sand dunes and for greening the desert through agro-forestry programmes;
7. Missing mechanisms of export promotion, adherence to sanitation and phyto-sanitation;
8. Lack of integrated farming approach;
9. Lack of up-scaling of farm-validated modern technologies and agricultural innovations; and
10. Proper institutional mechanisms and organizational management reforms and overcoming the felt constraints coming in the farm prosperity in the State.

It is important to note that despite several major challenges for agricultural sector in Rajasthan, the economy has undergone considerable changes in the recent past in terms of growing manufacturing and service sectors, with reducing share of agriculture and livestock. Over the last five years (i.e. during 2007-08 to 2011-12) the share of agriculture, industry and service sectors to NSDP has changed from 23.6 per cent, 30.0 per cent, and 46.4 per cent in 2007-08 to 23.4 per cent, 27.7 per cent and 49.0 per cent in 2011-12 respectively. Thus, there has been no significant changes in relative share of the agricultural sector in NSDP during last five years (Mrutyunjay Swain *et al.*, 2012). About 5.4 million households continue to be engaged in farming, of which roughly half are small and marginal farmers. Increasing agricultural productivity is critical for future of the agriculture sector in Rajasthan. Given the size of the agricultural sector, improved agriculture productivity is also one key element for further structural changes in the over all economy of the State (Government of Rajasthan, 2012 b).

It has been noted that the relative share of agriculture and allied sectors has not changed significantly during 2007-08 to 2011-12. The structural changes in Rajasthan agriculture have been in favour of more growing oilseeds, pulses and horticultural crops. Rajasthan is the India's largest producer of mustard, pear millet (*Bajra*), and three spices (coriander, cumin and fenugreek), cluster beans, *isabgol* and second largest producer of maize. Rajasthan has the second largest herd of livestock amongst Indian States. Contributing about 10 per cent of the country's milk and 30 per cent of mutton production (Government of Rajasthan, 2012b). Agriculture and livestock production take place in major parts of Rajasthan often in extreme climatic conditions. In most parts of rainfed areas of the state, only one crop can be grown during the year. Therefore, farmers need to make agricultural practices more resilient in the light of ever harsher and changing climatic conditions.

Drivers of Growth in Agriculture

Bhalla and Singh (2009) highlighted that agricultural growth in any region can seen because of (i) growth in crop output; (ii) diversification in agriculture towards high valued crops and livestock products; and (iii) increase in value of the given output examining these three aspects of agriculture growth in Rajasthan reveals that the over all growth in area, field and production of major crops in the State is quite impressive in recent years. The share of oilseeds, pulses and horticultural crops has increased substantially during the last two decades which proves that the process of diversification of agriculture towards high valued crops is in right direction.

Irrigation

Apart from several other inputs irrigation is also the important input in crop production. The discussion about variability of rainfall, rainfall and temperature trends, water resources and drought (natural calamity) have been discussed earlier, it is important to highlight some of the features of the irrigation system in Rajasthan. Out of the total gross cropped area in the State during 2009-10 (217.4 lakh hectare) about 33.6 per cent area was irrigated. The net irrigated area was about 58.5 lakh hectares consisting 34.5 per cent of the net sown area in 2009-10. The cropping intensity and irrigation intensity in the state has gradually declined. The irrigation intensity has been fluctuating from 1991-92 to 2009-10. It was 121.21 per cent in 1991-92 which attained highest as 125.01 per cent in 2000-01, 125.52 per cent in 2007-08 and 126.66 per cent in 2008-09. In 2009-10 it again declined to 124.94 per cent. The highest cropping intensity during 1991-92 to 2009-10 was 133.15 per cent in 1998-99 which declined to 128.10 per cent in 2009-10. The index of cropping intensity has also been fluctuating from 98.83 in 1991-92, 103.47 in 2002-03, 109.76 in 2008-09 and declining to 108.38 in 2009-10. The index of irrigation intensity was 98.95 in 2002-03, 98.98 in 2004-05. The index of irrigation intensity has been increasing after 2005-06 due to increasing amount of rainfall. The index of irrigation intensity was less than 99.00 before 1996-97 on account of consecutive drought years (Table 4.9).

Table 4.9: Cropping, Irrigation Intensity and Index of Cropping, Irrigation Intensity in Rajasthan (Base year 1991-92 to 1993-94=100)

Year	*Cropping Intensity*	*Irrigation Intensity*	*Index of Cropping Intensity*	*Index of Irrigation Intensity*
1991-92	116.81	121.21	98.83	99.45
1992-93	119.06	122.71	100.73	100.68
1993-94	118.62	121.71	100.36	99.86
1994-95	119.73	119.69	101.30	98.20
1995-96	118.69	121.57	100.41	99.75
1996-97	123.25	120.66	104.27	99.00
1997-98	130.75	123.14	110.62	101.04
1998-99	133.15	123.81	112.65	101.58
1999-00	124.35	123.56	105.21	101.38
2000-01	121.21	125.01	102.55	102.57
2001-02	124.06	124.44	104.96	102.10
2002-03	122.30	120.60	103.47	98.95
2003-04	124.55	122.03	105.37	100.13
2004-05	127.27	120.63	107.68	98.98
2005-06	128.88	124.22	109.04	101.92
2006-07	128.45	122.51	108.68	100.52
2007-08	129.91	125.52	109.91	102.99
2008-09	129.74	126.66	109.76	103.92
2009-10	128.10	124.94	108.38	102.51

Sources: Directorate of Economics and Statistics 2009-10.
Agricultural Statistics of Rajasthan (2009-10).

Area, Production and Productivity Index

The study of area index, production index and productivity index has been studied from 1980-81 to 2009-10. It has been seen that the area index has been showing fluctuating trend with minimum of 76.54 in 2002-03 and highest of 128.12 in 1997-98. The area index has been lowest in 1987-88 and 2002-03 due to severe drought condition in the State. The production index is also showing the same trend being lowest of 95.50 in 1987-88 on account of drought condition. The highest production index was 321.24 in 2003-04. The production index has been showing more or less increasing trend except 1987-88 and 2002-03. The productivity index is showing increasing trend from 1981-82 to 1983-84 and again a declining trend till 1987-88 with 172.27 in 1988-89. Again the declining trend of productivity index is seen from 1989-90 to 1995-96. The productivity index was higher in the year 2003-04 after than declining trend has been observed till 2009-10 (Table 4.10 and Fig. 4.10).

Table 4.10: Area Index, Production Index and Productivity Index (Rajasthan)

Year	*Area Index*	*Production Index*	*Productivity Index*
1980-81	100.01	100.01	110.02
1981-82	105.02	115.05	110.50
1982-83	104.24	130.26	115.56
1983-84	108.26	160.16	130.15
1984-85	100.01	140.43	125.52
1985-86	108.25	135.60	120.25
1986-87	105.10	120.20	120.12
1987-88	80.01	95.50	110.25
1988-89	108.08	190.23	172.27
1989-90	103.09	165.94	155.53
1990-91	114.38	211.43	165.06
1991-92	107.56	181.89	147.16
1992-93	117.03	216.67	157.22
1993-94	109.58	156.59	121.90
1994-95	117.66	228.80	161.30
1995-96	113.53	211.77	148.08
1996-97	121.09	265.68	175.95
1997-98	–128.12	267.27	157.52
1998-99	122.55	257.71	154.86
1999-00	104.97	222.52	158.89
2000-01	99.37	173.45	156.40
2001-02	112.07	244.94	192.66
2002-03	76.54	133.47	148.01
2003-04	121.71	321.24	–246.13
2004-05	119.39	–289.60	197.12
2005-06	117.45	153.84	130.98
2006-07	115.03	167.63	145.72
2007-08	119.03	166.66	140.01
2008-09	121.58	185.09	152.24
2009-10	116.25	138.83	119.43

Sources: Directorate of Economics and Statistics 1985-86 to 2009-10.
Agricultural Statistics of Rajasthan (2009-10).

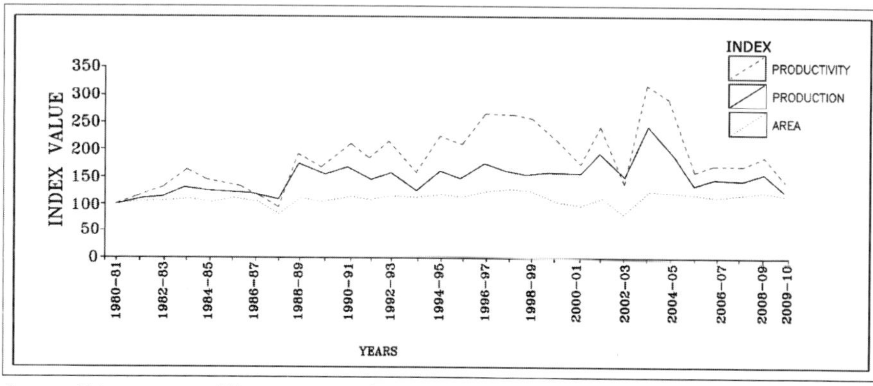

Source: Directorate of Economic and Statistics, Government of Rajasthan, Jaipur.
Fig. 4.10: Rajasthan: Index value, area, production and productivity

The analysis of land-use changes reveal the following trend in Rajasthan:

- The data reveals that the forest area is showing an increasing trend in Rajasthan from 6.50 per cent in 1985-86 to 7.98 per cent in 2009-10. However, the IRS data product of 2008 reported only 4.70 per cent area under forest cover in Rajasthan. There has been an increasing trend of forest cover in desert district from 1985-86 to 2002-10.
- The area not available for cultivation is showing a decreasing trend from 1985-86 to 2009-10. However, the land put to non-agricultural uses shows increasing trend.
- Other uncultivated land excluding fellow has been showing a declining trend during the study period (1985-86 to 2009-10).
- Fellow land has also been showing an increasing trend due to extension of irrigation facilities, fertilizers and other input in agriculture. Despite of irrigation facility in IGNP there has been declining trend in net sown area in Hanumangarh district.
- The same trend in total cropped area has been noticed except declining trend in Hanumangarh distict.
- The study further indicates that climate change will have less impact on land-use and agricultural production till the resource inputs are available. These resource inputs are in the form of soils, land, water availability, fertilizers and pesticides. The analysis of crop area, production and productivity indexes reveal that the rate of increase in productivity has shown a diminishing return. The impact of modern techniques of agricultural production is visible till 1980-90. But after that the impact of chemical fertilizers has failed to show any significant impact. It seems that despite of intensification of

agriculture, the agricultural productivity is increasing at a diminishing return. As a matter of fact it implies that climate change is playing more role. The analysis presented earlier on temperature and rainfall trends, indicate that the temperature has started increasing after 1970s. Since, the climate changes produce its impact after some period of time, the decreasing trend has begun after 1980-90.

5
Desertification and Climate Change

Introduction

Desertification is the greatest single environmental threat to the future well being of the earth. It has been estimated that about one-third of the land surface of the globe is affected by desertification. This area contains about 15 per cent of the total population of the world. The problem of desertification has become most acute in drought prone state of Rajasthan during last four five decades. The severe drought that extended from 1981 through 1987 focused the attention of the Government and scientists on the human aspects of land degradation and led to the convening of the seminars, discussions and relief works in drought affected areas of north-western part of India.

Defining Desertification: Changing Paradigm

Desertification is a term that has been in use since at least 1949 when Auberville, a perspective and well informed botanist and ecologist published a book on Climates, Forests, etc. desertification de L Afrique Tropical. He defined desertification as the change of productive land into a wasteland as the result of ruination by man-induced soil erosion. He associated it with the humid and sub-humid tropics where he worked. Desertification is a term employed rather loosely specially with reference to the deterioration of arid and semi-arid lands that are used beyond their capacity for sustained production by cultivators and herdsmen.

Despite the fact that a world conference has been held on the subject, there is no generally accepted definition of **'desertification'**. In the International Conference held at Nairobi in 1977, desertification has been defined as the "diminution or destruction of the biological potential of the land, which leads ultimately to desert like conditions and is an aspect of the widespread deterioration of eco-system under the combined pressures of adverse and fluctuating climate and excessive exploitation."

Unfortunately, there has been much confusion over the meaning of *desertification.* The largely invalid concept of expanding deserts and advancing sand dunes has become a more permanent image in the public eye than the less visible and much more serious phenomenon of land degradation in drylands which is addressed here this is the issue that affects so many people and is largely man-made. If fully recognized and tackled, it should be resoluable by man.

Desertification, as defined by UNEP, is land degradation in arid, semi-arid and dry sub-humid areas resulting mainly from human activities. This definition was modified by the 1992 UN Conference on Environment and Development (UNCED), to read as follows:

> "Desertification is land degradation in arid, semi-arid and dry sub-humid areas resulting from various factors, including climatic variations and human activities."

This definition has been internationally negotiated and approved a UNCED as the operational standard for Agenda 21. Unfortunately, however, it does include the three separate elements of short-term drought, long-term climate fluctuations and land degradation induced by human beings. Each of these is different and needs to be addressed in different ways, although there are interactions.

Droughts are natural phenomena which recur periodically. Severe and prolonged drought in degraded dry lands results in increased demands on the production of humid lands. In case of prolonged drought in certain key areas, global food security is already at a precarious state and this situation will become critical if desertification continues unabated.

Desertification and Droughts

A common misapprehension abut desertification is that it spreads from a desert core, like a ripple on a pond. The truth is that land degradation can and does occur far from any climatic desert; the presence or absence of nearby desert has no direct relation to desertification. The process being as a patch on the landscape where land abuse has become excessive. From that patch which might be around a watering point or in a cultivated field, land degradation spreads outward if the abuse continues. Ultimately, the patch may merge into homogeneous area, but that is unusual on a large scale.

The second misconception is that droughts are responsible for desertification. Ironically, association of the Sahelian drought with desertification had led to the common misconception that drought causes desertification. In truth drought do not cause desertification, but it only exaggerates the harmful effects of improper land management.

The researches all over the world indicate that desertification is caused by overgrazing, excessive woodcutting, land abuse, improper soil and water management and land disturbance; its effects appear as reduced productivity of land and environmental degradation. At UNCED the data of the conference were reviewed and thee possible causes of desertification have been defined.

1. Climatic change resulting from changes in global atmospheric circulation;
2. Mismanagement resulting from an inexorable build-up of pressure on resources reducing large areas to desert like conditions, i.e. the same effect on 1, but caused differently.
3. Biogeophysical feedback relating 2 to 1 by which means it is hypothesized that changes in land-use are able to affect changes in climate. Several theories have been proposed by scientists to seek a feedback mechanism between land-use and climate (The *dust theory*—inverse relation between dust and rainfall: The *albedo theory* direct relation between deforestation and reflectivity; the *moisture theory*—reduced vegetation and reduced soil moisture contribute diminution of rainfall; the *Carbondioxide Theory*—relates that high percentage of carbondioxide increases air temperature resulting global warming.

There are four views prevalent on the problem of desert expansion and desertification in Rajasthan. These are as follows:

Stationary Theory: The scientists of CAZRI held the view that Great Indian Desert is stationary since its formation some times in Mid-Miocene period. Some temporary modifications in the climate may, however, been experienced during the glacial period, but this did not appreciably affect the desert condition. According to them, the climate boundary of the desert is neither expanding nor deteriorating, but has been occupying essentially the same position. However, green revolution, introduction of irrigation, new tehnologies and better infrastructure have made drastic changes in agricultural land-use in post-independence era as a result farmers economy has improved. Contrary to it environmental issues like increased sand movement and atmospheric dust load have increased due to deep ploughing (Amal Kar, 2014).

Sand Spread Theory: Another alarming view is that desert has been spreading in the form of sand movement towards the north-eastern part of Rajasthan through the twelve Aravalli gaps (Dhabria, S.S., 1984).

Man-induced Sand Reactivation Theory: The third view holds that the existence of fossilized dunes and sand sheet are not result of current aeolian activity. They were deposited in the former arid phase and because of the increasing human and animal pressure in the form of cultivation and urbanization in Jaipur have been reactivised causing initiation and intensification of desertification hazard in this part (Sharma, 1988). This group also attributes desertification to the processes of modernization, as the expanding horizon of new science and technology aimed of maximum output from land or returns from these investments (Hooja, 1985).

Similar to other parts of the world, the process of desertification in Rajasthan is the result of a combination of several factors, some of which are natural and completely beyond human control (such as uncertainity of monsoon system and droughts), while others are the result of human and animal interaction with the environment.

Desertification in Rajasthan

Rajasthan has undergone a propound ecological change during the last two-three decades. The State embraces a huge area or approximately 3,42,262 square kilometers and of population of 3,41,08,292 (1981), 4,40,05,990 (1991) and 68.62 crores (2011). It is the first largest state of India in terms of area and occupies 10.41 per cent area of India. The Aravalli range divide the State into two halves and they have undergone massive deforestation during the last 30 years or so. The north-western twelve districts consist of the Thar Desert comprising various types of sand dunes and rocky plateau. The state is frequently drought prone and the occurrence of droughts has rendered the problem of desertification very acute.

Overgrazing, soil erosion and salinization in irrigated land are long standing problems of Rajasthan. Though waterlogging and salinization are of relatively recent origin, ameliorative measures to reduce land degradation have been undertaken by Government in the affected areas. The problem is an immense one in terms of the area involved and amount of degradation that has occurred.

Desertification Characteristics: Historical Perspective

Man first placed his imprint on the landscape of Rajasthan thousands years ago. Historical records and evidences make it clear that the destructive impacts of man was already apparent five thousand years ago in the Ghaggar and lost Saraswati plains. It was then what has been called the first period of accelerated man made erosion began. Wood cutting and cultivation of slopping lands of the Aravallis and sand dunes were responsible for most of the land degradation that occurred in the Aravallis and around former fertile plains of the Ghaggar, Chambal, Luni and Banas rivers of Rajasthan. Sedimentation of canals, dams, waterlogging and salinization combined of raise havoc in most of the river valleys of Rajasthan.

In recent years grazing pressure seems to have become a much greater problems than it was before. There has been an expansion of human and livestock population between 1971-1981 having higher growth rate than 2011. Cultivation has been extended into pasture lands. Wells have been constructed in arid and semi-arid parts and improved transportation facilities have made formerly remote areas accessible to livestock. As a consequence, cropping has become a high risk land-use in the climatically marginal areas that were formerly the best pasture lands. Destruction of vegetations on the Aravallies, sandy lands and upland has led to well spread soil erosion, depletion of soil fertility.

Location and Type : Areas of Desertification

The process and characteristics of desertification differ under different morpho-climatic zones of Rajasthan. Desertified lands are mainly distributed through western district which generally fall under arid and transitional between arid and semi-arid morphogenetic regions and the vast areas to the north-east of Aravallis, extending from the Sambhar inland basin to the border of Haryana and U.P. It is also found in relatively small areas of south-eastern uplands parts of Rajasthan and the eastern plain of the Banganga and Chambal and its tributaries. The desertified lands may be classified into four major areas according to origin and morphogenetic characteristics:

Desertification in Arid Region

The three hundred millimeter isohyet forms the boundary of this region. The districts of Barmer, Jaisalmer, Jodhpur, Bikaner, part of Jalore and Ganganagar fall under this region. The area is mostly dunal except parts of Jodhpur and Jalore. The very severe desertified lands are concentrated in western part of Jaisalmer and Barmer where 80 to 100 per cent area is under sand dunes of different heights and types. As we proceed towards north-east and south-east of this complex dunal country, strong to moderate sheet and gully erosion has rendered the rocky plateau under serious threat of desertification. In the Indira Gandhi Canal Command Area, the problem of Salinization and water logging has cropped-up. Uneconomic and intensive use of canal water in Ganganagar and Bikaner have developed some alarming problems. In these areas due to sandy terrain and seepage losses from the main canal, distributaries and field channels, the groundwater table is rising at an alarming rate of 0.3 to 0.83 meter per year. In the Ghaggar flood area round Suratgarh, Baropal and Manek-thar areas water table has risen by 5-6 meters within past 30 years. Due to this rise in water level 6920 MCM water is being added to the aquifer annually leading to the development of soil salinity (*kankar*) in many places around 18 reservoirs of the Ghaggar Diversion Scheme. A study conducted by Dhiman (1986) on the basis of remote sensing technique reveals that the critical marshy lands and salinized land are increasing at a faster rate since 1972-1986. The studies of CAZRI in Nachna area reveal that rise of water table has also caused seepage problem where canal water has only been introduced recently. It has been estimated if in the canal irrigated area, the present rate of groundwater is allowed to continue, then, more than 40 per cent of the canal command are is likely to develop drainage and salinity problem within 50-100 years. Canal irrigation and subsequent agricultural practices on stablished dunes and marginal area of Ganganagar, Bhadra and Bikaner are causing fresh reactivation of sand dunes and formation of new sand dunes.

In contrast to above mentioned areas of the canal command of arid morphogenetic region water table has gone down to 15 to 20 meters due to consecutive droughts from 1982-87, 2002-2003 and 2009 and human and animal consumption. The growth rate of human population and animal population during 1971-81 has been reported highest from 42 to 47 per cent which is highest so far as the deserts of the world are concerned. However, the growth rate of population during 2001-2011 is still higher in Barmer (32.55%), Jaisalmer (32.22%) and Jodhpur (27.69%) against the growth rate of Rajasthan which is 21.44 per cent.

Desertification in Transitional Region

The area between the Aravallis and 300 mm. isohyte is under severe desertification hazard. The area is mostly characterized by moderate wind, sheet and gully erosion. The strong wind and gully erosion hazards have been observed between Nagaur and Churu. Because of intense human and animal pressure the reactivation of dunes has set in, forming new barchan dunes in the vicinity of Churu, Jhunjhunu and Nagaur. Strong sheet erosion is perceptible in western slopes of the Aravallis due to largescale deforestation and cultivation on piedmont marginal areas. The growth rate of population in these areas was 27 to 32 per cent between 1971-81 which was highest in Churu and Jalore districts. The growth rate during 2001-2011 has shown a declining trend in these districts (Churu 20.35% and Jalore 26.31%), but it is still higher than the state. In some of the areas of this morphogenetic zone desertified lands are found around water bodies in the lower reaches of the water courses. This type of desertified land is mainly linked to the changes of river courses and to exhaution of water resources. The abandonment of Khadins in this region is usually a result of either decrease of water volume or drying-up of water resources due to consecutive droughts. Other causes of desertification in this region include the reactivation and shifting sand on the edge of inter-dunal flates due to destruction of vegetation on sand mounds through collection of firewood and overgrazing.

Desertification in Semi-arid Region

The areas adjacent to desert margin like semi-arid region of Rajasthan is vulnerable to desertification hazard. The area is characterised by widespread sheets and obstacle dunes on the windward sides of the isolated hogbacks and hills. Particularly large field occurs in the Sambhar area, west of Jaipur and extended up to Jaipur. Dunes extend a considerable distance east and north of Jaipur up to Kotputli. A large number of obstacle dunes are found at Banada, Jhir, Dausa, Lalsot, Karauli, Gangapur, Newai, Acharol and Tonk. This appears to mark the easternmost extension of the aeolian features in Rajasthan. The strong to very severe sheet and gully erosion has been noticed in this zone which has been accelerated during last 20-30 years as a result of large scale deforestation from the Aravallis, over cultivation and excessive biological stress. The intricate network of ravines and gullies along the lower Chambal river is evident which may be classified under severe desertification stage. The Jaipur district records highest growth rate of population (26.91%) between 2001-2011 in this semi-arid zone. In Dausa (23.75%), Dholpur (22.78%), and Alwar (22.75%) districts, the growth rate is still higher than Rajasthan. The consecutive droughts of previous years have also accelerated the effects of desertification. The surface reservoirs have dried up. The river course has been silted due to accelerated sheet erosion from the Aravallis. The introduction of canal irrigation in this semi-arid morphogenetic region especially in the Morel Command Area, Parbati Command Area, the saline soils are developing causing decline in productivity of soils. The systematic mapping of the Moral Dam Command Area on the basis of landsat imageries reveal nearly 1153 hectares of the command area is under threat of salinity which constitute nearly 2.91 per cent of the area.

Desertification in Sub-humid and Humid Region

The nature and processes of desertification in this morpho-climatic zone is different and desertified lands are confined in patches where deforestation has been maximum. The Aravalli hills, Chittor hills, uplands of tribal Banswara and Dungarpur areas are under threat of desertification. The cutting of forest for fire wood needs and clearance of forest for construction of Mahi, Bajaj Sagar Dam and series of dams along with Chambal resulted into strong to severe sheet erosion. The growth rate of population of these areas has been reported from 20 to 37 per cent. The parts of Kota and Banswara districts have reported 32 per cent growth rate during 1981-91 which has shown a declining trend during 2001-2011. However, Kota (24.39%) and Banswara (26.58%) still have a higher growth rate than state. Overgrazing is another cause of desertification. The nomads of eastern Rajasthan generally cross these areas for search of fodder and good pasture lands in adjoining districts of Madhya Pradesh Therefore, these areas are under the grip of overgrazing causing serious ecological and socio-economic problems in these areas.

Causes of Desertification in Rajasthan

The following arc thc main causes of desertification in Rajasthan:

Rainfall Trends in Rajasthan

The study of rainfall trends of the last 112 years and more in Rajasthan shows that certain stations have recently experienced apparent downward trend in rainfall, whereas others have shown a high variability from year to year, but no constant trend has been noticed. Now, the question arises, is there any widespread evidence that rainfall in the Indian desert and adjoining areas is decreasing or increasing? If so, is a fluctuation in progress, or is the decrease a lasting change? On a very long time scale there have certainly been great changes in climate specially within the Quarternary period. During the past 10,000 years Rajasthan has been alternating between arid and humid phases resulting in accumulation of sand dunes especially during the arid phase.

There is thus plenty of evidence of past-desiccation of climate that led to the spread of the desert surface in the north-eastern part of the present desert boundary. Is there evidence of a similar desiccation in progress today? Is the widespread desertification now in progress due to such a downward trend of rainfall? There is strong evidence to suspect that more and more humid and sub-humid zones of the country, which had earlier experienced no significant moisture stress, are now under the intensification of man-induced desertification process.

The unprecedented drought situation that has occurred during 1987, 2003 and 2009 years in Rajasthan is not a new thing. There are plenty of references of such a situation in the Vedic literature and in the recorded history of India. The droughts of 1877, 1900, 1918, 1940, 1965, 1972, 1979, 1984-85, 1986-87 and 2009 are well known because of their disastrous consequences. The most notable characteristics of such droughts are a random distribution in time, spatial coherence over large and in some regions, persistence for years at an end.

Climate Desertification Linkages

The prolonged drought for four consecutive years has prompted the question whether the climate is being affected by human action? There are two mechanisms whereby this might happen. First overgrazing, unwise use of land and stripping of forest cover, all tend to make the surface more reflective to solar radiation in this part of the country. Dynamic modelling indicates that such raised albedos tend to diminish rainfall and further potential disastrous positive feedback. Second, in regions like Rajasthan, far from the ocean, much of the rainfall is caused by re-evaporated soil moisture. Faulty land-use practices have reduced the storage capacity of enhanced evaporation.

Moisture Index (P/PE) and Desertification

Gore, P.G., *et al.* (2011) using rainfall normal for the periods from 1901 to 1950 and 1941-1990 based on moisture Index classification of the moisture index over India have calculated for arid, semi-arid and dry subhumid as P/PE=0.05-0.20, 0.21.05, and 0.51-0.65 respectively. In the state of Rajasthan, the districts falling in arid, semiarid and dry subhumid having moisture indexes in different regions is given in table 5.1.

Table 5.1: Moisture Index in Different Regions of Rajasthan

Arid (P/PE = 0.05-0.20)			*Semi-arid (P/PE = 0.21-0.50)*			*Dry Sub-humid (P/PE = 0.51-0.65)*		
	1901-1950	1941-1990		1901-1950	1941-1990		1901-1950	1941-1990
Rajasthan			*Rajasthan*			*Rajasthan*		
District	—	—	*District*	—	—	*District*	—	—
Sri Ganganagar	0.15	0.16	Jhunjhunu	0.27	0.28	Sawai Madhopur	0.51	0.57
Hanumangarh	0.11	0.14	Alwar	0.44	0.44	Bundi	0.52	0.53
Bikaner	0.15	0.17	Sikar	0.31	0.33	Kota	0.58	0.59
Barmer	0.15	0.15	Bharatpur	0.39	0.39	Chittorgarh	0.54	0.53
Jodhpur	0.17	0.20	Jaipur	0.31	0.38			
Jaisalmer	0.08	0.09	Ajmer	0.34	0.37			
			Tonk	0.44	0.46			
			Sirohi	0.26	0.37			
			Bhilawara	0.45	0.45			
			Pali	0.23	0.24			
			Jalore	0.22	0.23			
			Churu	0.21	0.24			
			Udaipur	0.45	0.45			

Source: Impact of climate change of land degradation over India:
National Climate Centre, RR No. 1/2011, May 2011
(NCC Research Report by P.G. Gore, B.A. Roy and H.R. Hatwar).

The analysis of Table 5.1 reveals that in Rajasthan the districts like Sri Ganganagar, Anupgarh, Bikaner, Jodhpur and Jaisalmer (Arid region), Jhunjhunu, Sikar, Jaipur, Ajmer, Tonk, Sirohi, Pali, Jalore and Churu (Semi-arid region) and Sawai Madhopur and Bundi (Dry sub-humid region) show increase in P/PE ratio. The increase is significant in Sirohi, Jaipur and Sawai Madhopur districts. The districts like Kota and Chittorgarh show decrease in P/PE ratio. The districts like Barmer, Alwar, Bharatpur, Bhilwara and Udaipur from semi-arid region show no change. However, the stations having significant change in P/PE ratio from 1901-1950 to 1941-1990 are Sirohi, Jaipur and Sawai Madhopur.

The decrease in P/PE ratio values for the two periods as above indicates land degradation to less soil moisture availability. There are 23 districts in Rajasthan from three climatic regions show land degradation of which three districts (Sirohi, Jaipur and Sawai Madhopur) show significant land degradation.

Recently, Ralnapriya and Padmini Pani (2014) understanding causes of land degradation in relation to climate and deforestation have indicated that in Rajasthan monsoon rainfall is moderately correlated with land degradation having 0.54, therefore, deforestation or less vegetative cover and population pressure are the main cause of land degradation.

Population Pressure and Management of Resources

Successive census reports have revealed that population of Rajasthan has been growing at an alarming rate. It has registered more than double over the past thirty years, having risen from 1.59 crores in 1951 to 3.41 crores in 1981, 68.62 crores in 2011. The highest growth rate, however, has been recorded during 1971-81. In absolute terms, population has increased by 238 lakhs during last 80 years. In 1981, Rajasthan has recorded a growth rate of 32.97 per cent as compared to the national growth rate of 24.66 per cent which has declined to 21.44 per cent, but still, higher than the national average (17.64 %). In particular, the western arid districts have registered an abnormal growth. The alarming growth rate led to an adverse environmental situation enumerated as follows:

1. Reduction in man-land ratio which is normally devoid of adequate natural sustenance. There is a corresponding shrinkage of cultivated land.
2. Rise in net shown area at the cost of grazing land and marginal land.
3. Over dependence on agriculture has led to land degradation due to non-availability of adequate cultivated lands.
4. Urbanization and industrialization in Kota, Jaipur, Banswara and Dungarpur districts have further aggravated the man land situation and reduction of forests for fuel and housing material.

Animal Pressure

Similar to increase in human population, there has been a rise in number of livestock population (Table 5.2). This has led to overgrazing, deforestation and exploitation of the woody bio-mass of the region.

Table 5.2: Growth of Livestock in Rajasthan

Sl. No.	*Type of Animals*	*1956*	*1961*	*1966*	*1972*	*1977*	*1983*	*Variation in 2007 over 2003 % increase (+) or decrease (—)*	*Variation in 2012 (P) over 2007 % increase (+) or decrease (—)*
1.	Cows	+12	+09	–09	–05	+03	+05	—	9.94
2.	Buffaloes	+13	+17	+05	+07	+11	+19	(+) 6.018	16.99
3.	Sheep	+37	–16	+20	–03	+16	+35	(+)11.111	(–)18.86
4.	Goat	+58	–08	+28	+18	+02	+25	(+)27.794	0.76
5.	Camel	+33	+25	+20	+16	+01	–53	(–)15.643	(–)22.79
6.	Pig	+32	–02	+16	+40	+11	+37	(–)38.120	(–)13.96
7.	Poultry	+39	+75	+14	+50	+25	+46	(–)20.123	
	Total	**+26**	**+02**	**+11**	**+04**	**+06**	**+23**	**(+)15.068**	**1.89**

Source: Some Facts About Rajasthan, Department of Planning Rajasthan, Jaipur, 2015.

Table 5.2 indicates that number of livestock in Rajasthan is showing an increasing trend except camal, pigs and poultry. It is evident that the number of camel are declining since 1977, but drastic reduction is noticed after 1983 (–53%). However, the variation in 2007 over 2003 is —15.643. The total livestock population in 2007 has been reported as 56663183 and the variation in 2007 over 2003 is 15.068 per cent. The percentage of goats and sheep are highest in the order 37.89 per cent and 20.26 per cent respectively. Cattle and buffaloes are 21.05 per cent and 19.57 per cent respectively (Census-2007) in rural areas. The number of sheep and camels have declined in 2012 our 2007 (Table 5.2).

Land-use Changes

Due to increased human and livestock population, vast area of Rajasthan have been brought under cultivation often extending to marginal lands, piedmont slopes of the Aravallis and other fringe areas, where the physical conditions do not warrant farming. The detailed analysis of land-use changes has already been discussed in previous chapter.

The official data shows an increase of forest area from 6.50 per cent in 1985-86 to 7.98 in 2009-10, but the satellite data 2008 reveals that the forest area in Rajasthan is 4.70 per cent (India State Forest Report 2011). The field study reveales that degradation of forests of the Aravalli and other parts of Rajasthan is taking place at an alarming rate. The total forest area in the state is 32,639 sq. km. which is 9.54 per cent (2011) of the total geographical area. If this rate of depletion will continue, the Aravallis will be barren by the mid of twenty-first century.

Massive investments on various irrigation projects in Rajasthan have been made, but the returns are not commensurate. Introduction of irrigation farming without adequate drainage measures has given rise to accumulation of minerals and salts in the soil and near surface formation in a tract having naturally impeded drainage and heavily charged with minerals and salts. All these factors have led to diminishing returns from land and intensification of man-induced desertification processes.

Development of new settlements on stabilized dunes, urbanization and construction of roads without regard to drainage channels (as in case of Jaipur city), canal without regard to physical constraints have led to rapid mobilization of stable sand belts and consequent desertification by removal of vegetation cover and wind action.

Status of Desertification in Rajasthan

No scientific data was available about the status of desertification in India and Rajasthan till 2002-2003 and 2005-06. In the year 2000, the Regional Action Programme (RAP) to combat desertification was formulated by UN Convention to Combat Desertification (UNCCD) and accordingly RAP for Asian region had been initiated as collective effort by the member countries. Under RAP Thematic Programme Network (TPN-I) is on desertification Monitoring and Assessment. In India Space Application Centre (ISAC), Ahmedabad has been identified as the national focal institution to coordinate TPN-I activities and establish the national network for desertification monitoring and assessment. To implement TPN-I activities in India, a working group was constituted by SAC in which 17 institutions were involved. Accordingly the mapping of status of desertification was worked out on the scale of RF.I: 5,00,000 and an Atlas on Desertification/land degradation has been published in 2007 by Government of India (Ajai, 2007). A separate Atlas has also been published for pilot projects at district-level on 1:50,000 scale in the same year.

As per earliar analysis done by SAC on the basis of AWiFS data 2005-06 about 67 per cent of area was affected due to desertification/land degradation in Rajasthan. Out of this figure the most significant process of desertification in Rajasthan is wind erosion which accounts for 44.42 per cent of area, followed by water erosion (11.22%), vegetal degradation (6.25%), and salinization (1.07%) in canal command areas,water logging (0.01%), man made (0.09%) and others (4.36%) only 32.58 per cent area is unaffected by the process of desertification in the state (Ajai, 2007).

However, in 2013 a fresh study on spatio-temporal status of desertification was initiated by Space Application Centre, Ahmedabad under the heading

Second Cycle of desertification mapping. The study is based on multi-temporal digital IRS AWiFS data (2003-05 and 2011-13), ancillary information and aided data of Forest Survey of India to bring out the changes in desertification and land degradation status of country including Rajasthan (SAC, 2016).

The present study conducted by SAC, Ahmedabad reveals that Rajasthan is the state with highest area under desertification/land degradation with respect to country TGA and second highest area under desertification/land degradation with respect to state TGA. The state is observed with 62.90 per cent of the total geographical area under desertification/land degradation for the period of 2011-13. The desertification/land degradation area in Rajasthan has decreased about 0.29 per cent since 2003-05 (Figs. 5.1 and 5.2).

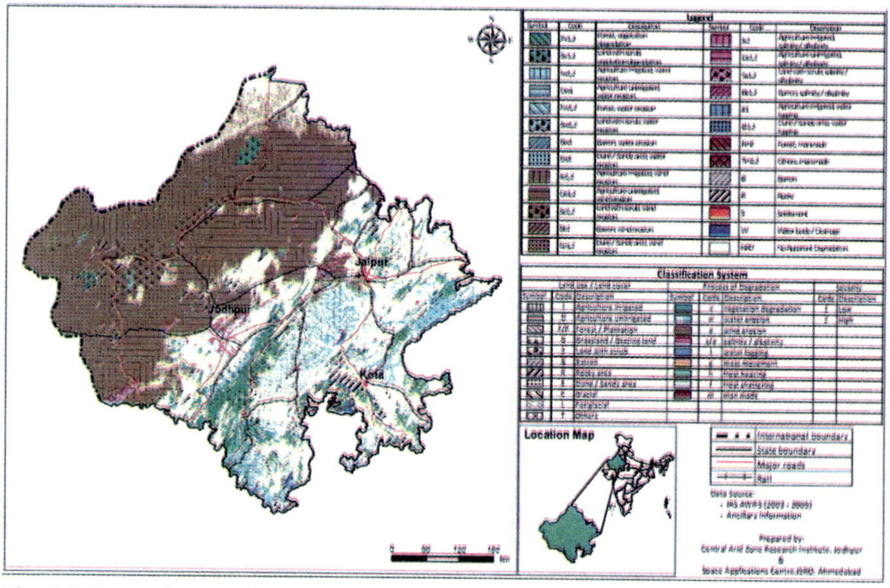

Fig. 5.1: Desertification/Land Degradation Status Map of Rajasthan (2003-05).

Source: Desertification and Land Degradation Atlas of India, Space Application Centre, ISRO, Government of India, 2016.

Desertification Processes in Rajasthan

The most significant process of desertification/land degradation in Rajasthan is Wind Erosion (44.41% in 2011-13 and 44.80% in 2003-05) followed by Vegetation Degradation (7.62% in 2011-13 and 7.59% in 2003-05) and Water Erosion (6.18% in both 2011-13 and 2003-05) (Fig. 5.3 and Table 5.3).

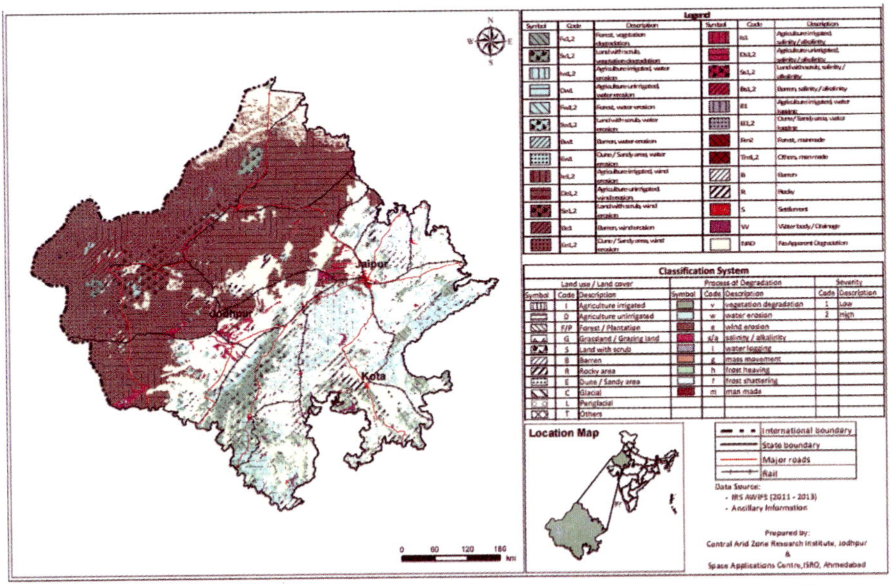

Fig. 5.2: Desertification/Land Degradation Status Map of Rajasthan (2011-13).

Source: Desertification and Land Degradation Atlas of India, Space Application Centre, ISRO, Government of India, 2016.

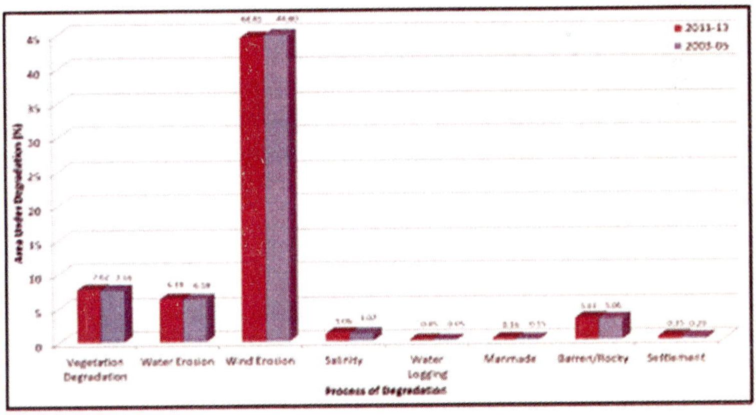

Fig. 5.3: Desertification Processes in Rajasthan.

Source: Desertification and Land Degradation Atlas of India, Space Application Centre, ISRO, Government of India, 2016.

Table 5.3: Processes of Desertification in Rajasthan

Process of Desertification/Land Degradation	*2011-13*		*2003-05*		*Change (ha)*
	Area (ha)	*Area (%)*	*Area (ha)*	*Area (%)*	*(2011-13) - (2003-05)*
Vegetation Degradation	26062121	7.62	2596003	7.59	10218
Water Erosion	2116314	6.18	2116082	6.18	233
Wind Erosion	15197874	44.41	15332054	44.80	–134180
Salinity	363768	1.06	365666	1.07	–1898
Water Logging	18421	0.05	18421	0.05	0
Manmade	53058	0.16	50865	0.15	2193
Barren/Rocky	1052374	3.07	1047818	3.06	4556
Settlement	118482	0.35	98696	0.29	19786
Total Area under Desertification	**21526512**	**62.90**	**21625604**	**63.19**	**–99092**
No Apparent Degradation	**12546925**	**36.66**	**12448140**	**36.37**	**98785**
Total Geographical Area (ha)			**34223900**		

Source: Desertification and Land Degradation Atlas of India, Space Application Centre, ISRO, Government of India, 2016.

The spatio-temporal analysis of desertification indicates that the area under desertification in Rajasthan has reduced by 0.29 per cent or 9992 hectares. This reduction in desertification and land degradation is as a result of combating desertification programmes of the Central and State Governments such as DDP, DPAP, CDP and watershed schemes. The increase in water bodies is because of increasing trend of rainfall after 2003. Table 5.3 further reveals that the area under vegetal degradation, man made activities and settlements have increasd. However, the wind erosion and problem of salinity have shown a declining trend in contraction to the study conducted by Amal Kar (2012) where he stated that "there is likely hood of aridity spreading beyond the present eastern boundary .... wind erosivity is expected to increase .... destabilizing the sandy landscapes contributing more of the atmospheric dust load".

Monitoring of Desertification (Change Detection)

In order to work out the changes in the status of desertification the SAC has started II cycle mapping with the help of LISS III data of 2002-2003 and 2012-2013. The analysis is still under progress. However, the analysis of two districts—Ajmer and Dausa is under consideration to understand the changes in the status of desertification on LISS data of 2002-2003 and 2012-2013. The details of these districts are being discussed below:

Ajmer district

Ajmer district is situated in the centre of Rajasthan having an area of 8,481 sq. kms and population of 2,180,528 in 2001 and 2,584,913 in 2011. The population growth rate over the decade 2001-2011 was 18.48 per cent which is less than the State's growth rate 21.44 per cent. The density of population is 305 person per square kilometers, which is higher than 2001 (257). The climate is semi-arid. The district has a hot dry summer and a cold bracing winter. The average annual rainfall in the district as a whole is 454.9 mm. On an average there are 21 rainy days in a year in the district. The year 2002-03 was the drought year and rainfall is showing an increasing trend after 2009.

Dausa district

The area of district is 3432 sq.kms. The total population of the district in 2001 was 1,317,063 which increased to 1,63,4409 in 2011. The growth rate of population between 2001 and 2011 was 24.09 per cent which is higher than national growth rate and even the state (21.44%). The density of population is 476 persons per sq. km. as per 2011which has increased over 2001 (384/sq.km.) The climate of the district is semi-arid. The district has a dry climate except during the south-west monsoon season. The average rainfall in the district is 679.8 mm. Rainfall is uniform throughout the district. On an average there are 34 rainy days in a year. The year 2002-03 was the drought year and after 2009 rainfall is showing an increasing trend.

Geomorphologically, the district may be divided into landform units of fluvial origin, denudational origin and hills mainly structural. Terrain is mainly undulating produced by extensive deposition of alluvium. Ravines are confined to major and small tributaries of the Moral. These narrow, deep depressions are generally carved out by the fluvial action. Structural hills are linear arcuate hill showing definite trend lines with varying lithology with faulting and folding. Pediments are broad gently sloping rock flooring erosion surface of low relief between hill and plain.

Soils are greyish brown to brown and yellowish brown with wide variations in texture from sandy loam to clay loam having a clear upper boundary of argillic horizon.The forest area of the district was 7.67 per cent in 2009-2010.

Data Used

IRS LISS-III data of two different time series (2002-03 and 2012-13) have been used. During each of these years, three season viz. *kharif, rabi* and summer have been used. The data set have been used for vulnerable areas desertification status mapping at 1:50,000 scale.

The Classification System

The classification system adopted is a three level hierarchical classification system. The following steps were followed in standardizing the Classification System for DSM:

Table 5.4: Level 1: Land-Use/Land Cover.

The Following categories have been indentified as below:

Agriculture-Unirrigated	(D)
Agriculture- Irrigated	(I)
Forest/ Plantation	(F)/P*
Grassland/Grazing Land	(G)
Land with Scrub	(S)**
Barren/Rocky Area	(B/R)# [B(Sc) indicating Scree area in cold Deserts
Dune/Sandy Area	(E)
Water body/Drainage	(W)
Glacial/Periglacial	C/L
Others	(T)

Source: Classification of Land-use as provided by SAC, Ahmedabad.

Table 5.5: Level 2: Processes of Degradation

Type of processes resulting in degradation:

Vegetation Degradation	(v)
Water Erosion	(w)*
Wind Erosion	(e)
Water Logging	(i)
Salinization/Alkalinization	(s/a)**
Mass Movement (in cold areas)	(g)
Frost Heaving (in cold areas)	(h)
Frost Shattering (in cold areas)	(f)
Man made	(M)

Source: Classification of Land-use as provided by SAC, Ahmedabad.

Table 5.6: Level 3: Severity of Degradation

This level represents the degree and severity of the degradation:

Slight	1
Moderate	2
Severe	3

Source: Severity Classes as provided by SAC, Ahmedabad.

The Methodology followed as per flow chart given below (Fig. 5.4) :—

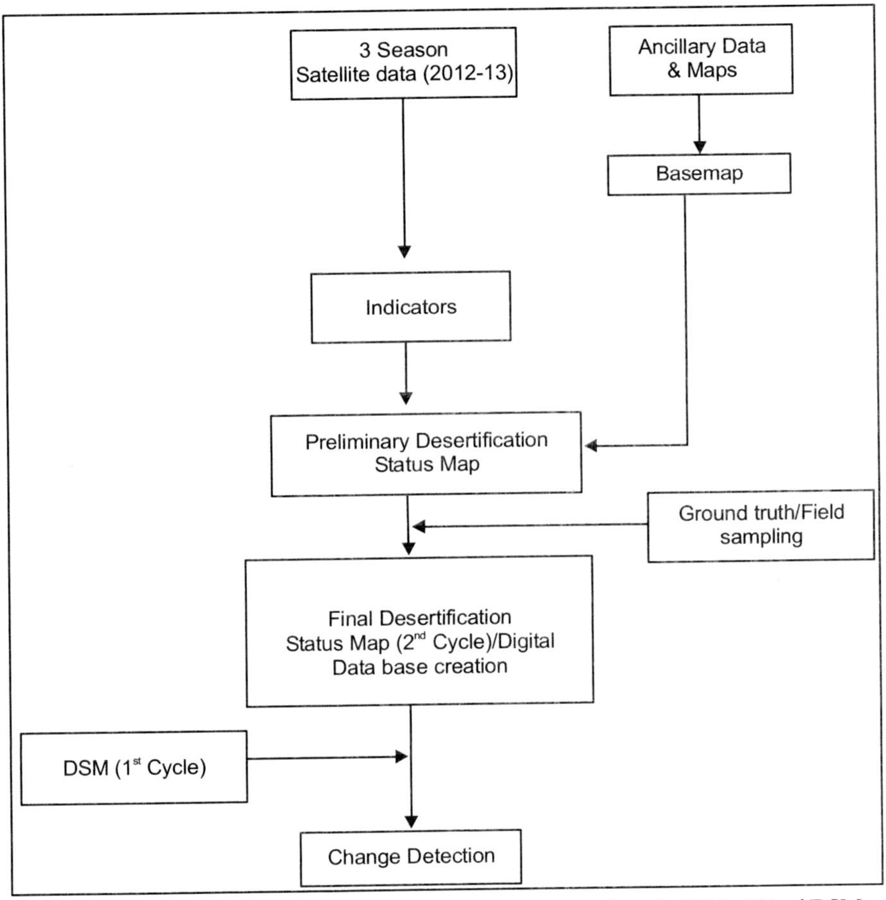

Fig. 5.4: Methodology for First LISS-III (2002-03) & second cycle (2012-13) of DSM

Change Detection in Ajmer District

Table 5.7 indicates the status of desertification in 2002-03 in Ajmer district. The total area under desertification in the district was 1211.853 sq. kms. The maximum area was affected by the process of vegetal degradation which accounts for nearly 7.00 per cent. Nearly 2.00 per cent area was under water erosion. The saline area accounted for nearly 3.00 per cent. The comparative study of tables 5.7 and 5.8 make it evident that there has been a decrease of 45.85 sq. kms area under the process of desertification/land degradation. It is further evident that the area under water bodies has increased in 2012-13 because of increasing amount of rainfall. The district suffered severe drought in the year 2002-03 which accelerated the process of desertification in the absence of good vegetal cover. The man made activities have also increased from 2002-03 to 2012-13 (Figs. 5.5, 5.6, 5.7 and 5.8).

Table 5.7: Desertification Status Mapping of Ajmer District Category-wise Area in sq. km. with percentage (2003-04)

	DSMI (2003-04)		
Raw Labels		*Area sq.km*	*% in ref. total geographical area*
Bs	(Barren)	29.56835	0.336805825
Bs1	(Barren saline slight)	102.236593	1.16455196
Bs2	(Barren saline moderate)	173.917451	1.98105104
Bv2	(Barren vegetal moderate)	2.601663	0.02963490
Bw1	(Barren water erosion slight)	54.345239	0.61903329
Bw2	(Barren water erosion moderate	4.64264	0.05288317
Ee1	(Dune/sandy wind erosion slight)	23.391396	0.26644565
Ee2	(Dune/sandy wind erosion moderate)	23.62935	0.26915613
Ee3	(Dune/sandy wind erosion severe)	8.600877	0.09797048
Fv1	(Forest vegetal degradation slight)	103.741413	1.18169
Fv2	(Forest vegetal degradation moderate)	27.170624	0.30949391
NAD	(No Apparent Degradation)	7138.448491	81.3123165
NAD		1.957359	0.02229579
NAD			
NAD		0.037356	0.00042551
NAD		7140.443206	81.3350378
R	(Rocky Area)	52.001647	0.59233801
Settlement		82.913638	0.94444892
Sv1	(Land with scrub vegetal degradation slight)	342.872704	3.90557890
Sv2	(Land with scrub vegetal degradation moderate)	249.466637	2.84161329
W	(Water body)	3.64119	0.04147590
Tm2	(others)	1.478772	0.01684432
Tm3	(others)	5.608865	0.06388920
W	(Water body)	343.839522	3.9165916
We2	(Water wind erosion moderate)	2.937692	0.03346252
Grand Total		**8779.049469**	**100**
		603.068168	
		553.845596	
		8786.55058	
		2.937692	
DSMI		1211.853103	
	45.844267 Difference		
		1166.008836	
%charge		3.931725523	

Source: IRS Data—SAC, Ahmedabad (2003-04).

Table 5.8: Desertification status Mapping of Ajmer District Category-wise Area in sq. km. with percentage (2011-12)

	DSMI (2012-13)		
Row Labels	*Row Labels*	*Sum of area_sqkm*	*% In ref.total geographical area*
B (Barren)	B	5.782385	0.0658095
Bs1 (Barren saline slight)	Bs 1	61.276497	0.69738968
Bs 2 (Barren saline moderate)	Bs 2	186.023893	2.117143597
Bs 3 (Barren saline severe)	Bs 3	0.388941	0.004426549
Bv 1 (Barren vegetal slight)	Bv 1	0.824724	0.009386209
Bv 2 (Barren vegetal moderate)	Bv 2	23.013757	0.26192027
Bw 1 (Barren water erosion slight)	Bw 1	54.345239	0.618504822
Bw 2 (Barren water erosion moderate)	Bw 2	4.64264	0.052838027
Ee 1 (Dune/sandy wind erosion slight)	Ee 1	49.19732	0.559916199
Ee 2 (Dune/sandy wind erosion moderate)	Ee 2	5.546306	0.063122678
Ee 3 (Dune/sand wind erosion savere)	Ee 3	35.579548	0.40493192
Fv 1 (Forest vegetal degradation slight)	Fv 1	111.199638	1.26556647
Fv 2 (Forest vegetal degradation moderate)	Fv 2	18.601706	0.211706583
Fv 2 (Forest vegetal degradation severe)	Fv 2	0.470748	0.005357597
NAD (No Apparent Degradation)	NAD	7156.508074	81.44843655
R (Rocky Area)	R	52.001647	0.59183233
Settlement	Settlement	84.560854	0.962389657
Sv 1 (Land with scrub vegetal degradation slight)	Sv 1	286.848547	3.264632058
Sv 2 (Land with scrub vegetal degradation moderate)	Sv 2	231.202791	2.631326012
Sv 3 (Land with scrub vegetal degradation slight)	Sv 3	22.746353	0.258876937
Tm 1 (others)	Tm 1	2.384756	0.027140981
Tm 2 (others)	Tm 2	6.244626	0.071070279
Tm 3 (others)	Tm 3	7.686774	0.087483409
W (Water body)	W	379.472816	4.318791687
	Grand Total	**8786.55058**	100
		557.113847	
		556.893342	
		52.001647	
	DSM2	1166.008836	

Source: IRS Data—SAC, Ahmedabad (2012-13).

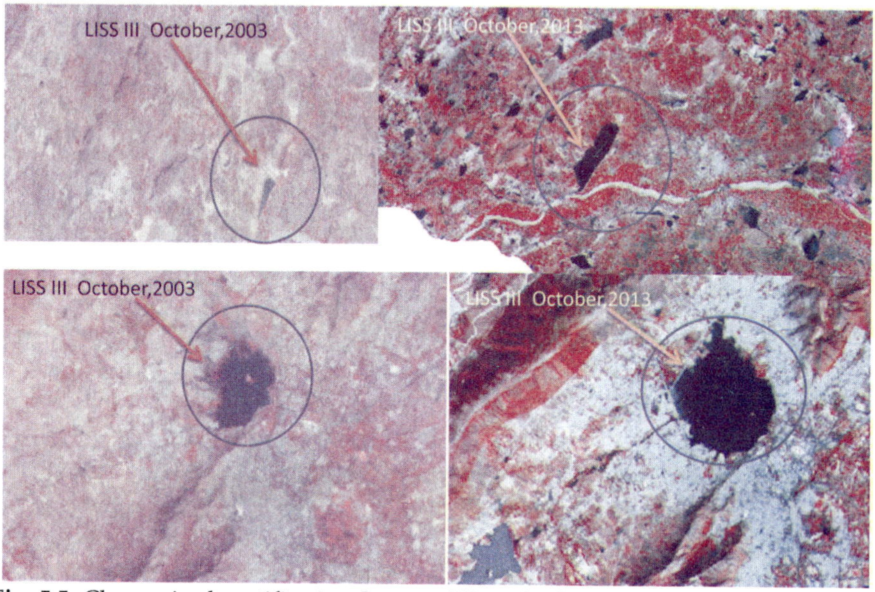

Fig. 5.5: Change in desertification Status – Water body (2002-03 to 2012-13), Ajmer Rajasthan, India

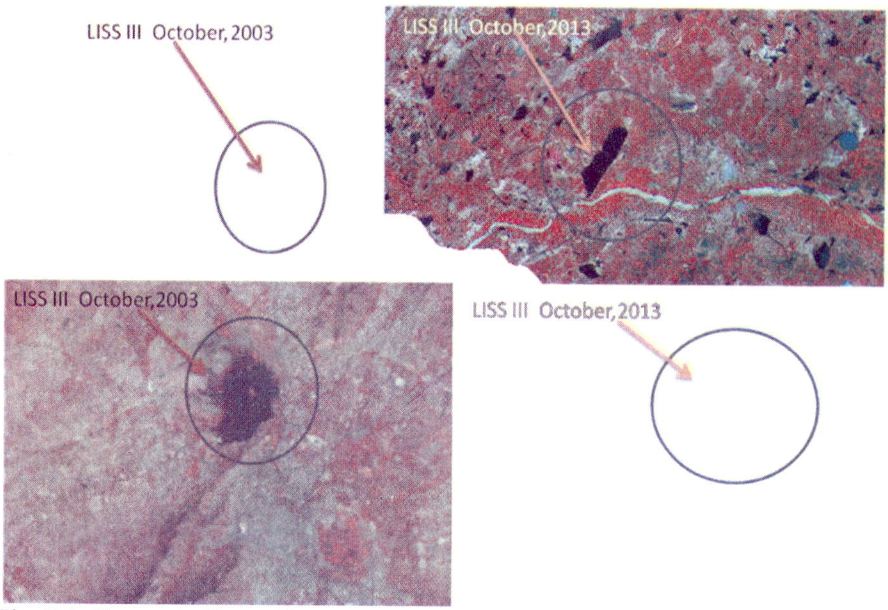

Fig. 5.6: Change in Desertification Status—Water Body (2002-03 to 2012-13), Ajmer Rajasthan, India

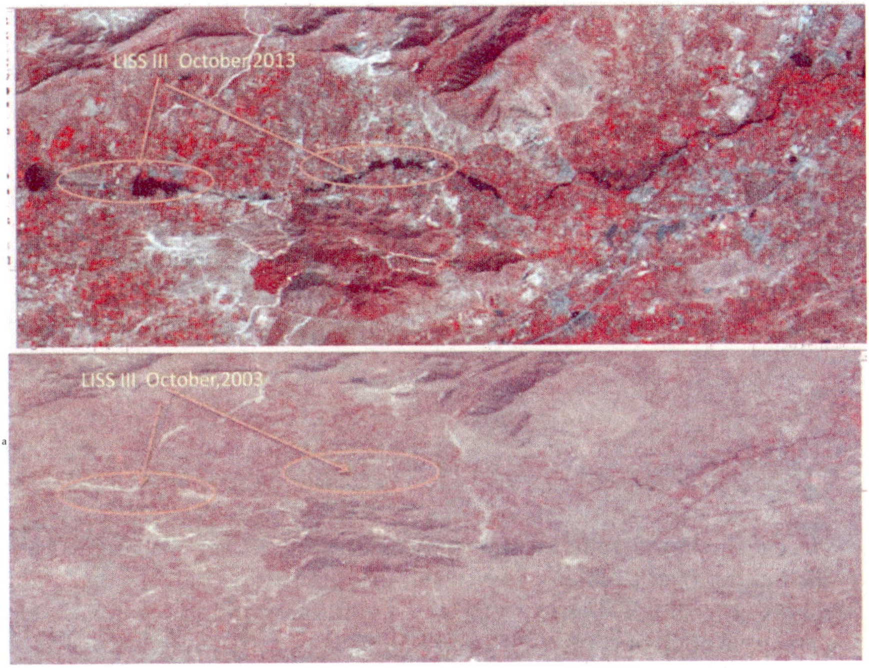

Fig. 5.7: Change in Drainage System

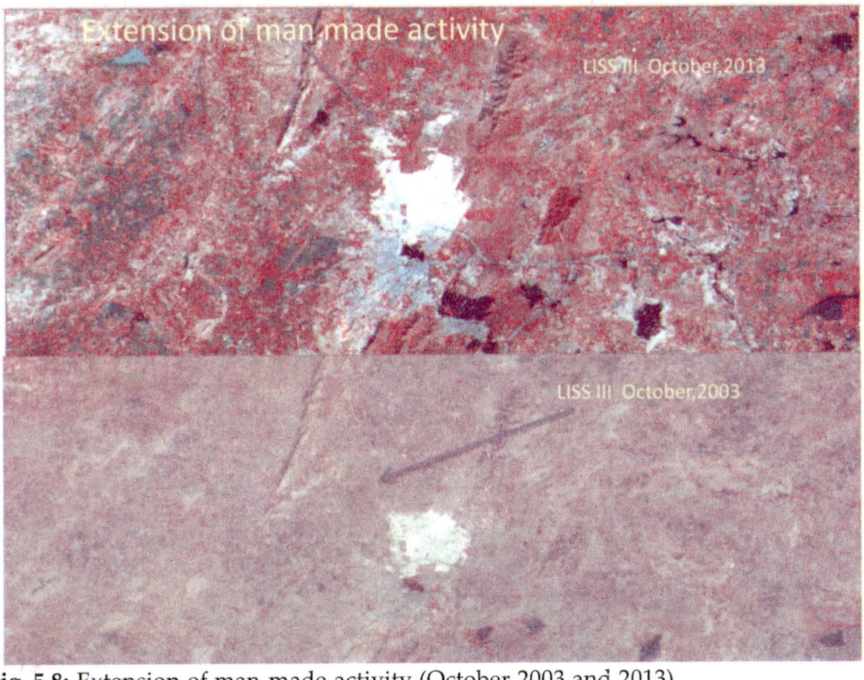

Fig. 5.8: Extension of man-made activity (October 2003 and 2013)

Change Detection in Dausa District

The data related to land-use/ land cover, processes of desertification with their severity have been tabulated in table 5.9. The land under agriculture unirrigated and agriculture irrigated is widely distributed. Nearly 69.07 per cent or 2352.49 sq. kms area of the district is under this category in which no apparent process of desertification may be observed. However, 26.97 per cent of area of the district is under the process of desertification particularly vegetal degradation and water erosion to accelerate gully erosion in the foot hill dunes. The area under these processes account for nearly 25.9 per cent (Fig. 5.9).

The study of tables 5.9 and 5.10 make it clear the changes in various categories of degradation processes. The area under no apparent degradation (NAD) has been reduced to 55.97 per cent in 2012-13 against 69.07 per cent of 2002-03, meaning thereby that the process of degradation has shown an increase of 9.10 per cent area in the district. The vegetal degradation in 2012-13 has shown an increasing trend (33%) as evident from tables 5.9, 5.10 and fig. 5.9.

In conclusion, it may be stated that desertification is recognized as an important national problem. The strategy of combating desertification must be based on ecological characterization of different morphogenetic regions and an application of control measures suitable to each natural zone by taking into account the degrees of desertification involved. Prevention of further deterioration should always be the first consideration whether the land is in good or poor condition, because prevention is always cheaper than reclamation. If priorities are to be established on an economic basis it is better to concentrate first on protecting and improving the good land rather than the poor land. For example, most of the Indira Gandhi Canal Command Area, Chambal Command Area, Morel Command Area, Jawai Canal Area, etc. are not now salinized or waterlogged; much of it can be kept that way with only the application of good field water management practices. The strongly developing desertified land and seriously desertified lands should be the focal point of desertification control as they constitute a great threat to agricultural production: effective measures include building system of shelter belts, planting forest in the depression among dunes and growing sand, fixing plants and combining forestry with agriculture. In the developing desertified lands which have not yet reached the serious stage and the land can still yield certain amount of biological production, this type of land is often not given adequate consideration. Therefore, the measures should be to :

1. Prompt action to reorient cropping methods and control dry farming techniques.
2. Improvement of appropriate control measures, such as establishment of wind break networks, reducing the pressure of grazing, closing off

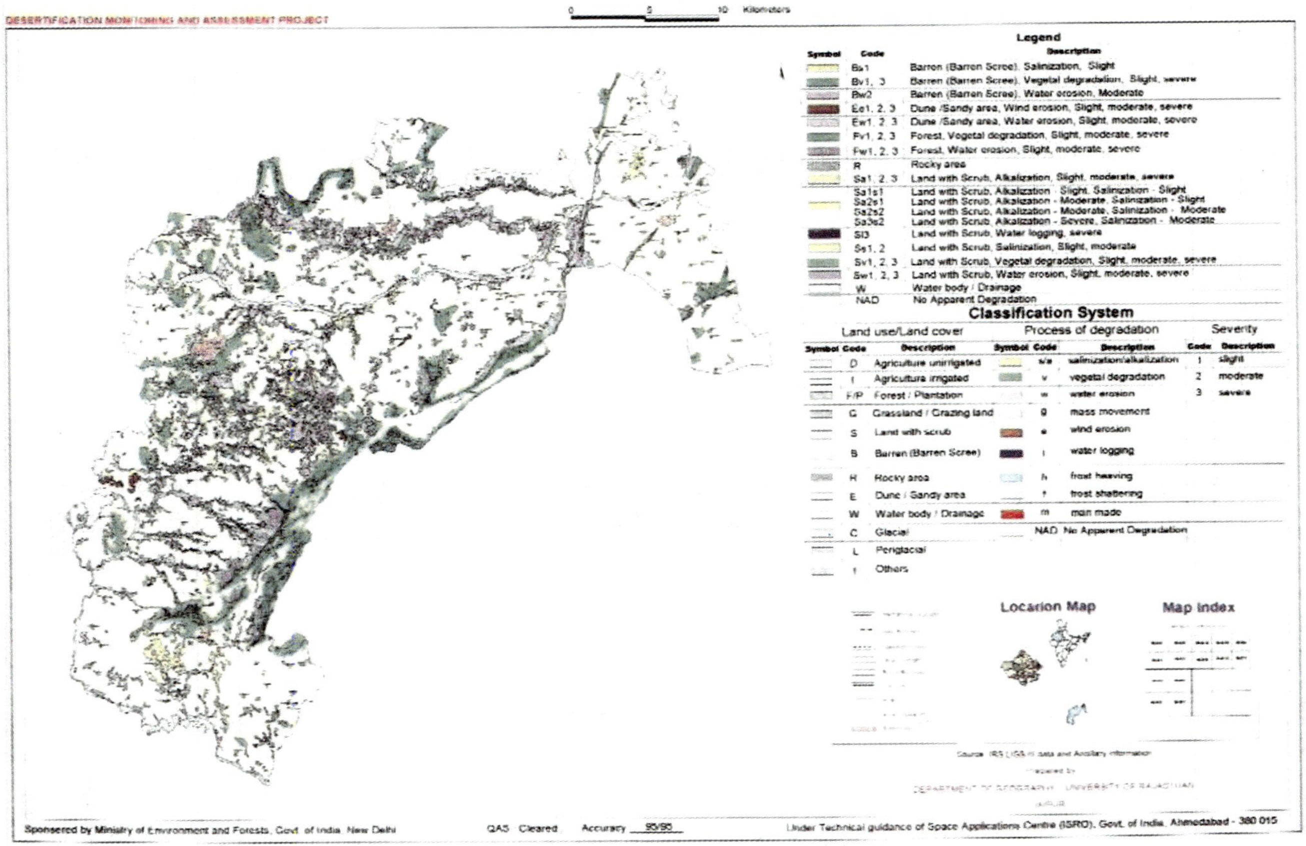

Fig. 5.9: Desertification Status Map (Dausa District, Rajasthan State)

Table 5.9: Desertification Status Mapping of Dausa District (category-wise area in sq. km. with percentage 2002-03).

	District	
	Area	*%*
Bs1	0.06	0.01
Bv1	0.29	0.03
Bv3	0.43	0.05
Bw2	0.11	0.01
Ee1	7.49	0.82
Ee2	0.37	0.04
Ee3	0.18	0.02
Ew1	0.16	0.02
Ew2	1.91	0.21
Ew3	1.70	0.19
Fv1	80.25	8.73
Fv2	58.54	6.37
Fv3	42.79	4.66
Fw1	2.52	0.27
Fw2	4.72	0.51
Fw3	26.54	2.89
Sal	7.94	0.86
Sa2	11.50	1.25
Sa3	12.35	1.34
Sals1	15.48	1.69
Sa2s1	1.95	0.21
Sa2s2	3.01	0.33
Sa3s2	0.61	0.07
Sl3	0.24	0.03
Ss1	6.83	0.74
Ss2	2.84	0.31
Sv1	94.47	10.28
Sv2	100.97	10.99
Sv3	32.18	3.50
Sw1	118.62	12.91
Sw2	148.89	16.21
Sw3	132.84	14.46
Total		**100.00**
	918.76	26.97
Water-bodies	5.88	0.17
River	97.25	2.86
Settlement	31.76	0.93
NAD	2352.49	69.07
Out area		
Total	**3406.14**	**100.00**

Source: Based on IRS Data (LISS-III 2002-03).

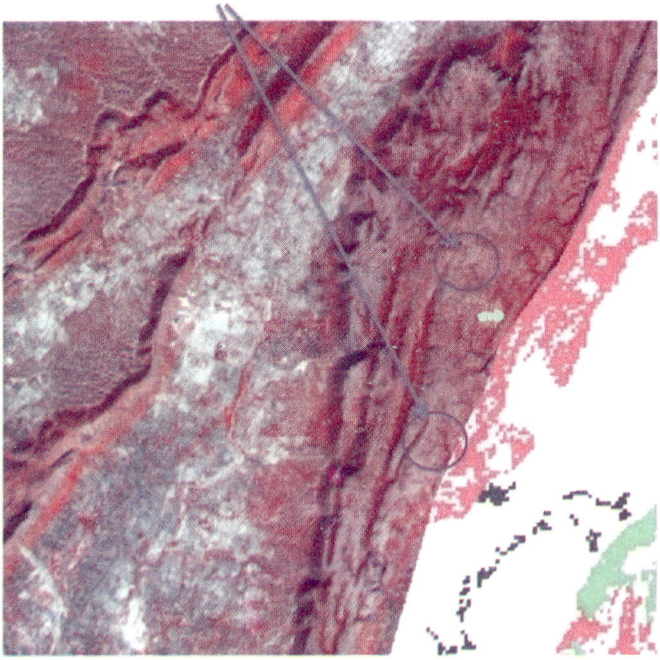

Fig. 5.10: Forest Vegetation Degradation Year 2012-13

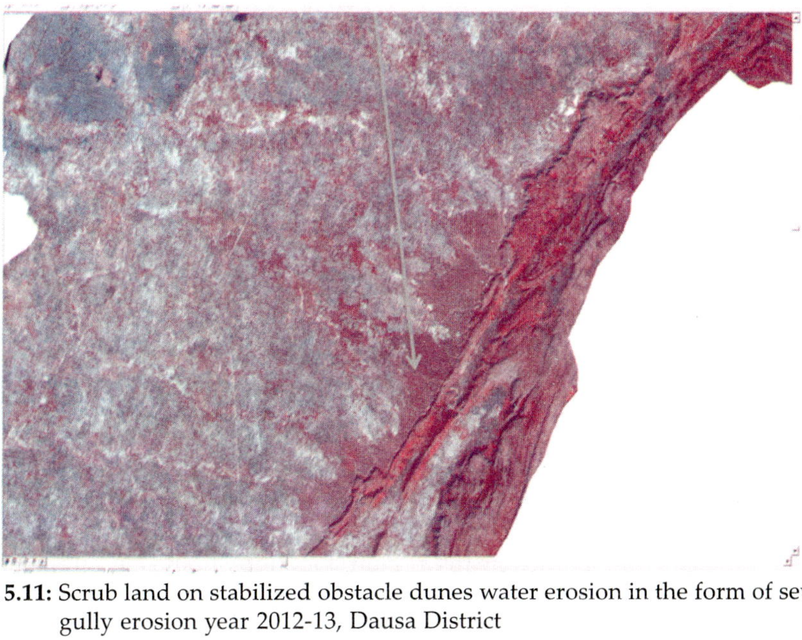

Fig. 5.11: Scrub land on stabilized obstacle dunes water erosion in the form of severe gully erosion year 2012-13, Dausa District

Table 5.10: Desertification Processes in Dausa District—2011-13

DSM-2 Process	2011-2013 Area (sq. km.)	Dausa % Degradation
Salinisation	22.713935	0.491974186
Vegetation	1531.255089	33.16633493
Man Made	89.955771	1.948403797
Water Erosion	394.17344	8.537629312
Barren+Rocky	2.893531	0.062672653
NAD	2575.903927	55.79298512
Total	**4616.895693**	**100**
Degradation	**2038.098235**	

Source: Based IRS Data (USS-III 2012-13).

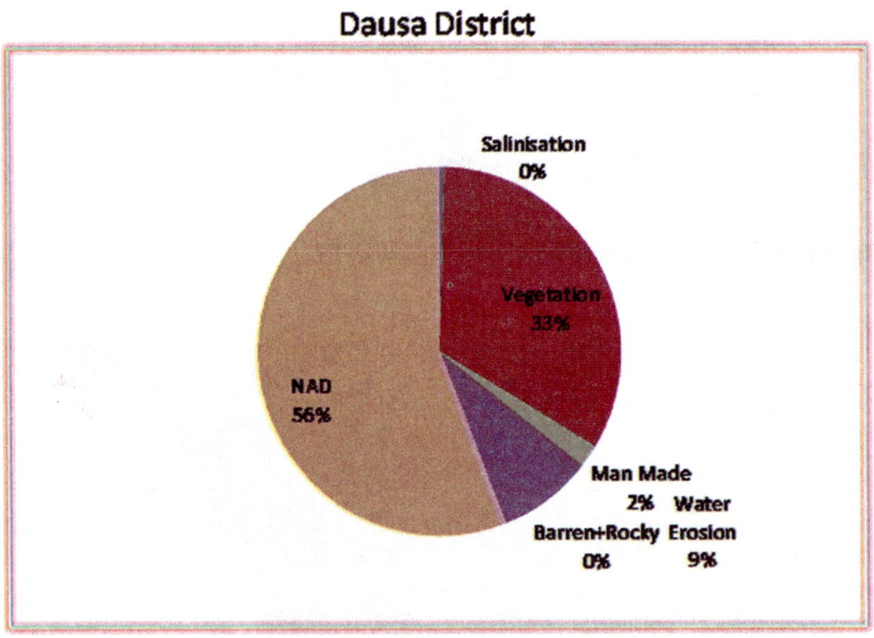

Fig. 5.12: Dausa District—Desertification Processes

natural pasture lands and protecting natural vegetation. The slightly desertified lands could easily become developing desertified land under incorrect utilization. It is important that adequate consideration should be given to this type of land and necessary preventive measures be taken. These should be complemented by planned land-use based on the land capabilities in the region. Grazing should be regulated in accordance with the carrying capacity of pastures.

3. There has to be comprehensive water harvesting management distribution policy for Rajasthan. Otherwise, the state will stagger from crisis-to-crisis. Such an approach to utilization would stop the advance of desertification, reverse the process and therefore, ensure sustained production.

6
People's Perception and Response

It is significant to evaluate the perception, experiences and response of the people living in different agro-climatic regions of Rajasthan in regard to climate change issue. The perception of people on climate change is complex and responses often unpredictable. The expected impact of climate change and responses are significantly related to both age and experience of farmers. In order to understand the diversity of coping strategies, it is necessary to explore the social, political and institutional factors that provide contexts for these individual perception. Perceptions, however, are not static and may shift over time or are expressed differently under altering circumstances. Thus, perception of climate change and associated risks are crucial to formulate appropriate relief and mitigation policies. The problem of climate change is further compounded by the fact that government perceives natural hazard/drought/floods etc. as *a crisis situation* and a short-term problem and manages them as an *isolated event*. At the household level, people perceive such events as a natural hazard, beyond human control. Both lead to different kinds of approaches and solutions. The survey of 407 households of 20 sample villages will certainly provide an insiders' view about climate change and their perception towards natural hazards. The primary survey conducted through structured schedule contain 32 questions related to climate change issues, water resources, land-use changes, agricultural productivity, and adaptation measures at an individual level or at government level. The profiles of sampled villages are described in the following pages.

Profiles of Sample Village

4 C Village (Arid Region)

4 C village is in Sri Ganganagar tehsil of Sri Ganganagar district having 361 hectares of land and 423 population (2011). The total households are 86. People are dependent on agriculture. Wells, taps, tank, tube-wells, hand-pump and

canals are sources of water. This village is also witnessing a climate change as seen in other villages.

Dabla Village (Arid Region)

Dabla village is in Jaisalmer tehsil of Jaisalmer district. The total area of the village is 2653 hectares of land having 2607 population (2011). The total number of households are 414. Agriculture is the main source of livelihood. The village is also witnessing the rising trend of temperature and changing rainfall pattern. Droughts are common at an interval of 3 to 4 years. Groundwater is very deep and is showing declining trend. Rainfed crops are grown.

Keroo Village (Arid Region)

Keroo village is a big size village in Jodhpur tehsil of Jodhpur district having 5612 hectares of land and 9523 population (2011). The number of households are 1296. Drinking water source is hand-pump and tankas in houses. Agriculture is the main occupation dependent on vagaries of monsoon rainfall. Groundwater level is showing declining trend due to overdraft and changing rainfall pattern. Changes in land-use are evident.

Sanavada Village (Arid Region)

Sanavada village is located in Barmer tehsil of Barmer district having 800 hectares of land and 1341 population (2011). Traditional water harvesting system in the form of Tanka is still prevalent. Groundwater in some areas is showing increasing trend due to change in rainfall pattern. But it is a localized phenomenon. Temperature in the area is also showing an increasing trend. Heat waves are common during summer. Agriculture is the main source of occupation. However, people are also engaged in tertiary activities.

Takhatpura Village (Arid Region)

Takhatpura village is a moderate size village having 2185 hectares of area and 1616 population (2011). The total households are 231. The village is located in Chhatargarh tehsil of Bikaner district. Agriculture is the main occupation. The source of drinking water is canal. Other sources are wells, taps, tanks, hand-pumps etc. The survey reveals that due to canal irrigation the productivity is fairly high. However, the signs of climate change are visible on account of changing rainfall pattern and temperature rise.

Dhamana Village (Transitional Zone)

Dhamana village is located in Sanchore tehsil of Jalore district. The total area is 1542 hectares. The total population is 3854 (2011). Agriculture is the main occupation of villagers. The survey of this village indicates that the water table in the area is declining due to overuse and changing rainfall pattern. Temperature are increasing as reported by the respondents of the village. The productivity is also declining. The total households are 483.

Gharat Village (Transitional Zone)

Gharat is located in Pindwara tehsil of Sirohi district having 962 hectares of area and 2708 population (2011). Agriculture is the main occupation. The field study of the village brought out that temperature in village is showing increasing trend and declining trend in rainfall. However, no change in cropping pattern has been noticed. Groundwater is depleting. There is no evidence of occurrence of cold waves. Forest cover is showing a declining trend due to wood cutting by inhabitants by the tribal population for fuel requirement.

Kitalsar Village (Transitional Zone)

Kitalsar village is located in Degana tehsil of Nagaur district having 1668 hectares of area. The total population of the village is 2946 (2011). Taps, wells, tanks, tube-wells, hand-pumps, canal and springs are the sources of water. Tube-well is the only source of drinking water during summer. The interview of respondents of the village reveals that most of the persons perceived climate change, declining water table, rising temperature and land-use change.

Beri Village (Semi-Arid Region)

Beri is a small village of Weir tehsil of Bharatpur district having 399 hectares of land. The total households are 293 and population is 1335 (2011). Agriculture is the main occupation which is dependent on rainfall. Agricultural productivity is declining in the village. Hand-pump is the main source of drinking water. Other sources of drinking water are wells, taps, tube-wells, canal, lakes and springs. The observations reveal that the impact of climate is visible in the changing land-use, productivity and declining groundwater levels.

Bhandana Village (Semi-Arid Region)

Bhandana village is located in Dausa tehsil of Dausa district having 871 hectares of area and 2485 population (2011). The total households are 273. Agriculture is the main source of occupation with animal husbandry as secondary source of income. The survey of this village also supports that climate change is visible in the form of declining rainfall and increasing temperature, declining groundwater level and decline in productivity. Groundwater is very deep. Wells and hand-pumps are sources of water.

Chainpura Village (Semi-Arid Region)

Chainpura village is in Newai tehsil of Tonk district having 597 hectares of area and 1777 population (2011). The households are 245. Agriculture is the main occupation. The analysis of data and personal interview of cultivators reveal that there has been a declining trend in rainfall and increasing trend of temperature. Nearly 50 years ago, ample groundwater was available due to good monsoon rainfall which encouraged farmers to grow cotton and sugarcane. Now, the cropping pattern has been changed and farmers are growing only commercial crops such as groundnut and mustard. Groundwater level has drastically declined which has been responsible for declining productivity.

Chhapli Village (Semi-Arid Region)

Chhapli village is located in Bhim tehsil of Rajsamand district. The total village area is 312 hectares. The total population is 999 (2011). The number of households are 185. The occupation of the people is agriculture. The main working and marginal workers are 138 only. Workers population is less than non-workers. The sources of drinking water are hand-pump, wells, tanks, tube-wells, canals, spring and rivers. The cultivators indicated that the rainfall is declining in the area and temperature is increasing due to climate change. Decline in productivity has also been noticed during field work. Due to government efforts there is good cover of forest in adjoining hills. The area is generally put to food crops. People do not encourage commercial crops. However, the groundwater is in a good condition due to other natural sources of surface water.

Dhodhakari Village (Semi-Arid Region)

Dhodhakari village is in Behror tehsil of Alwar district having 256 hectares of land and 405 population (2011). The total number of households are 56. Agriculture is the main occupation dependent on monsoon rainfall. The study reveals that rainfall is showing a decreasing trend while temperature is increasing. Hailstorms are generally experienced during winter. The productivity of land has increased due to electricity, tube-well irrigation and fertilizers. Land prices have enhanced due to proximity of Nimrana industrial area.

Kardala Village (Semi-Arid Region)

Kardala village is located in Kishangarh tehsil of Ajmer district having 669 hectares of area and 793 population (2011). The total number of households are 119. 30 per cent of the population is Scheduled Caste. Agriculture is the main occupation. The field survey indicates that the area is witnessing declining trend of rainfall and increasing trend of temperature. The productivity is dependent on good and bad monsoon, if the rainfall is good productivity is also good and the *viceversa*. Drought is frequent after 3-4 years period. Hailstorms are on increase during winter. Groundwater level has drastically declined during the last 30-40 years. The quality of water is also a problem (fluoride). Tank is an indigenous water conservation practice in the village. Hence, every house has this provision.

Lilee Village (Semi-Arid Region)

Lilee village is in Lachhmangarh tehsil of Alwar district having 398 hectares of area and 2218 population (2011). The total number of households is 275. Agriculture is the main occupation. The analysis of the survey conducted in the village indicates that the area is witnessing climate change. The productivity of land is showing declining trend. However, due to use of fertilizers and irrigation facilities, increase in productivity has also been noticed. Drinking water through hand-pumps, wells, and tube-wells is available. However, groundwater is showing a declining trend on account of overdraft of water. Changes in land-use are perceptible.

Bagdar (Sub-Humid Region)

Bagdar is a small village of Jhalarapatan tehsil of Jhalawar district. The village area is 337 hectares. The total population is only 356 (2011). The number of households is 84. Agriculture is the main occupation. Hand-pumps are source of drinking water. Other sources are wells, tanks, tube-wells and other natural sources like rivers, lakes and spring. The temperature is rising and rainfall is declining as revealed by field survey.

Kanwas Village (Sub-Humid Region)

Kanwas village is located in Sangod tehsil of Kota district. The total village area is 8273 hectares. The total population is 8011 (2011). Nearly 17 per cent of the population belongs to Scheduled Caste and 10.5 per cent belong to Scheduled Tribes. The number of households is 1143.

Agriculture is the main occupation which is dependent on monsoon rainfall. Out of total population, 2022 people are working as main workers and marginal workers. Nearest town is Sangod. Sources of drinking water are hand-pumps, tanks, river, canal and wells. The water table is stable in the village. However, there is great pressure on groundwater resources because of over-exploitation. The productivity has declined due to rising temperature and uncertainty of rainfall.

Ghasiyar Village (Humid Region)

Ghasiyar village is in Girwa tehsil of Udaipur district having 292 hectares of area and only 273 population (2011). It is a tribal village having more than 99 per cent population of Scheduled Tribes. Agriculture is the main occupation of tribals. However, the fields are small and narrow for cultivation. The survey of village reveals that temperature in area is showing an increasing trend. Cold waves are negligible. Rainfall has been trendless neither showing increase nor decrease. The groundwater is rich except seasonal change in water level. Forest depletion has been noticed in the area.

Nandor Village (Humid Region)

Nandor village is located in Sagwara tehsil of Dungarpur district having 653 hectares of area and 2906 population (2011). The major population is of Scheduled Tribes (704). Agriculture is the main occupation dependent on monsoon rainfall. The personal survey reveals that the area is witnessing increasing trend of temperature and declining trend of rainfall. Forests are depleting due to scarcity of rainfall and availability of water. It is interesting to note that despite climatic changes groundwater is moderate. Productivity of land is showing declining trend because of low input in the field in the form of fertilizers and scarcity of water.

Sadri Village (Humid Region)

Sadri village is a small village of Ghatol tehsil of Banswara district having an area of 203 hectares. The total population is 482 (2011). Agriculture is the main occupation of this tribal village. It has been noted during fieldwork that temperature is showing increasing trend, rainfall indicating declining trend. Heat waves are frequent. Productivity of food crops is declining. Wells and hand-pumps are the sources of drinking water. Groundwater is declining rapidly. So, is the case of forest depletion.

Perception and Responses on Climate Change Issues

A sample survey of 20 villages (5 in arid region, 3 transitional zone between arid and semi-arid; 7 semi-arid region, 2 sub-humid region and 3 from humid region) including 407 households was conducted to know the perceptions and responses of individuals on climate change issues. Table 6.1 indicates the responses of people in different agro-climatic regions. Figure 6.1 and Table 6.1 clearly reveal that more than 75 per cent people perceive climate change at a world level phenomenon. The people from arid region responded very strongly on climate change. However, 15 per cent people strongly disagreed on the issue of climate change. Ten per cent people did not express their views on the issue.

Table 6.1: Perception on Climate Change

The world climate is changing	*Arid Region*	*Transitional Zone*	*Semi-Arid Zone*	*Sub-Humid Zone*	*Humid Zone*
Strongly Agree	11	20	22	3	0
Agree	58	40	104	32	51
Abstain	31	0	0	5	9
Disagree	1	0	0	0	0
Strongly Disagree	0	0	0	0	0

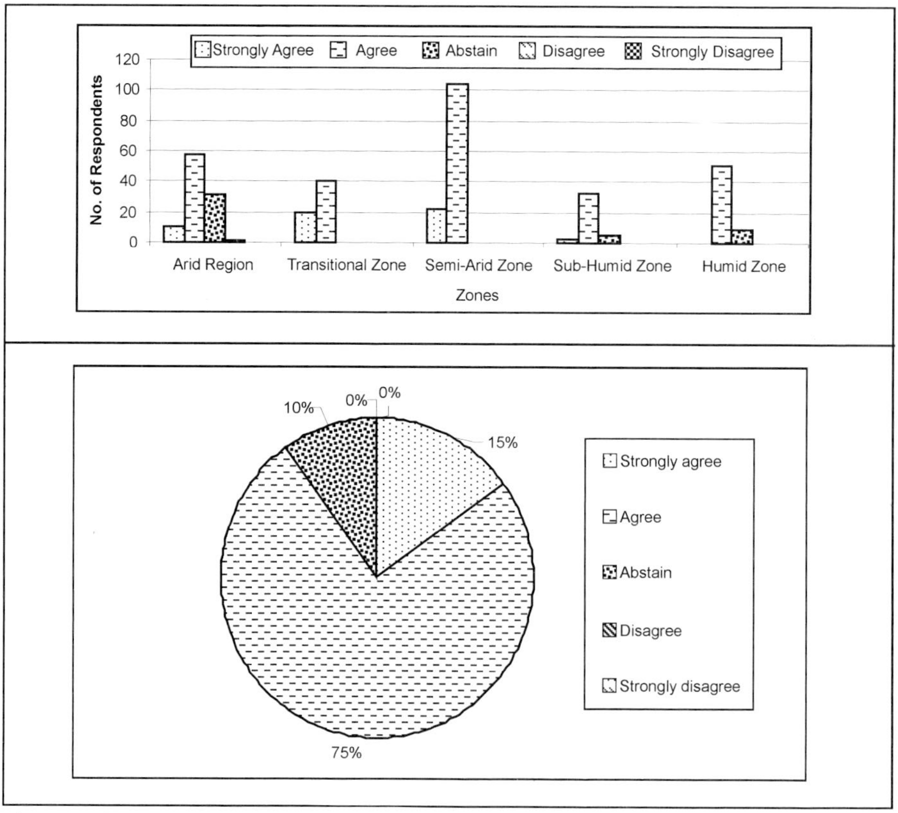

Fig. 6.1: The World climate is changing

The climate change is due to both natural and human causes. Table 6.2 provides the responses of the people. Except sub-humid region most of the people agreed that climate change is due to natural cause. Out of total responses, more than 83 per cent agreed and 8 per cent strongly agreed that it is due to natural causes beyond human control (Fig. 6.2).

The cause of climate change was further put before the people to know whether human activities are the cause of it. Table 6.3 brings out the fact that people from different regions agreed but several people expressed

Table 6.2: Perception on Climate Change

Climate change is natural occurrence	*Arid Region*	*Transitional Zone*	*Semi-Arid Zone*	*Sub-Humid Zone*	*Humid Zone*
Strongly Agree	7	0	0	25	0
Agree	84	60	127	11	51
Abstain	8	0	19	3	9
Disagree	1	0	1	1	0
Strongly Disagree	1	0	0	0	0

Source: Based on personal village survey.

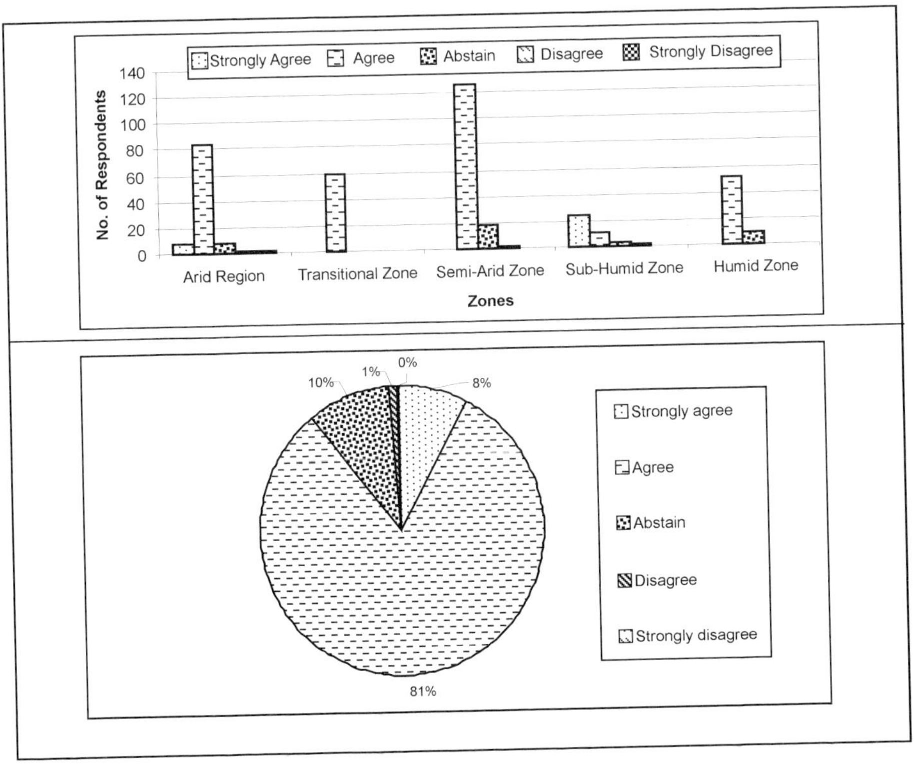

Fig. 6.2: Climate change is Natural Occurrence

disagreement and majority of the people abstain themselves from the issue. However, nearly 70 per cent perceived that human activities are also responsible for climate change (Fig. 6.3). The respondents of transitional zone 100 per cent attributed climate change to human activities.

Table 6.3: Perception on Climate Change

Climate change is the result of human activities	*Arid Region*	*Transitional Zone*	*Semi-Arid Zone*	*Sub-Humid Zone*	*Humid Zone*
Strongly Agree	3	0	0	1	0
Agree	68	60	73	26	51
Abstain	20	0	55	13	9
Disagree	6	0	19	0	0
Strongly Disagree	4	0	0	0	0

Source: Based on personal village survey.

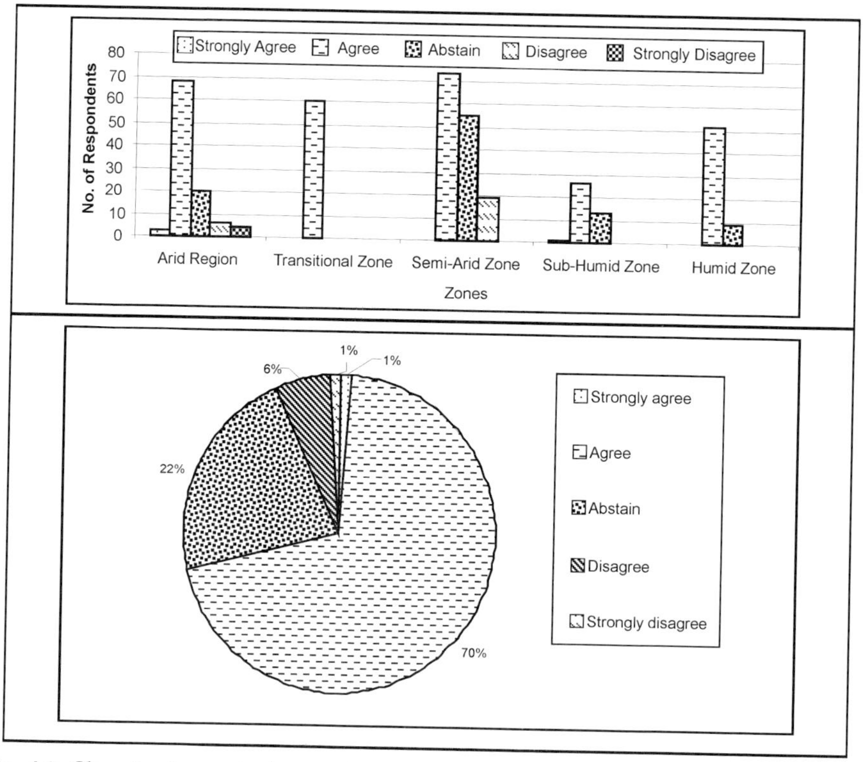

Fig. 6.3: Climate change is the result of human activities

The households of all the regions are of the view that both natural and human activities are responsible for climate change. 74 per cent people perceived that climate change is due to both natural and human activities (Table 6.4 and Fig. 6.4).

Table 6.4: Perception on climate change

Climate change is to both (natural and men's activity)	*Arid Region*	*Transitional Zone*	*Semi-Arid Zone*	*Sub-Humid Zone*	*Humid Zone*
Strongly Agree	6	2	20	5	0
Agree	84	56	74	28	51
Abstain	7	2	52	3	9
Disagree	1	0	1	4	0
Strongly Disagree	2	0	0	0	0

Source: Based on personal village survey.

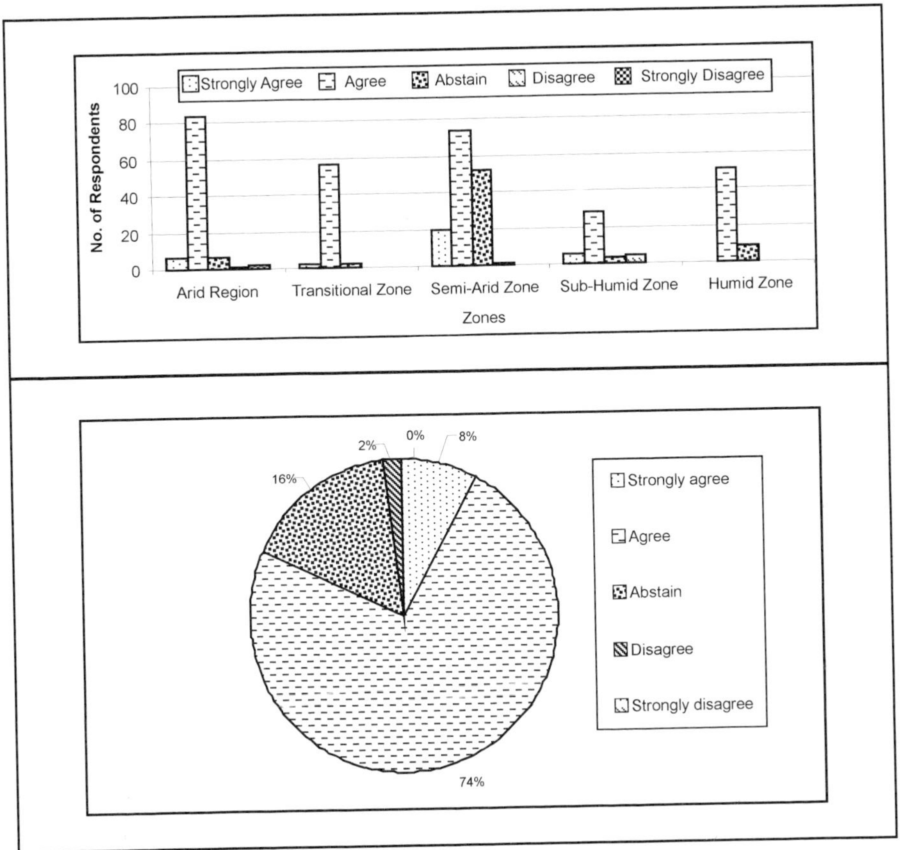

Fig. 6.4: Climate change is due to both natural and human activity

The respondents of 20 villages of different regions feel that the rainfall pattern during two to four decades has changed. More than 96 per cent expressed their views in favour of change in rainfall pattern (Table 6.5 and Fig. 6.5).

Table 6.5: Perception on Climate Change

Do you think the pattern of rainfall has changed during last 20-40 years?	*Arid Region*	*Transitional Zone*	*Semi-Arid Zone*	*Sub-Humid Zone*	*Humid Zone*
Yes	92	39	120	39	51
No	7	1	6	1	9

Source: Based on personal village survey.

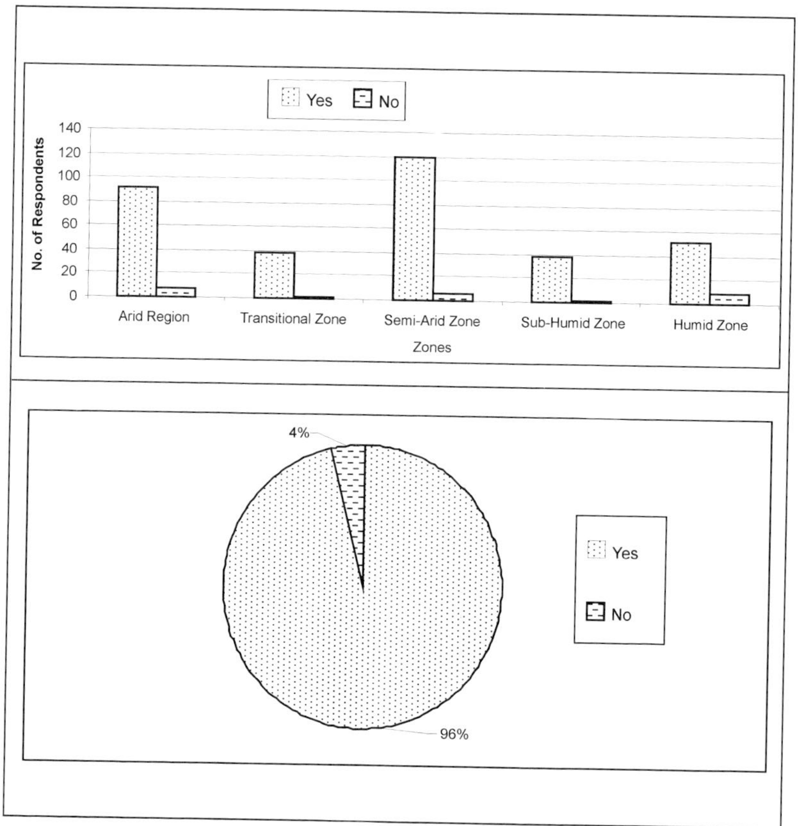

Fig. 6.5: Do you think the pattern of rainfall has changed during last 20-40 years?

The people of villages were also interrogated about change in temperature in the last 40 years. The analysis of data (Table 6.6) reveals that the temperature has risen in the last 40 years. As much as 41 per cent revealed that the intensity of temperature increase is of medium level while 47 per cent opined that it is of low level. The response of arid and humid regions is towards low rise in temperature (Fig. 6.6). However, 5 per cent people and 7 per cent felt that it was high rise and no rise in temperature, respectively.

Table 6.6: Perception on Climate Change

Does the temperature have risen in the last 40 years?	*Arid Region*	*Transitional Zone*	*Semi-Arid Zone*	*Sub-Humid Zone*	*Humid Zone*
High Rise	1	0	20	1	0
Medium Rise	19	29	85	31	6
Low Rise	71	31	42	8	45
No Rise	31	0	0	0	0

Source: Based on personal village survey.

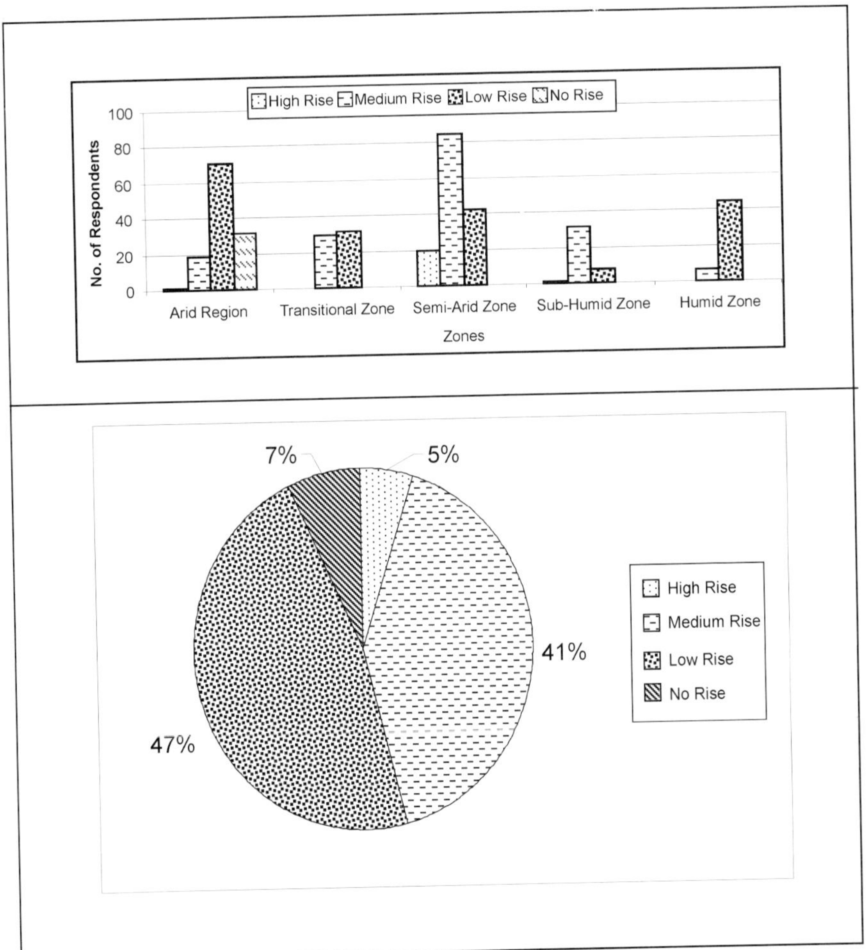

Fig. 6.6: Does the temperature has risen in the last 40 years?

It is amazing that a large number of respondents of all villages have expressed that temperature will continue to rise in the next 10-15 years. As much as 81 per cent revealed that it will continue to rise in future. However, 12 per cent disagreed to such increase in temperature (Table 6.7 and Fig. 6.7).

Table 6.7: Perception on Climate Change

Whether temperature will continue to rise in next 10-15 years	*Arid Region*	*Transitional Zone*	*Semi-Arid Zone*	*Sub-Humid Zone*	*Humid Zone*
Strongly Agree	3	20	21	2	0
Agree	65	40	125	38	48
Abstain	15	0	0	0	12
Disagree	13	0	0	0	0
Strongly Disagree	1	0	0	0	0

Source: Based on personal village survey.

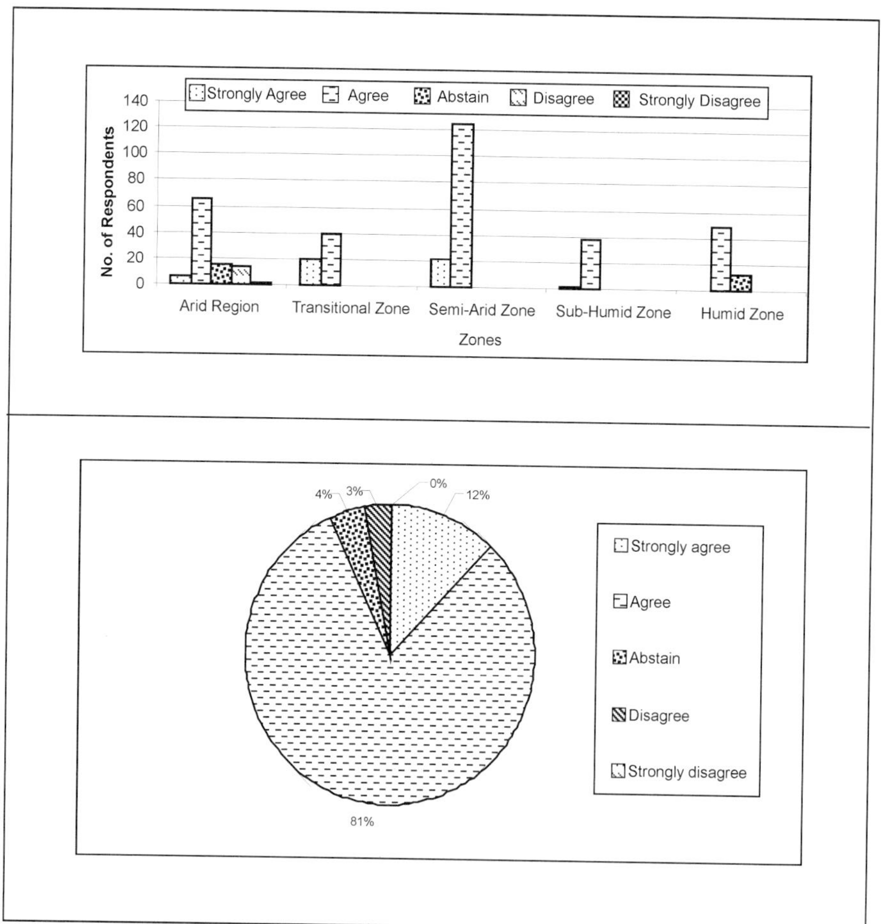

Fig. 6.7: Whether temperature will continue to rise in next 10-15 years

The natural hazards such as heat waves, cold waves, drought, and floods are common in Rajasthan. When the question was posed before the respondents regarding increase in cold wave, their response was mixed. 52 per cent expressed that there is no increase in the cold waves. However, the people of arid and humid regions strongly disagreed (Table 6.8 and Fig. 6.8).

Table 6.8: Perception on Climate Change

Whether intensity of cold wave increased in this area in the last 20-40 years	*Arid Region*	*Transitional Zone*	*Semi-Arid Zone*	*Sub-Humid Zone*	*Humid Zone*
Yes	13	39	118	40	9
No	88	41	49	0	51

Source: Based on personal village survey.

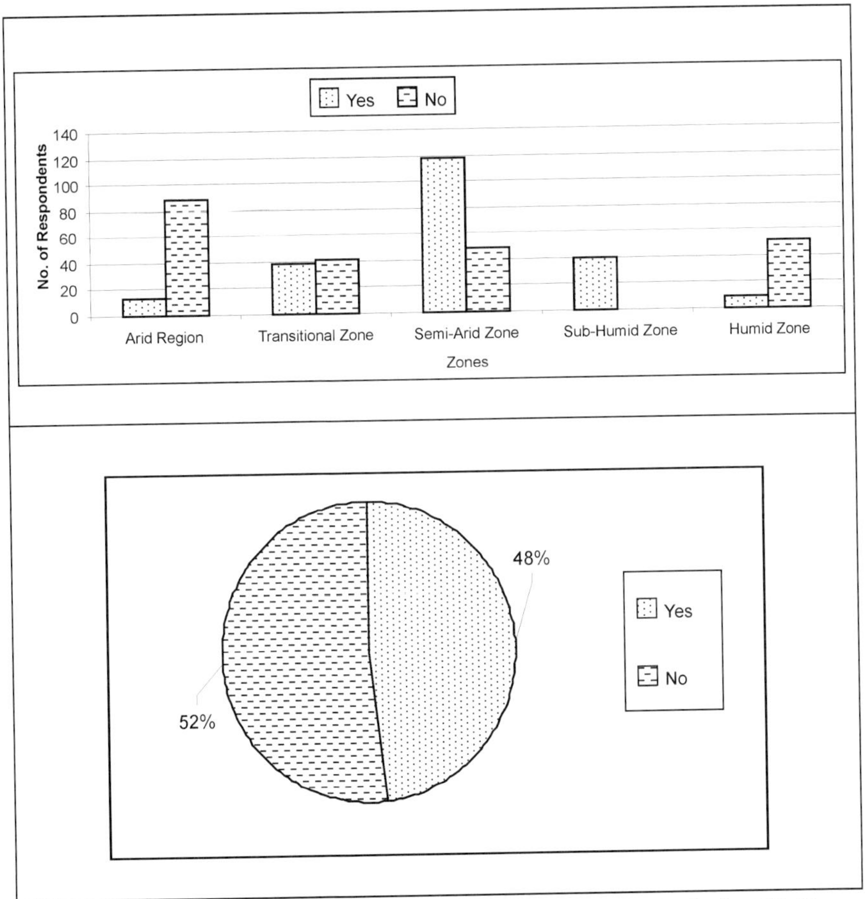

Fig. 6.8: Wether the intensity of cold wave increased in this area in the last 20-40 years.

The responses of the people towards increase in heat wave conditions during summer in the last 20-40 years were very strong. Except arid region all the respondents felt that the heat waves have increased too much during the last 20-40 years. Over all 90 per cent felt that there has been an increase in heat waves (Table 6.9 and Fig. 6.9).

Table 6.9: Perception on Climate Change

What about the heat waves during the summer, did they increase too in last 20-40 years?	*Arid Region*	*Transitional Zone*	*Semi-Arid Zone*	*Sub-Humid Zone*	*Humid Zone*
Yes	60	60	147	40	51
No	41	0	0	0	9

Source: Based on personal village survey.

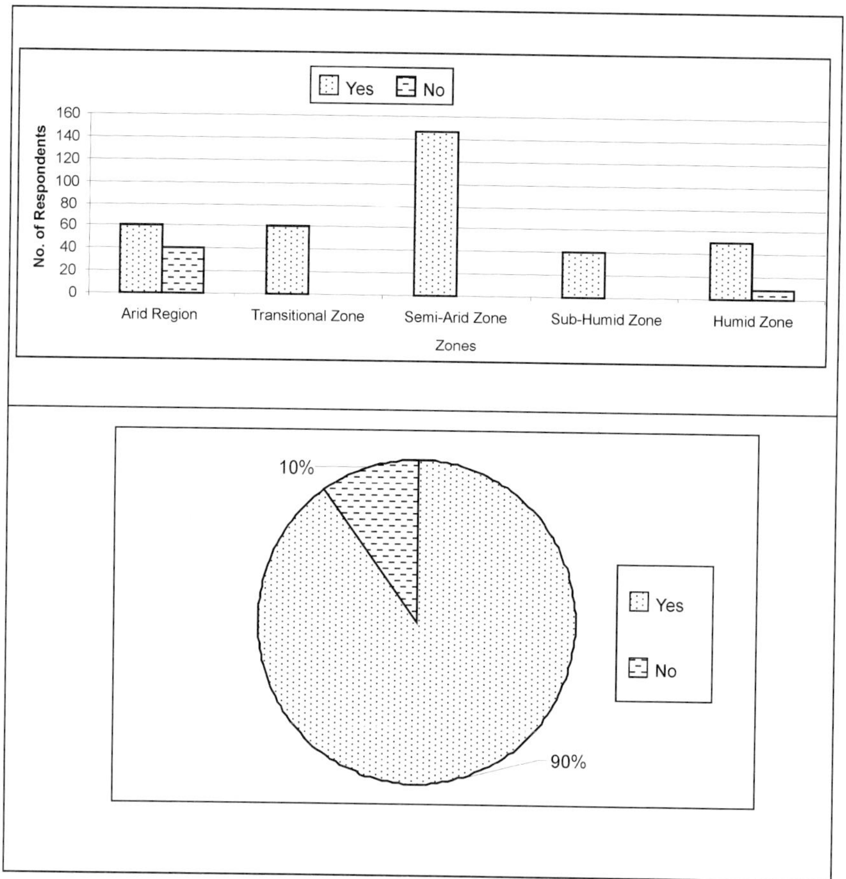

Fig. 6.9: What about the heat waves during the summer, did they increased too in last 20-40 years?

The analysis of physical damage due to extreme events (cold wave, heat wave and other conditions—flood etc.) reveals that in comparison to cold waves the heat waves (48%) and freak weather conditions (38%) have caused more damage to the state (Table 6.10 and Fig. 6.10).

Table 6.10: Perception on Climate Change

Whether the region suffers physical damage due to extreme events?	*Arid Region*	*Transitional Zone*	*Semi-Arid Zone*	*Sub-Humid Zone*	*Humid Zone*
Cold wave	30	20	40	1	9
Heat wave	68	40	127	37	51
Other condition	39	40	125	1	51

Source: Based on personal village survey.

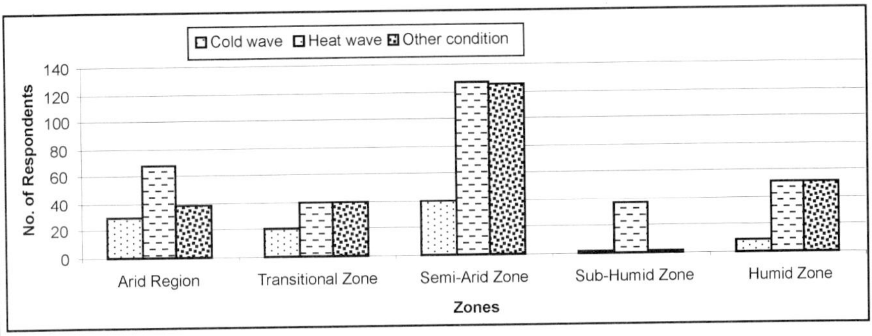

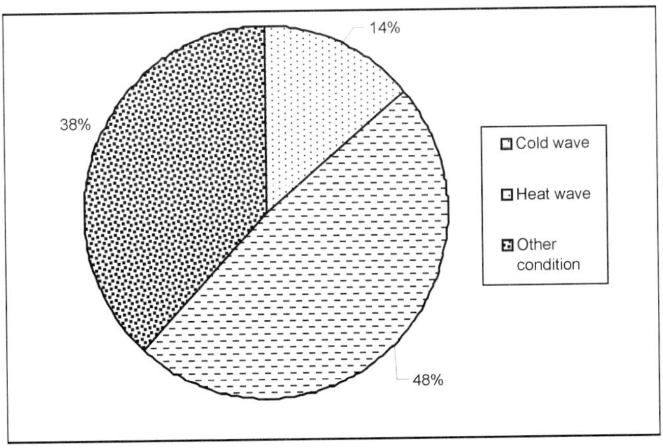

Fig. 6.10: Whether the region suffers physical damage due to extreme events

The study of intensity of climate hazard assessment indicates that 65 per cent people perceived it as low level, 21 per cent felt it as medium level, while 12 per cent felt it of very high level. The respondents of the transitional zone felt that intensity is of very high level (Table 6.11 and Fig. 6.11).

Table 6.11: Perception on Climate Change

What is the intensity of climate hazard assessment?	*Arid Region*	*Transitional Zone*	*Semi-Arid Zone*	*Sub-Humid Zone*	*Humid Zone*
Low	64	16	26	0	40
Medium	20	11	42	0	7
High	2	7	40	0	0
Very high	12	21	20	0	0

Source: Based on personal village survey.

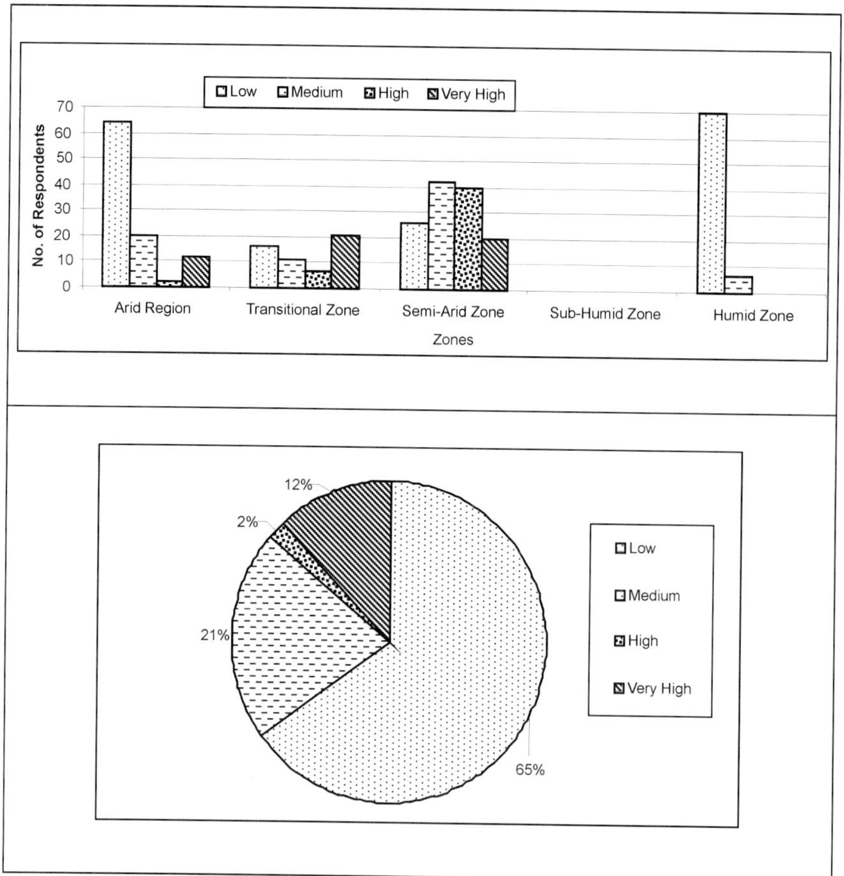

Fig. 6.11: What is the intensity of climate hazard assessment?

The analysis of assessment of intensity hazard-wise reveals that the major hazard in the state are drought (29% people perceived), heat waves (23% people perceived), hailstorms (20%), cold waves (11%), frost (7%), and cyclones (5%). The drought is a major hazard followed by heat waves in different regions of Rajasthan (Table 6.12 and Fig. 6.12).

Table 6.12: Perception on Climate Change

What is the intensity of climate hazard assessment?	*Arid Region*	*Transitional Zone*	*Semi-Arid Zone*	*Sub-Humid Zone*	*Humid Zone*
Drought	76	60	126	0	31
Flood	2	26	20	0	0
Cloud bursts	10	0	1	0	0
Heat waves	60	40	121	0	51
Cold waves	30	21	73	0	1
Cyclone	12	0	1	0	0
Hailstorm	51	42	60	0	0
Frost	19	20	95	0	0

Source: Based on personal village survey.

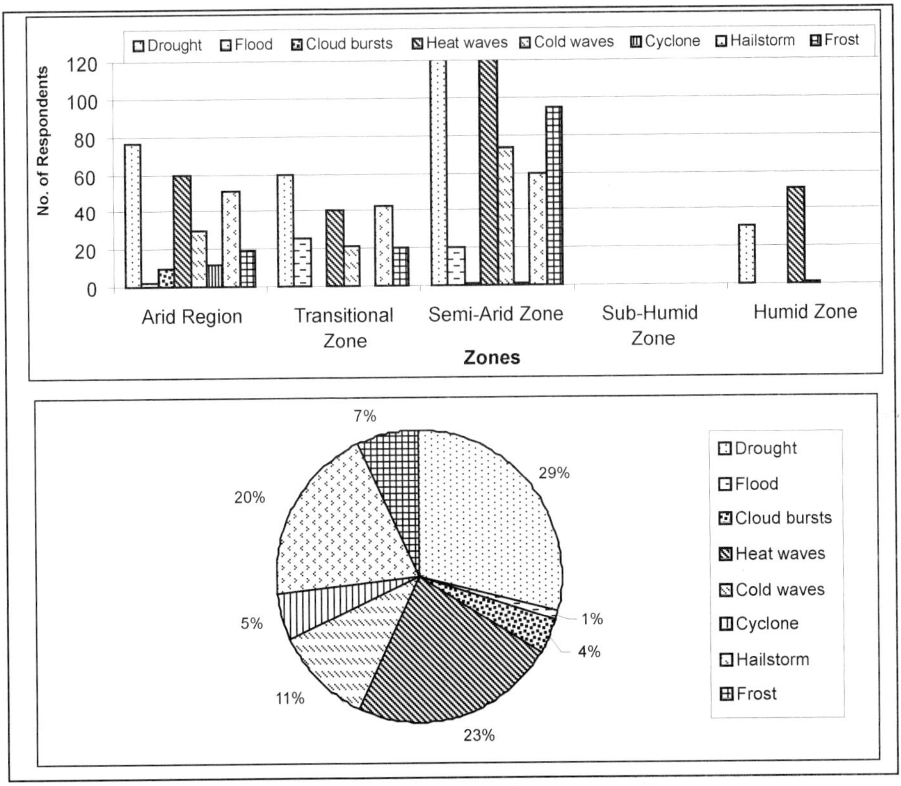

Fig. 6.12: What is the intensity of climate hazard assessment?

Perception and Response on Water Resource

The survey in respect of water resources indicates that major sources of drinking water in the villages are tanks, ponds, reservoirs, canals and other sources like wells and hand-pumps. As much as 28 per cent people responded that ponds are the major source of drinking water, 23 per cent people opined that tanks are the source of drinking water, while 16 per cent and 15 per cent people expressed that other sources like wells and canals are also source of drinking water (Table 6.13 and Fig. 6.13).

Table 6.13: Perception on Water Resources

What are the sources of water including drinking water?	*Arid Region*	*Transitional Zone*	*Semi-Arid Zone*	*Sub-Humid Zone*	*Humid Zone*
Pond	59	36	89	25	26
Reservoir	58	21	39	6	21
Tank	68	41	79	0	2
Canal	44	21	40	0	20
Other	53	1	64	5	7

Source: Based on personal village survey.

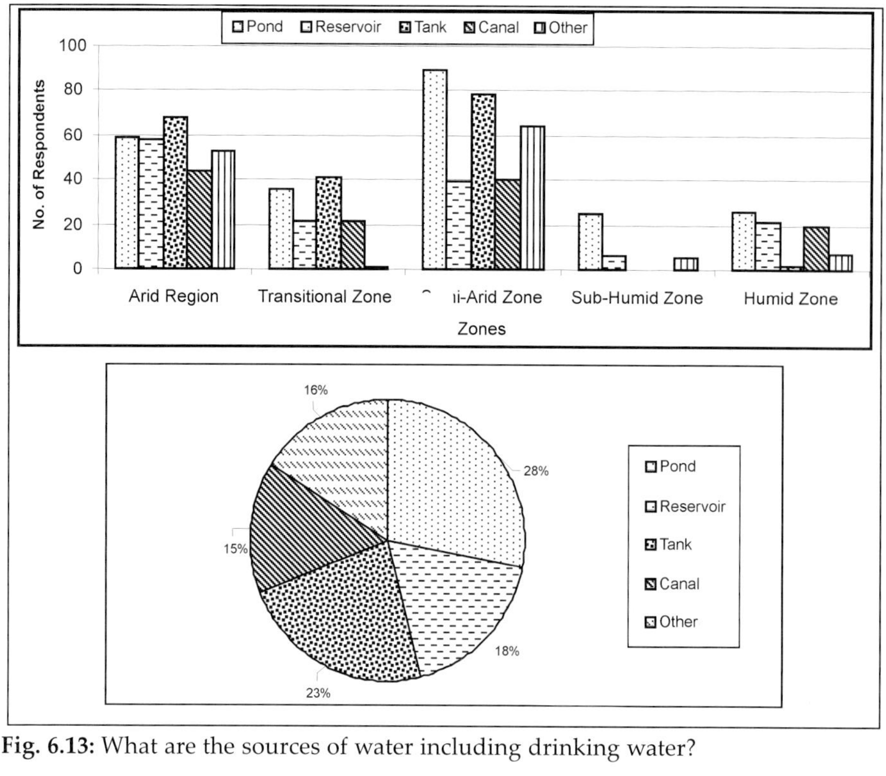

Fig. 6.13: What are the sources of water including drinking water?

The major traditional sources of water conservation in different regions of the state are Bavari, Tanka, Khadin, Kundi and village ponds. As much as 42 per cent interviewers found village ponds as the main source of conservation. This is followed by Tanka (28%), Kundi (11%), Khadin (9%) and Bavari (9%) (Table 6.14 and Fig. 6.14).

Table 6.14: Perception on Water Resources

What are the traditional sources of water conservation?	*Arid Region*	*Transitional Zone*	*Semi-Arid Zone*	*Sub-Humid Zone*	*Humid Zone*
Bavari	23	0	26	5	1
Tanka	60	39	80	2	0
Khadin	59	0	0	0	0
Kundi	29	0	39	0	0
Village pond	100	9	80	33	51

Source: Based on personal village survey.

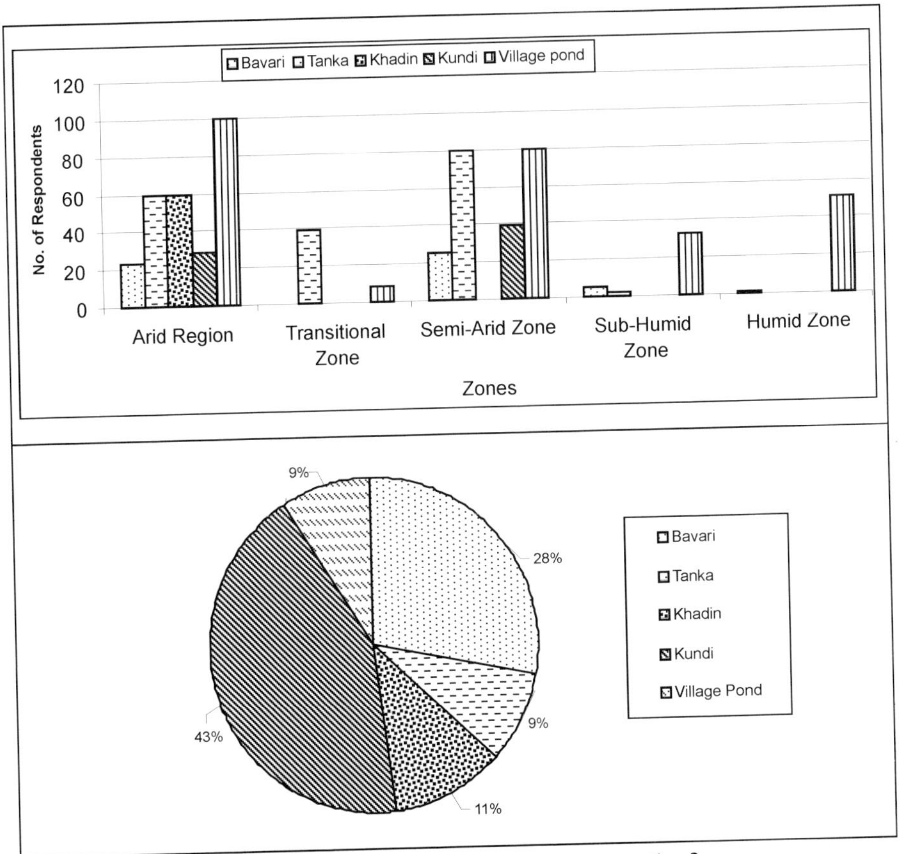

Fig. 6.14: What are the traditional sources of water conservation?

During primary survey of sample villages it has been noted that the undergroundwater level in all the regions has receded drastically and partially in humid region. Nearly 78 per cent people responded that groundwater level has declined during the last 3 to 4 decades because of changing climatic conditions and overuse of groundwater for irrigation and other industrial uses (Table 6.15 and Fig. 6.15).

Table 6.15: Perception on Water Resources

Has the underground water level receded?	*Arid Region*	*Transitional Zone*	*Semi-Arid Zone*	*Sub-Humid Zone*	*Humid Zone*
Yes	60	60	130	40	20
No	39	0	16	0	31

Source: Based on personal village survey.

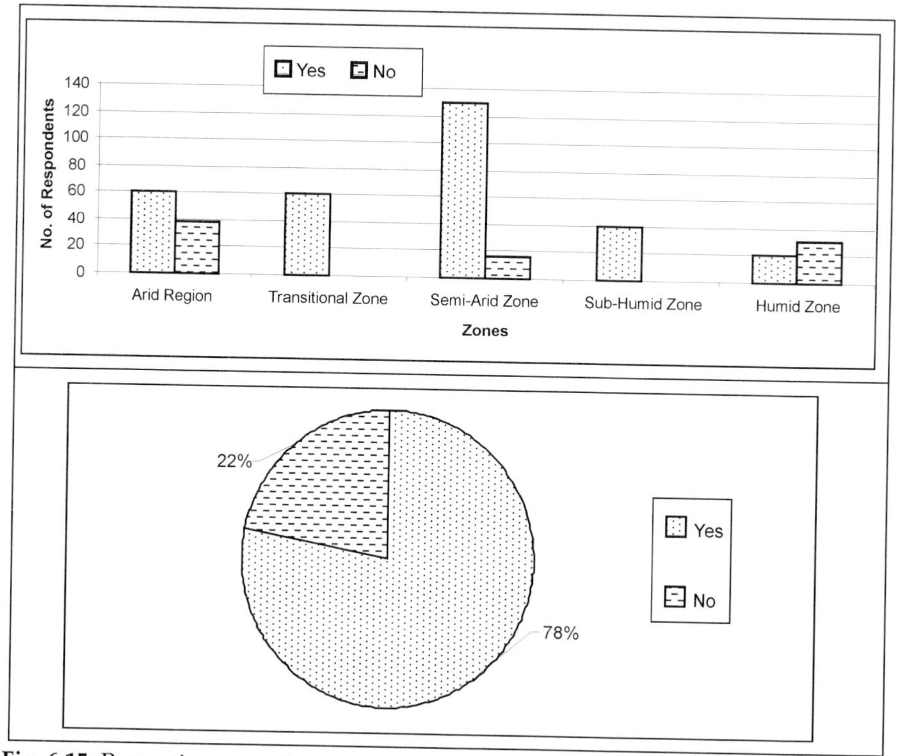

Fig. 6.15: Perception on water resources.

Perception and Response on Land Use Change and Productivity

The impact of climatic vagaries on land-use are of great significance, particularly in the ecologically fragile and economically developing state of

Rajasthan. In order to bring out the inner views of the people some questions were asked related to land-use change and productivity. Table 6.16 reveals that in all the regions majority of the respondents indicated changes in the land-use in the last 40 years except a few in transitional and semi-arid regions. Fig. 6.16 brings out the fact that 90 per cent of the respondents are in favour of change while 10 per cent expressed views against it.

Table 6.16: Perception on Land Use/Land Productivity

Do you think pattern of land use has changed in last 20-40 years?	*Arid Region*	*Transitional Zone*	*Semi-Arid Zone*	*Sub-Humid Zone*	*Humid Zone*
Yes	99	40	126	40	51
No	1	20	20	0	9

Source: Based on personal village survey.

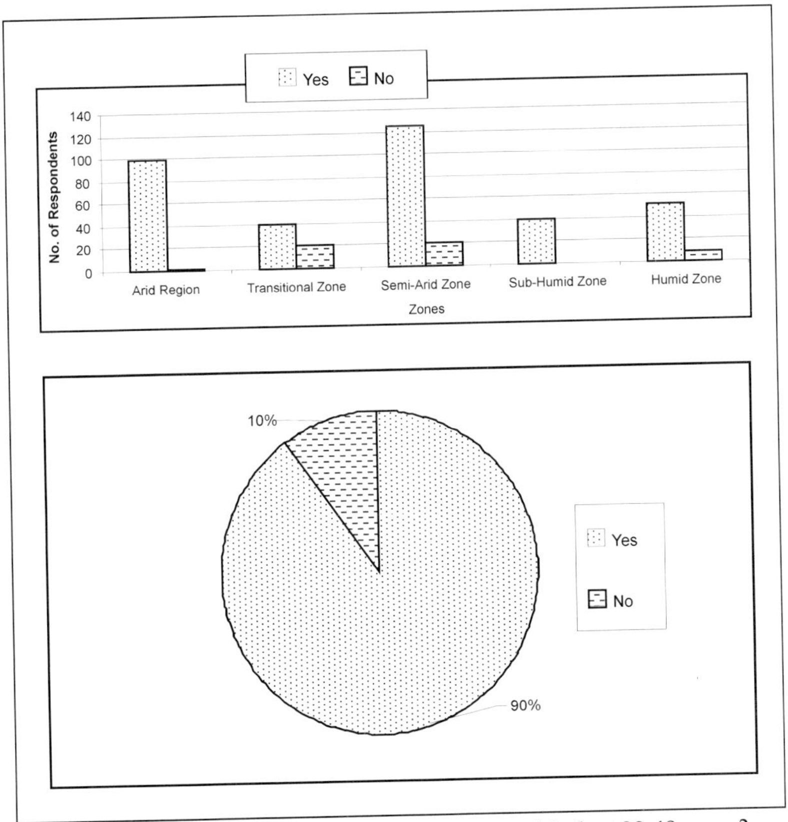

Fig. 6.16: Do you think pattern of land-use has changed in last 20-40 years?

The analysis was further carried to know whether land-use changes are due to rainfall variability. 92 per cent respondents felt that changes are due to rainfall variability (Fig. 6.17 and Table 6.17).

Table 6.17: Perception on Land Use/Land Productivity

Whether land use change are due to rainfall variability?	*Arid Region*	*Transitional Zone*	*Semi-Arid Zone*	*Sub-Humid Zone*	*Humid Zone*
Yes	71	60	146	40	51
No	29	0	0	0	9

Source: Based on personal village survey.

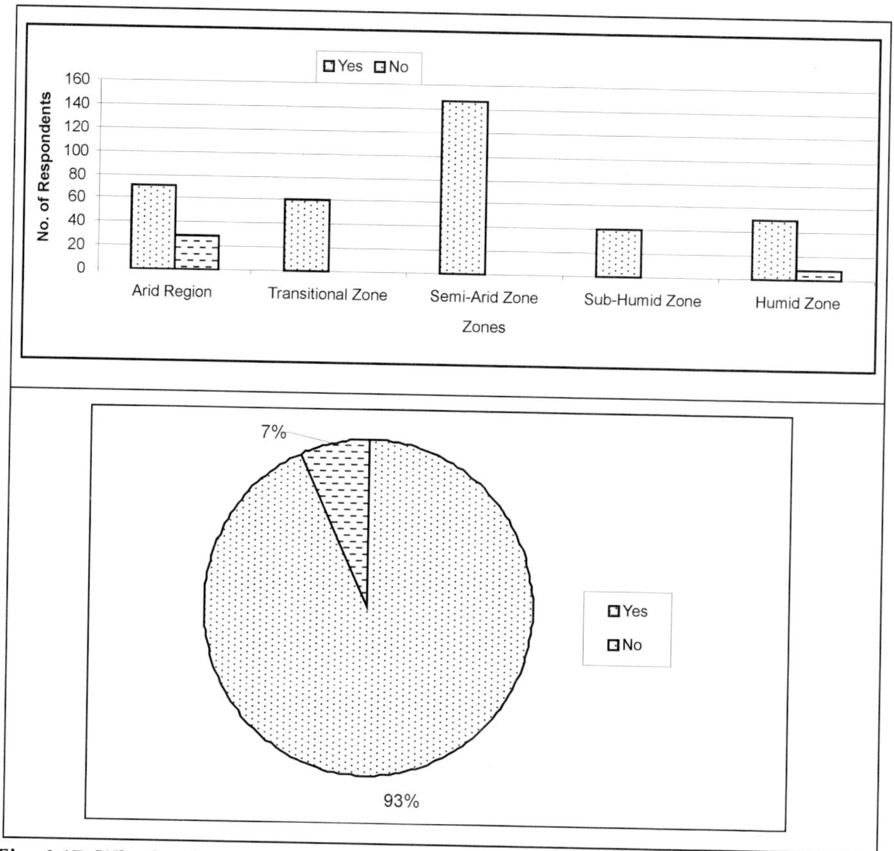

Fig. 6.17: Whether land-use changes are due to rainfall variability?

The changes in land-use may be due to institutional, social and economic factors than rainfall variability. The response of people on this issue was mixed.

Social and economic factors also play a great role in land-use changes. Hence, 29 per cent respondents attributed this change to economic return and 15 per cent and 19 per cent to social and institutional factors, respectively. However, 37 per cent people again attributed to climatic factors (Table 6.18 and Fig. 6.18).

Table 6.18: Perception on Land Use/Land Productivity

What are reason for the change in land use?	*Arid Region*	*Transitional Zone*	*Semi-Arid Zone*	*Sub-Humid Zone*	*Humid Zone*
Climate	27	20	86	22	51
Institutional factors	15	20	67	6	0
Social	2	20	53	10	0
Economic return	94	39	20	1	11

Source: Based on personal village survey.

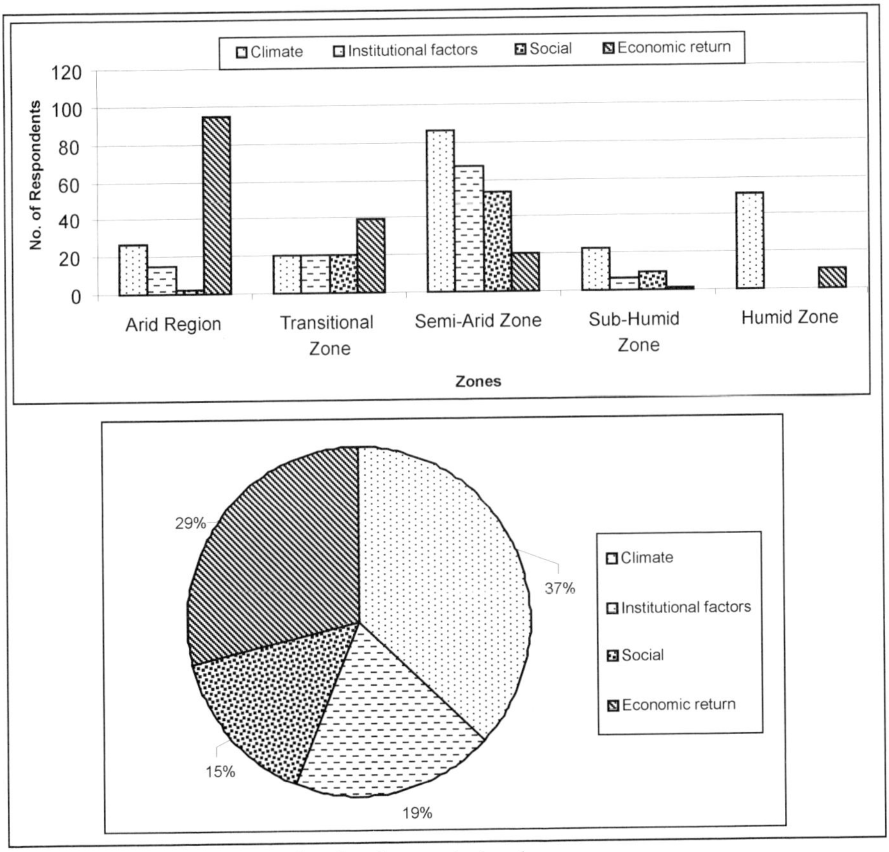

Fig. 6.18: What are reasons for the change in land-use

It is interesting to note that land degradation and climate change are interrelated and the anslysis of responses of people on this issue indicates that

76 per cent people felt land degradation will accelerate in response to climate change in the next 10-15 years. In arid and semi-arid regions people expressed mixed feelings (Table 6.19 and Fig. 6.19).

Table 6.19: Perception on Land Use/Land Productivity

Do you think that land degradation will increase in response to climate change in the next 10-15 years?	*Arid Region*	*Transitional Zone*	*Semi-Arid Zone*	*Sub-Humid Zone*	*Humid Zone*
Strongly Agree	3	0	36	2	0
Agree	69	59	83	38	51
Abstain	16	0	26	0	9
Disagree	11	0	0	0	0
Strongly Disagree	0	0	1	0	0

Source: Based on personal village survey.

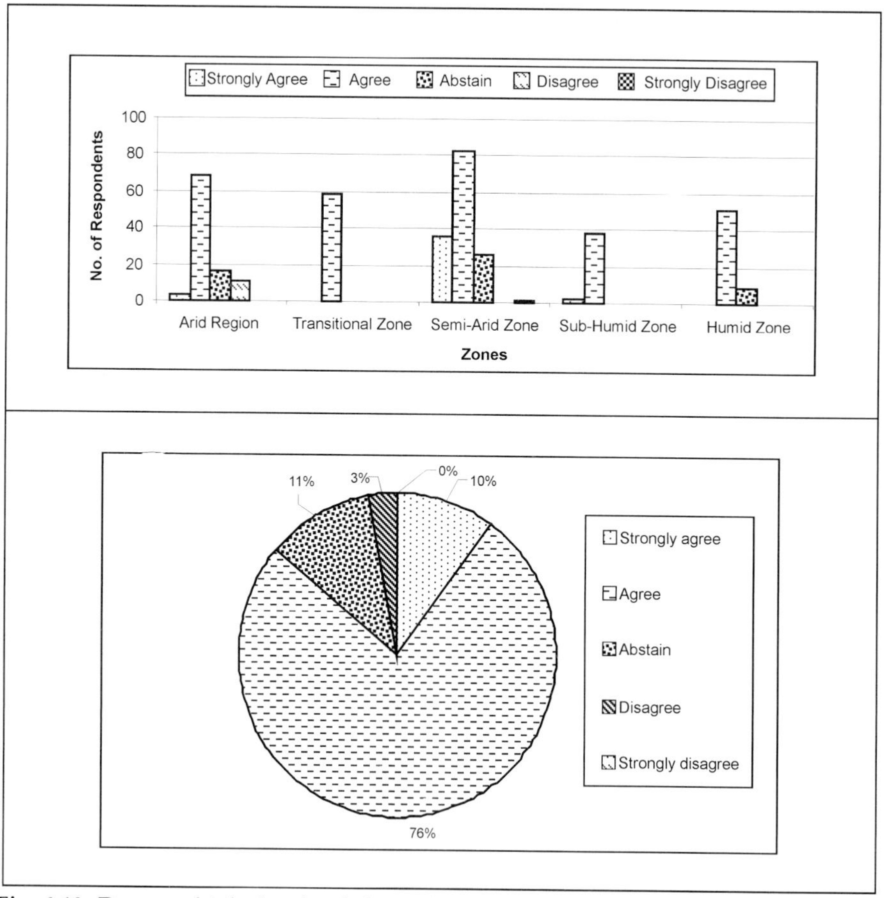

Fig. 6.19: Do you think that land degradation will increase in response to climate change in the next 10-15 years?

The same responses have been expressed by the people on the issue of wasteland increase in response to climate change in the coming 10-20 years. 76 per cent people felt that it will accelerate in future while 10 per cent strongly felt for the change (Table 6.20 and Fig. 6.20).

Table 6.20: Perception on Land Use/Land Productivity

Do you that wastelands will increase in response to climate change in the next 10-20 years?	*Arid Region*	*Transitional Zone*	*Semi-Arid Zone*	*Sub-Humid Zone*	*Humid Zone*
Strongly Agree	1	0	37	2	0
Agree	71	59	83	38	51
Abstain	16	0	26	0	0
Disagree	13	0	0	0	0
Strongly Disagree	0	0	0	0	0

Source: Based on personal village survey.

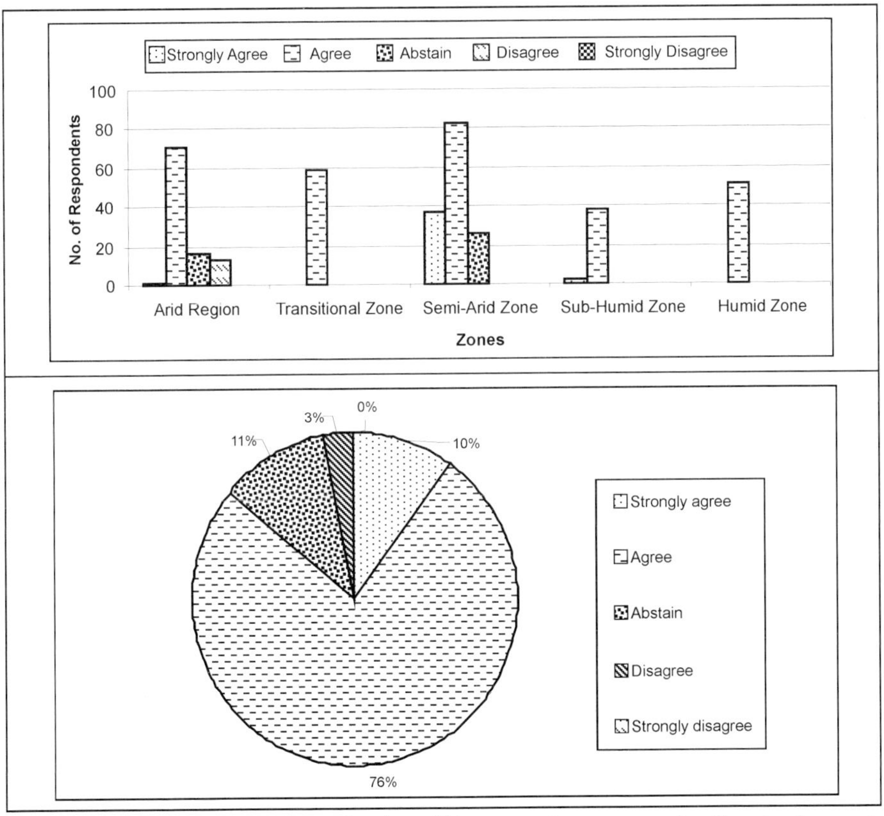

Fig. 6.20: Do you think that wastlands will increase in response to climate change in the next 10-20 years?

It is interesting to note that nearly 40 per cent strongly felt that forest will decrease in response to water stress due to climate change in the next 10-20 years. 39 per cent of the respondents agreed that denudation of forest will continue in the next few years due to vagaries of climatic factors and human activities (Table 6.21 and Fig. 6.21).

Table 6.21: Perception on Land Use/Land Productivity

Do you think that forest will decrease in response to water stress due to climate change in the next 10-20 years?	*Arid Region*	*Transitional Zone*	*Semi-Arid Zone*	*Sub-Humid Zone*	*Humid Zone*
Strongly Agree	18	37	58	2	42
Agree	39	22	48	38	9
Abstain	29	0	20	0	9
Disagree	15	0	20	0	0
Strongly Disagree	0	0	0	0	0

Source: Based on personal village survey.

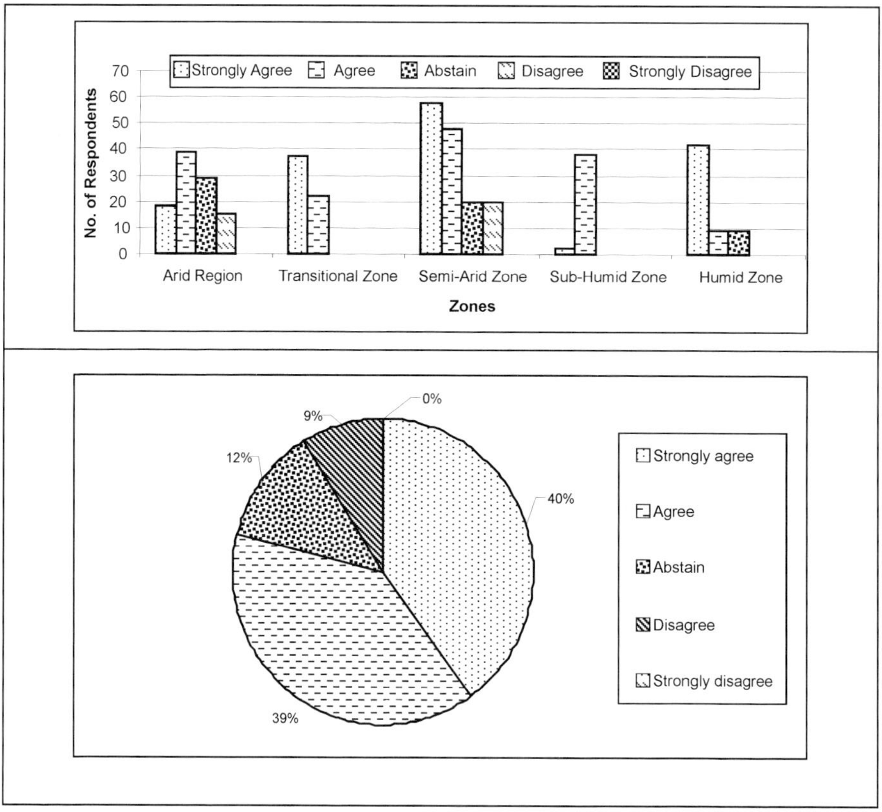

Fig. 6.21: Do you think that forest will decrease in response to water strees due to climate change in the next 10-20 years?

The land-use changes in the fragile ecosystem like Rajasthan may be attributed to climate change. Almost 87 per cent of the interviewers agreed that most of the changes in land-use are due to climatic factors (Table 6.22 and Fig. 6.22).

Table 6.22: Perception on Land Use/Land Productivity

Do you think climate change has caused changes in land use?	*Arid Region*	*Transitional Zone*	*Semi-Arid Zone*	*Sub-Humid Zone*	*Humid Zone*
Strongly Agree	3	0	2	2	0
Agree	72	59	124	38	51
Abstain	12	0	0	0	9
Disagree	12	0	20	0	0
Strongly Disagree	1	0	0	0	0

Source: Based on personal village survey.

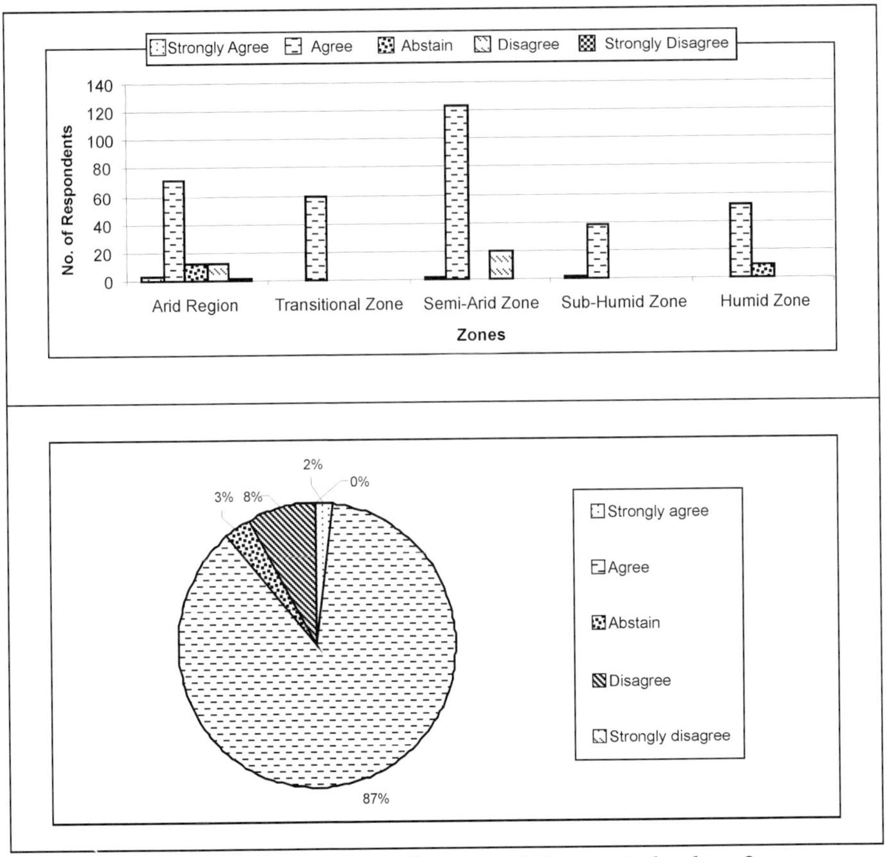

Fig. 6.22: Do you think climate change has caused changes in land-use?

The study was also conducted to know the changes in the cropping pattern during last 20-40 years. It has been noted that majority of respondents felt that food crops received more attention particularly in semi-arid and humid region. In arid, semi-arid and sub-humid regions there has been more emphasis on oilseeds. However, in arid region more and more attention has been made to grow commercial crops (Table 6.23). The share of various crops may be arranged in order of food crops (40%), oilseeds (40%) and commercial crops (Fig. 6.23).

Table 6.23: Perception on Land Use/Land Productivity

What are the changes in cropping during last 20-40 years?	*Arid Region*	*Transitional Zone*	*Semi-Arid Zone*	*Sub-Humid Zone*	*Humid Zone*
Food crops	20	40	86	0	51
Oilseeds	28	0	80	40	9
Commercial crops	87	19	40	0	0

Source: Based on personal village survey.

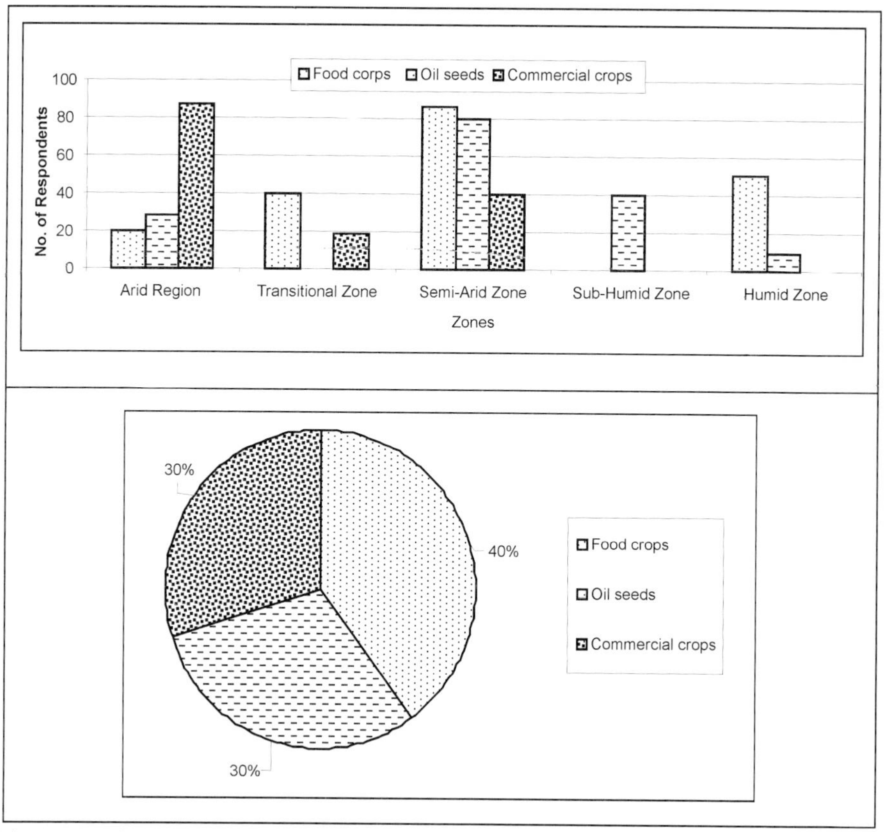

Fig. 6.23: What are the changes in cropping during last 20-40 years?

The productivity of crops is very much related to climate change, land-use and water resources. In view of changing climatic conditions the people of the region perceived that there has been changes in the productivity of land. Except 15 per cent, most of the people 85 per cent agreed that productivity has changed (Fig. 6.24 and Table 6.24).

Table 6.24: Perception on Land Use/Land Productivity

Are there changes in productivity due to climate change (rainfed) land and water resources?	*Arid Region*	*Transitional Zone*	*Semi-Arid Zone*	*Sub-Humid Zone*	*Humid Zone*
Yes	80	59	126	40	31
No	21	0	20	0	29

Source: Based on personal village survey.

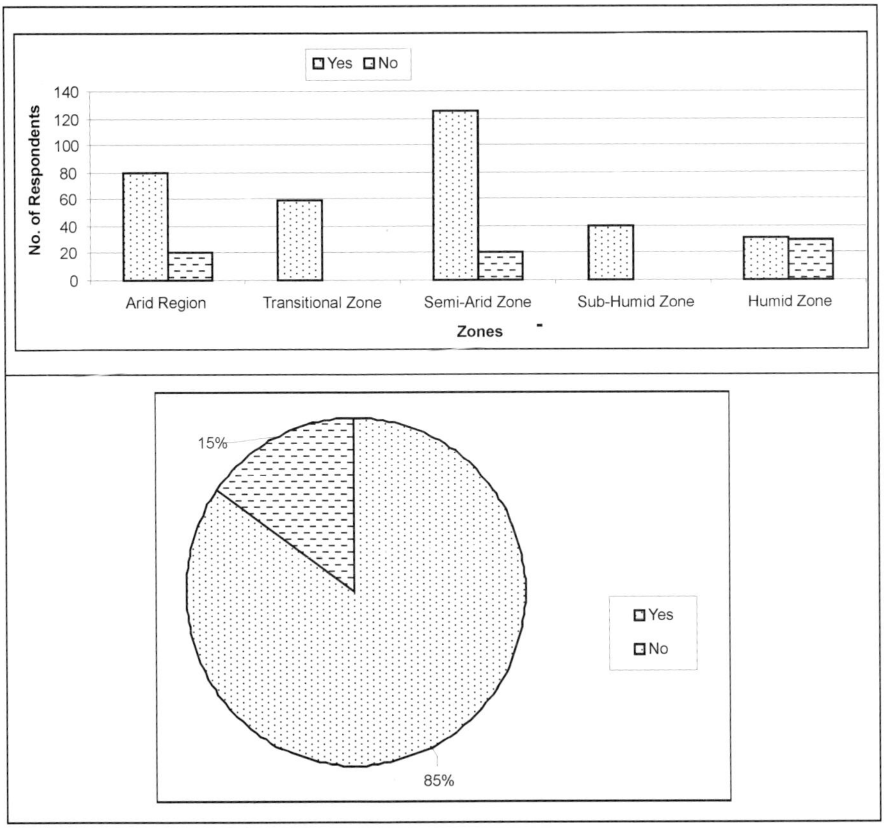

Fig. 6.24: Are there changes in productivity due to climate change (rainfed) land and water resources?

When the question related to increase in productivity in the last 20-40 years was asked, the people expressed mixed feeling, though 65 per cent agreed that productivity has increased in the state despite of harsh geographical conditions. However, 35 per cent people opined that productivity has declined in the last 2 to 4 decades (Table 6.25 and Fig. 6.25).

Table 6.25: Perception on Land Use/Land Productivity

Productivity has increased in last 20-40 years	*Arid Region*	*Transitional Zone*	*Semi-Arid Zone*	*Sub-Humid Zone*	*Humid Zone*
Yes	54	20	62	32	51
No	48	40	28	4	9

Source: Based on personal village survey.

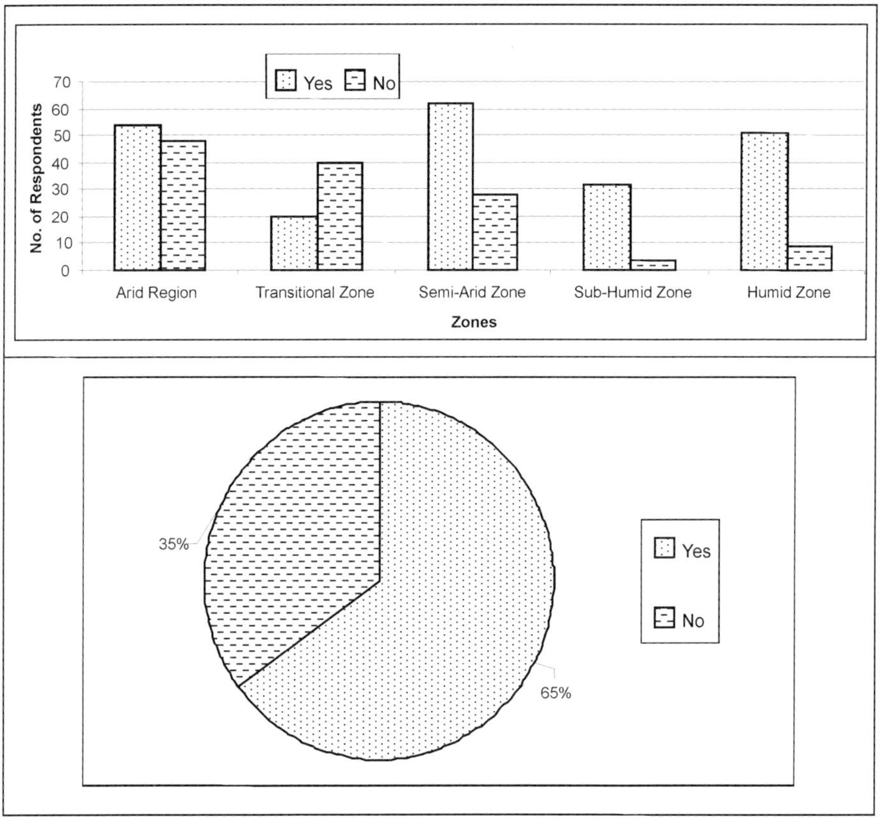

Fig. 6.25: Productvity has increased in last 20-40 years

People's Perception Related to Adaptation Measures

Climate change is happening and something should be done to mitigate this problem. In this context 71 per cent people agreed while 10 per cent strongly agreed that something can be done on personal basis to reduce the impact of climate change in different agro-climatic regions (Table 6.26 and Fig. 6.26).

Table 6.26: Perception on Adaptation Measures

Do you feel that something can be done on personal basis to reduce the impact of climate change?	*Arid Region*	*Transitional Zone*	*Semi-Arid Zone*	*Sub-Humid Zone*	*Humid Zone*
Strongly Agree	1	20	20	0	0
Agree	50	39	100	40	51
Abstain	47	0	26	0	9
Disagree	2	0	0	0	0
Strongly Disagree	1	0	0	0	0

Source: Based on personal village survey.

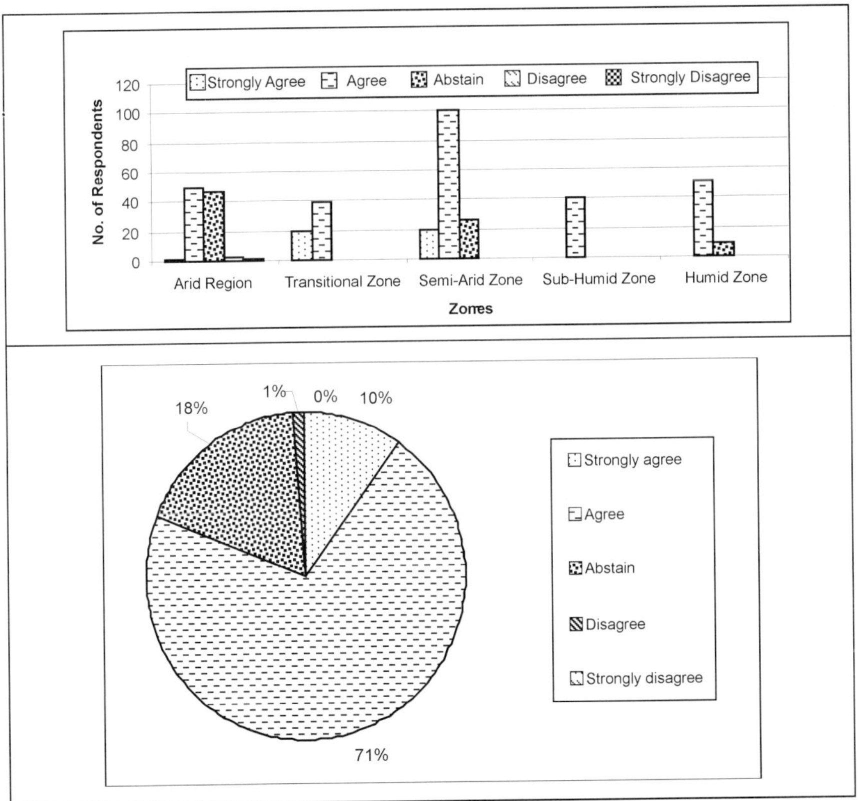

Fig. 6.26: Do you feel that something can be done on personal basis to reduce the impact of climate change?

It is amazing to find that the public participation in adaptation/mitigation of climate change is of low level (85%) while 10 per cent of respondents felt it is of medium level (Table 6.27 and Fig. 6.27).

Table 6.27: Perception on Adaptation Measures

What is level of public participation in adaptation/mitigations of climate change?	*Arid Region*	*Transitional Zone*	*Semi-Arid Zone*	*Sub-Humid Zone*	*Humid Zone*
Low	79	59	126	21	51
Medium	17	0	20	19	5
High	4	0	0	0	4

Source: Based on personal village survey.

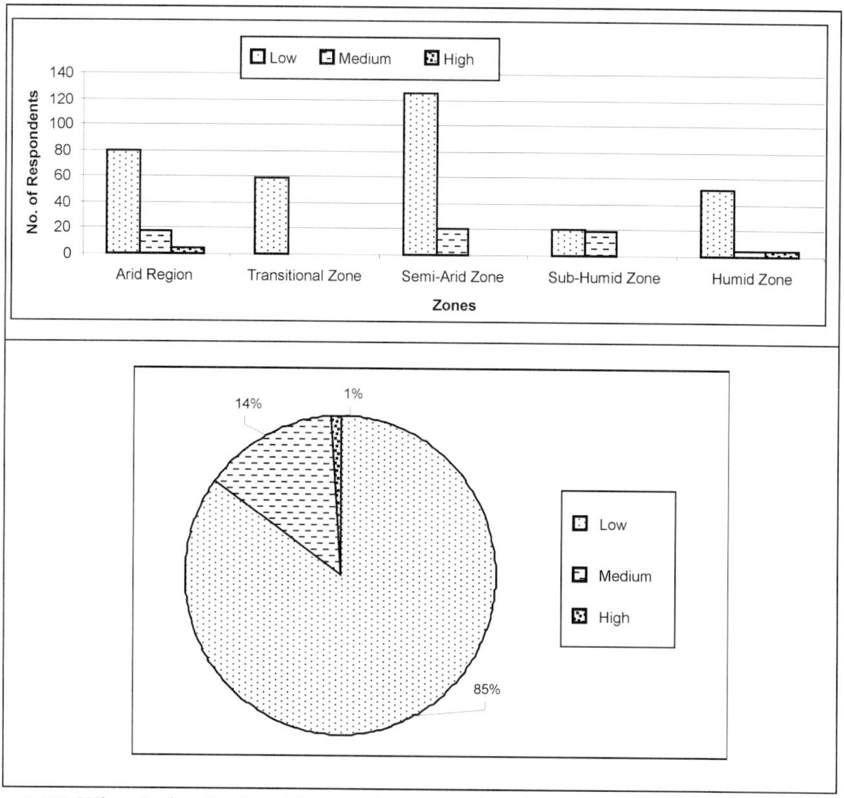

Fig. 6.27: What is level of public participation in adaptation/mitigations of climate change?

The awareness of government functionaries is also low as indicted by Table 6.28 and Fig. 6.28.

Table 6.28: Perception on Adaptation Measures

What is overall awareness of Government functionaries about climate change?	*Arid Region*	*Transitional Zone*	*Semi-Arid Zone*	*Sub-Humid Zone*	*Humid Zone*
Low	83	59	146	19	51
Medium	15	1	1	21	5
High	3	0	0	0	9

Source: Based on personal village survey.

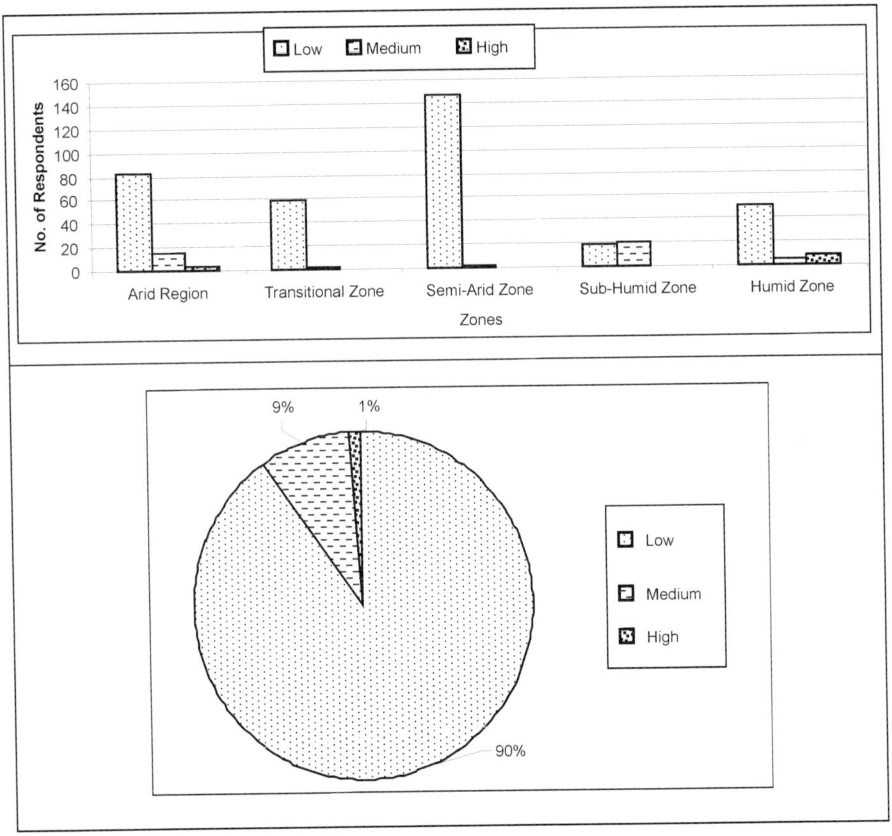

Fig. 6.28: What is over all awareness of Government functionaries about climate change?

So, is the case of awareness of elected representatives. 95 per cent people perceived that the awareness of elected representatives to adaptation is low (Table 6.29 and Fig. 6.29).

Table 6.29: Perception on Adaptation Measures

What is level of awareness of elected representatives?	*Arid Region*	*Transitional Zone*	*Semi-Arid Zone*	*Sub-Humid Zone*	*Humid Zone*
Low	82	59	146	37	51
Medium	16	1	1	3	9
High	2	0	0	0	0

Source: Based on personal village survey.

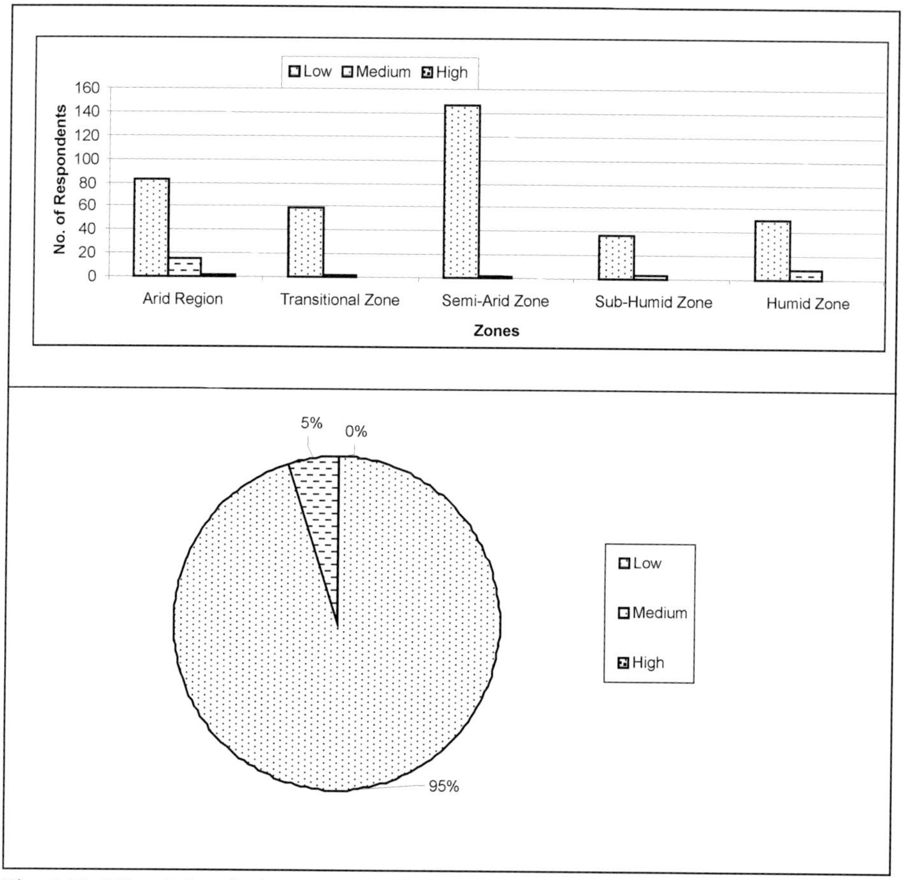

Fig. 6.29: What is level of awareness of elacted representatives?

As has been observed that during drought period people generally migrate to other areas with livestock. The analysis of data reveals that 68 per cent respondents felt that they do not migrate due to climate change particularly drought. However, 32 per cent people felt that they do migrate when drought occur (Tables 6.30, 6.31 and Figs. 6.30, 6.31).

Table 6.30: Perception on Adaptation Measures

Do you think that people have left village due to climate change, particularly drought?	*Arid Region*	*Transitional Zone*	*Semi-Arid Zone*	*Sub-Humid Zone*	*Humid Zone*
Yes	53	21	36	19	9
No	48	39	111	21	51

Source: Based on personal village survey.

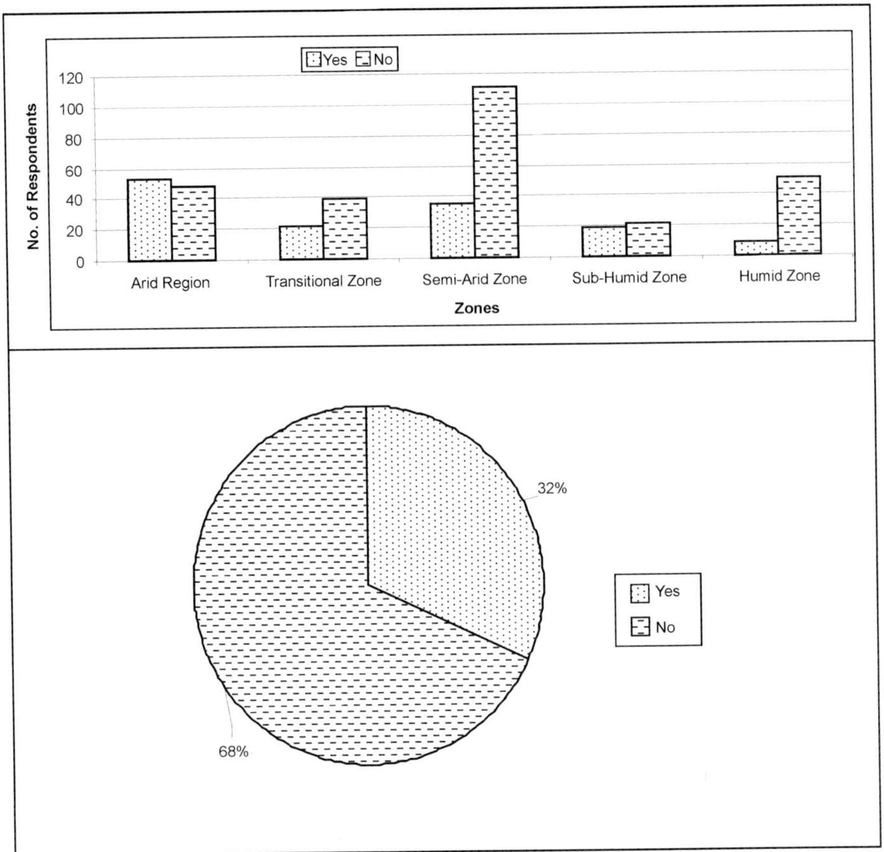

Fig. 6.30: Do you think that people have left village due to climate change, particularly drought?

Table 6.31: Perception on Adaptation Measures

If they migrate, do they take domestic animals with them?	*Arid Region*	*Transitional Zone*	*Semi-Arid Zone*	*Sub-Humid Zone*	*Humid Zone*
Yes	53	39	21	17	9
No	48	21	126	23	51

Source: Based on personal village survey.

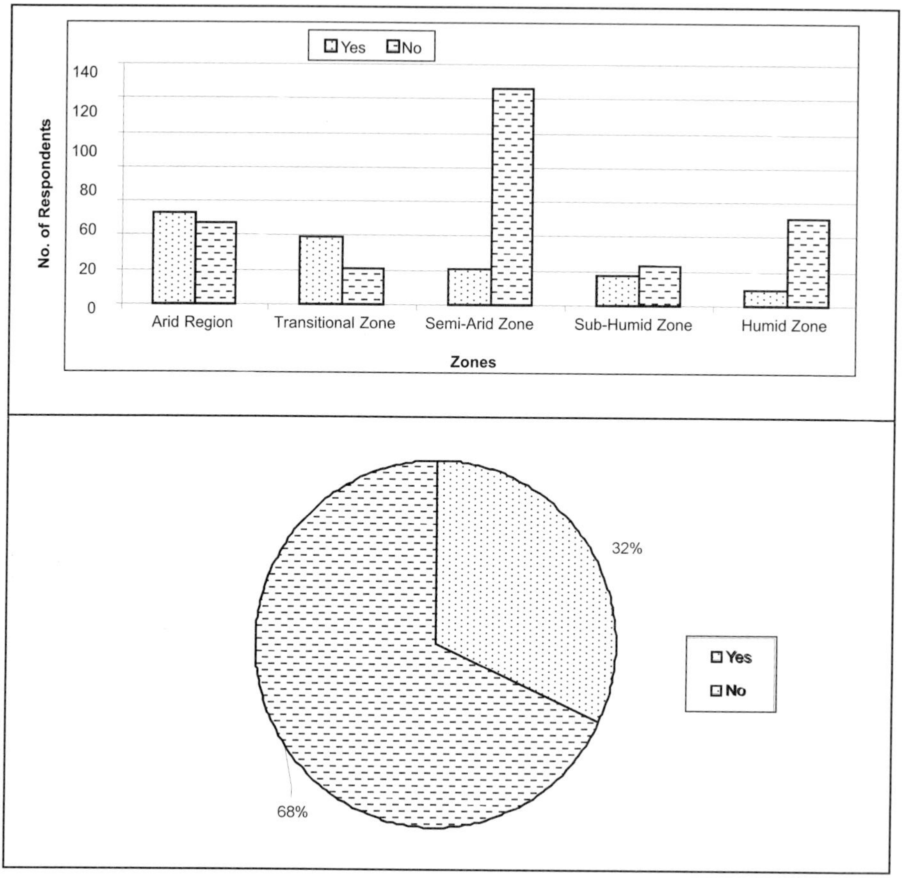

Fig. 6.31: If they migrate, do they take domestic animals with them?

The last question posed before the people during field survey and interview, the response was encouraging. 76 per cent respondents felt that they can personally limit the effect of climate change using indigenous technology and skill. However, 21 per cent people abstained and did not give any answer on the issue (Table 6.32 and Fig. 6.32).

Table 6.32: Percentage on Adaptation Measures

I personality can help to limit the effect of climate change	*Arid Region*	*Transitional Zone*	*Semi-Arid Zone*	*Sub-Humid Zone*	*Humid Zone*
Strongly Agree	7	8	21	7	0
Agree	37	51	98	23	51
Abstain	47	0	28	9	9
Disagree	5	1	0	0	0
Strongly Disagree	5	0	0	1	0

Source: Based on personal village survey.

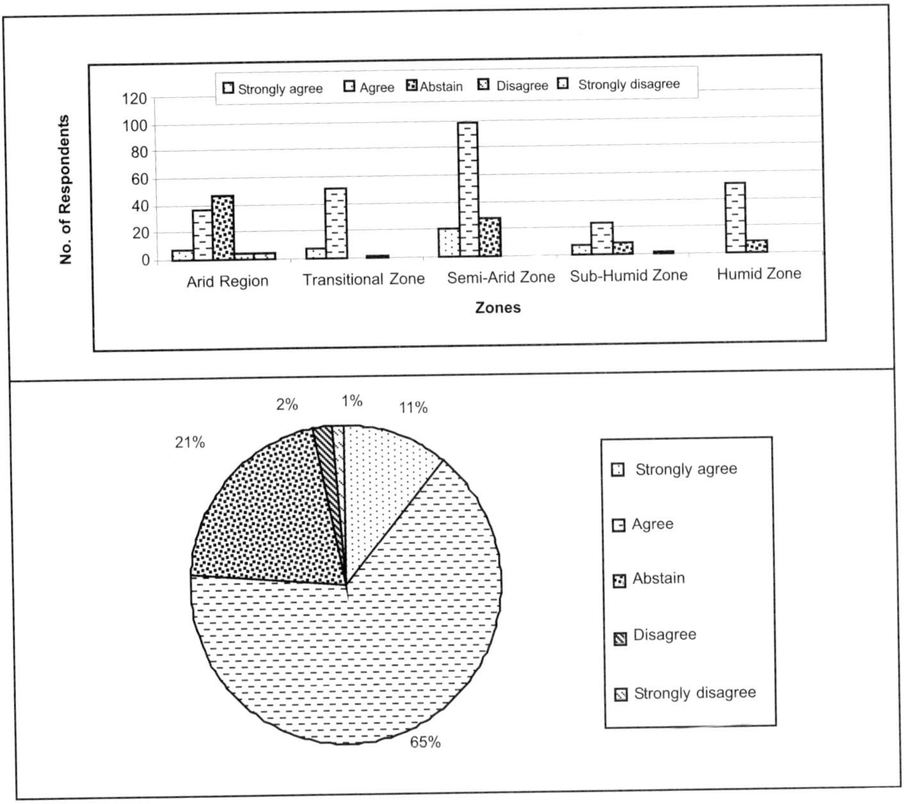

Fig. 6.32: I personally can help to limit the effect of climate change

The extensive survey conducted in 20 villages interviewing 407 respondents during 2013 and 2014 brings out the following conclusions:

- The majority of people are aware of the problem of climate change in Rajasthan.
- Most of the people attributed climate change to both natural and human factors.
- The survey also brings out that changes in rainfall is occurring in Rajasthan having slightly more rainfall in western Rajasthan and declining trend of rainfall in the eastern part of the Aravallis.
- The study well indicated that temperature is increasing in all the regions of Rajasthan which supports the linear regression analysis of temperature trend (1901-2012) analyzed in second chapter.
- The analysis further brings out that groundwater levels are declining in most the parts of Rajasthan except in some areas of humid region.

- The land-use changes in Rajasthan have been occurring and are likely to occur due to climatic vagaries and changing rainfall and temperature pattern.
- The productivity of land has declined due to climatic factors. However, wherever productivity is increasing, it is due to other social, economic and technological factors (fertilizers etc.)
- The decline in forest cover and accelerated land degradation is evident in the analysis of surveyed data.
- The responsibility of people, government functionaries and elected representatives to mitigate climate change is very low and need to make awareness amongst people to do personally for adaptation and preparedness well in advance when the change occurs.

7
Conclusions and Adaptive Strategies for Climate Change

In the present study climate change and its impacts on land and water resources have been examined. Climatically, Rajasthan is a very sensitive state having scanty rainfall, high air temperature, intense solar radiation and high wind velocity. In an era of climate change Rajasthan is likely to suffer further more shortage due to uncertainty of rainfall and increasing number of weather related hazards and disasters. The analysis of excessive rainfall, droughts, changing nature of number of rainy days and trend of temperature and rainfall during last 112 years or so make it clear that Rajasthan has been witnessing climate change during past and present centuries. The frequency of weather related hazards have increased both in western arid region as well as in the eastern part.

The analysis of rainfall data from 1877 to 2012 reveals that over all the monsoonal rainfall in Rajasthan shows both an increasing and decreasing trend. The decade from 1962-1972 was the decade of declining trend of rainfall while the period from 1972 to 1978 showed an increasing trend of rainfall. Again, the declining trend has been noticed from 1979 to 1996 with a severe drought of 1987 when –55 per cent deficient rainfall was recorded. After 1987 onward a slight increase in rainfall has been noticed. From 2000 onward till 2009 again a declining trend of rainfall has been observed with severe drought of 2002-03. However, from 2009 onward the current decade is indicating an increasing trend of rainfall.

The centennial rainfall trend (1901-2012) in Rajasthan shows that Banswara, Dungarpur, Jhalawar, Sirohi and Udaipur have recorded less rainfall than normal while rest of 27 districts of Rajasthan have received more rainfall, although all correlation values are insignificant. Annual total rainfall shows a decreasing trend in eastern Rajasthan while annual total rainfall in western Rajasthan shows either no trend or slight increasing trend.

Correlation analysis of annual average temperature with time series shows that fifteen districts of Rajasthan—Jaisalmer, Bikaner, Jodhpur, Banswara,

Dungarpur, Barmer, Udaipur, Jalore, Jhalawar, Sirohi, Chittorgarh, Rajsamand, Baran, Pali and Kota have recorded very significant warming while two districts —Bhilwara and Bundi have witnessed significant warming trend at centennial time scale. The districts like Sikar and Dausa have shown a cooling trend although this trend is insignificant. Over all the conclusion is that the temperatures have been showing a rising trend after 1981. Both mean maximum and minimum temperatures other than mean annual temperatures are showing an increasing trend over Rajasthan. Mount Abu, the only hill station of Rajasthan is showing a decreasing trend in minimum temperature while no trend in the maximum temperature has been observed (IMD, 2010).

The study further indicates that the number of rainy days in the state shows a declining trend and extreme rainfall events are more in frequency. The study further brings out the fact that there has been a shift of 15 days in timing of seasons in the state. The snowfall in Jammu & Kashmir, March 2015 are warning signs of climate change, a more cooling period in coming years.

The analysis of causes of climate change indicates that both natural and anthropogenic factors are playing a role. However, the short-term analysis of both temperature and rainfall (1901-2012) brings out clearly that changes are generally attributed to anthropogenic factors such as energy generation, transport, industrialization, urbanization, agriculture, changes in land-use, forestry, increase in livestock and population explosion resulting into increased emission of green houses gases in the atmosphere. However, there is a need of more data on GHG emission to prove the hypothesis.

The analysis of data reveals that there has been a clear warming trend in the past temperature records supports the hypothesis of a increasing trend of mean annual temperature over time found to be true. The analysis of rainfall data during 1901-2012 infect climate change on water further validate the hypothesis that rainfall in Rajasthan has been variable and trendless.

Impact of Climate Change on Water Resources

The preceding analysis has brought out the fact that the water resources of Rajasthan are deteriorating at the fast rate. The climate change poses uncertainties to the supply and management of water, the most precious resource of ecologically vulnerable state like Rajasthan. The critical situation of water resources in the light of above analysis is highlighted below and how climate change will act as an additional stressor in near future?

- Rajasthan is one of the most water deprived state with respect to rainfall per capita water availability.
- Surface water resources are limited and in a precarious situation. The rivers are rainfed and there is no perennial river in the state except the Chambal and Mahi.

- The per capita available water is about 0.000780 BCM against minimum requirement of 0.001 BCM. It is anticipated that the water availability would further fall in future.
- The state of groundwater exploitation which was 35 per cent in 1984 has reached to 166 in 2009 out of 239 blocks and 172 in 2011 out of 243 blocks.
- The scanty and irregular rainfall trend from 1901 to 2012 has resulted in poor recharge to groundwater. Climate change projections in declining rainfall and increasing mean temperature in future will further limit the recharge of groundwater resources.
- Indiscriminate use of groundwater has brought out changes in the quality making groundwater unsafe for health. Climate change will further increase groundwater extraction due to less availability of surface water and rising demand which could further deteriorate groundwater quality and have a serious implications for human health.
- The efficiency of water use in the state is low and unallocated for water loss is very high. About 2/3 of the water used for irrigation is affected by inefficiency. Losses in surface water irrigation are much higher as compared to those of groundwater irrigation.
- The deficit between demand and supply is 8 BCM at present and likely to increase to 9 BCM by 2015. Non-agricultural demand is expected to increase implying the state will slip from scarcity to an absolute scarcity zone.
- With the rising demand for water for both agricultural and non-agricultural purposes the share of water for agriculture is to set to 70 per cent from 83 per cent by 2050. This will have serious implications for agriculture.

In view of the above findings, Rajasthan needs more holistic and integrated management of its water resources. Collective action is necessary to bring together all sectors and stakeholders to work together towards integrated and participating solutions for water resource planning and management in view of climate change.

Climate Change and Land Use and Land Use Changes

The analysis of land-use changes brings out the following trend in Rajasthan:

- The official data reveals that the forest area is showing an increasing trend in Rajasthan from 6.50 per cent in 1985-86 to 7.98 per cent in 2009-10. However, the IRS data of 2008 reported only 4.70 per cent area under forest cover. The vegetal degradation of 6.25 per cent as per ISRO report on desertification (2007) clearly indicates indiscriminate degradation of the Aravallis and other parts of the state. The village-level survey also upholds this view. However, there has been an increasing trend of vegetal cover in desert districts from 1985-86 to 2009-10 due to slight increasing trend of rainfall and resultant moisture in soils. The degradation of forest has been more in southern districts like Banswara, Dungarpur, Udaipur and Sirohi as a result the surface reflectivity is on increase and warming trend is witnessing.

- The area not available for cultivation is showing a decreasing tend, but the land put to non-agricultural land shows increasing trend. Other cultivated land excluding fellow has been showing a declining trend.
- Fallow land has also been showing an increasing trend due to extension of irrigation facilities use of fertilizers and other inputs in agriculture. Despite irrigation facilities in IGNP there has been declining trend of net sown area and total cropped area in Hanumangarh district.
- The study further indicates that climate change will have less impact on land-use and agricultural production till the resource inputs are available. The analysis of crop area, production and productivity indexes reveal that the rate of increase in productivity has shown a diminishing return. The impact of modern techniques of agricultural production is visible in Rajasthan till 1980-1990. But after that the impact of chemical fertilizers has failed to show any significant impact. It seems that despite of agricultural intensification, the agricultural productivity is increasing at a diminishing return. As a matter of fact, it implies that climate change is playing more role. The analysis of temperature and rainfall trend from 1901-2012, brings out the fact that the temperature has started increasing after 1970s. Since the climate changes produce its impact after some time, the decreasing trend has begun after 1980-90. Hence, the impact of climate change is now visible in the state.

The earlier studies and the present study brings out the fact that the performance of agriculture in Rajasthan has been satisfactory despite of adverse climatic conditions. However, there are many challenges to be addressed so as to accelerate the agricultural growth in the event of climate change. Some of the challenges include the expansion of irrigation facilities, efficient water management, low productivity, practically very little value addition, distress sales, rising cost of cultivation, greening the desert through agro-forestry programme, upscaling of farm-validated modern technologies and agricultural innovations. Though challenges are many, there are also opportunities which are to be harnessed for future agricultural growth in Rajasthan.

People's Perception on Climate Change

The extensive survey conducted in 20 villages interviewing 407 respondents during 2013 and 2014 brings out the following conclusions:

- The majority of people are aware of the problem of climate change in Rajasthan.
- Most of the people attributed climate change to both natural and human factors.
- The survey also brings out that changes in rainfall is occurring in Rajasthan having slightly more rainfall in western Rajasthan and declining trend of rainfall in the eastern part of the Aravallis.
- The study well indicated that temperature is increasing in all the regions of Rajasthan which supports the linear regression analysis of temperature trend (1901-2012) analyzed in second chapter.
- The analysis further brings out that groundwater levels are declining in most of the parts of Rajasthan except in some areas of humid region.
- The land-use changes in Rajasthan have been occurring and are likely to occur due to climatic vagaries and changing rainfall and temperature pattern.
- The productivity of land has declined due to climatic factors. However, wherever productivity is increasing, it is due to other social, economic and technological factors (fertilizers etc.)
- The decline in forest cover and accelerated land degradation is evident in the analysis of surveyed data.
- The responsibility of people, government functionaries and elected representatives to mitigate climate change is very low and need to make awareness amongst people to do personally for adaptation and preparedness well in advance when do the change occur.

Adaptive Strategies and Adoption of Green Economy for Sustainable Development

Action to reduce environmental deterioration and mitigate the impact of climate change is the responsibility of central and State governments and individuals. At the same time it is widely believed that some impacts arising from industrial pollution over the past 100 years are evident in India as well as in Rajasthan. For this reason adaptation to climate change is also a necessity. However, the mitigation of desert state like Rajasthan will be less of a priority than adaptation. Both adaptation and risk reduction strategies must focus on raising the adaptive capacity of more vulnerable sections of society. It is often stated that India belongs to the category of large emitters which must take on carbon reduction commitments in order to mitigate global climate change. India is describe as fourth in the list of 20 countries in terms of total carbondioxide emissions from the consumption of energy after China, USA and Russia. India emits 1775.76 million metric tons of CO_2 (2011). However, in terms of per capita carbondioxide emission from the consumption of energy it ranks lowest, i.e. 1.45 metric tons of CO_2 per person. In this regard USA (17.62%) stands first in the list of 20 countries. Central Government is committed to address the issue of climate change as a national priority.

In view of this, the country has already taken serious steps and in 2008, a National Action Plan on Climate Change (NAPCC) for India was released by the former Prime Minister. In view of critically addressing the challenges posed by climate change along with the imperatives of poverty alleviation and economy growth for India, the NAPCC encomphases a range of measures. The focus of NAPCC is to improve the understanding of climate science, adaptation, mitigation, energy efficiency and natural resources management and conservation. It focus on eight priority missions to reduce climate change; these include National Missions on Solar Energy, Enhanced Energy Efficiency Sustaining, Habitat, Water the Himalayan Ecosystem, Greening India, Sustainable Agriculture and Strategic knowledge for Climate Change covering a range of actions including adaptation and mitigation.

In order to achieve synergy between national priorities and state of Rajasthan specific strategies, the state has already prepared the State Action Plan in accordance with the common framework set by the Ministry of Environment and Forest in 2010. To begin with, the State of Rajasthan established a climate change and CDM cell in the State Pollution Control Board to act as a nodal agency to deal with all the issues related with climate in Rajasthan. State specific missions for Rajasthan were developed highlighting research gaps and needs along with relevant policy measures in the light of vulnerabilities and capacities of the state.

The agenda led to setting-up of task forces for water resources, agriculture and animal husbandry, human health, enhanced energy efficiency and solar energy, urban governance, sustainable habitats and strategic knowledge for climate change. The RCCA identified above seven sectors and list out certain strategies. Hence, in view of present research findings and on the basis of NAPCC and SAPCC, the following adaptive measures are recommended in the sectors of water resources, agriculture and land-use changes, desertification and forest resources.

Adaptation to Depleting Water Resources

I. Ground water management with focused attention on over exploited areas:
   - Comprehensive assessment of groundwater resources including groundwater level, yield, recharge potential up to block/taluka level

and entering the data collected in water resources information system. This action would enable effective planning and management of the fast depleting groundwater resources of the state.

- Evaluation of groundwater recharge potential by conducting GIS based aquifer studies and assessing the feasibility and viability of groundwater recharging by rain water harvesting or artificial recharging.
- Conducting groundwater exploration studies to decipher the potential of deeper fresh water aquifer up to 1000 to 1500 metres, which would operate as a water bank for water stressed regions during extreme events like droughts for addressing regular water demand.
- To better plan and manage water resources in the event of climate change there should be a regulation on drilling bore well for groundwater use along with policy on mandatory groundwater metreing for bulk consumers in domestic, industrial and agricultural sector.
- Mass awareness programme is essential to enhance understanding of groundwater as a finite resource.
- Setting up research facilities for groundwater assessment and flow modelling with respect to climate change and climatic variables.

II. Enhancing preparedness for drought monitoring, drought mitigation and development of early warning system:

- Traditional water harvesting systems like *khadins, tankas, bawries, johads*, village ponds, *nadis, kundis, paar* system etc. be encouraged.
- Conducting feasibility studies for *in situ* water and soil conservation practices like contour bunding, vegetative barriers and percolation ponds in drought prone areas to reduce evaporation losses are needed.
- Developing on integrating drought monitoring systems which include climate, land and socio-economic indicators to fully characterize the spatial extent and potential impact of drought situation enabling proper planning of adaptation strategies for Rajasthan.
- Developing a drought management and mitigation policy which would integrate climate change concerns like frequency of drought events.
- Educating farmers about matching land-use systems with water availability by adopting efficient practices, low water requirement crops for agriculture.

III. Enhancing water conservation measures:

- Mass construction of rooftop rainwater harvesting system in urban areas be encouraged.

- Conducting pilot studies to explore augmentation of water resources by transforming the surplus flood water into utilizable water from flood prone districts like Sri Ganganagar, Hanumangarh, Bharatpur, Alwar, Dholpur and Jaipur.
- Promoting wastewater recycle and reuse in all sectors.
- Public awareness campaign to encourage rain water harvesting in rural and urban areas.

IV. Improving water use:
- Conducting studies on efficient crop water application and utilization by promoting water saving techniques like pressure irrigation methods, drip and sprinkler irrigation.
- Conducting feasibility studies for assessing the effectiveness of technologies (conjunctive use, sub-surface drainage system) in managing water logging in the Chambal and IGNP command areas.
- Mandatory water audit for all sectors.
- Rationalizing water pricing for domestic, industrial and irrigation water usage by introducing increasing block tariff.

V. Developing a Comprehensive water data base for assessment of impacts of climate change on water resources:
- Setting up a real time dynamic data base for water resources along with the development of a web enabled water resources information system.
- Review of availability of scale (spatial and temporal) of hydrological data like surface and groundwater resource.
- Review of network of automatic weather stations and rain-gauge stations.
- Setting up dedicated facilities with advanced computing systems and customize information on impacts of climate change for regional water basins.
- Conducting training programmes for government officials to facilitate the understanding of impacts of climate change on water resources.

Adaptation to Land Use and Agriculture

Climate change is likely to further exacerbate the already strained conditions for land-use and agriculture in Rajasthan. The present study has further predicted that due to climate change there may be significant increase in desertification processes in the next 100 years with disproportionate impact of global warming on coupled human and natural systems. Manifestation of a changing climate are being observed as the duration of seasons change. It is observed that summers are expanding and the winters are shrinking.

Temperatures in early March have been on the rise which makes the cultivation of wheat more challenging. Decreasing productivity of crops and livestock animals is also being observed. Decline in pasture lands is also being experienced. As a result of these changes, the real income of cultivators is also declined.

Therefore, there is a need to have a suitable scientific strategies to cope with the situation. The following strategies in land-use and agricultural sectors may be adopted:

I. **Enhancing productivity of crops and livestock:**
   - Development of such crops which are tolerant to droughts and thermal extremes, alkalinity and pests and less water consumptive be encouraged.
   - Programmes like macro management of agriculture be adopted. Emphasis should be on increasing and improving quality of produce of coarse cereals like *bajra* that are traditionally suited to the region.

II. **Restoration and development of wastelands:**
   - The state has nearly 1.194 million hectares as wastelands. Trees like *khejri* and subabool can be grown on large scale on wastelands in phased manner.
   - Reclamation of alkaline soils in Bhilwara, Chittorgarh, Alwar, Bikaner, Pali, Nagaur, Bharatpur, Ajmer, Jaipur, Sikar, Dholpur and Jalore be undertaken.

III. **Research and Assessment of Specific Climatic Risks to Agriculture:** Installation of automatic weather stations in the State be planned to get better picture of climatic variability. In the light of climate change impacts rigorous data generation for conduction of vulnerability assessments using crop growth models should be undertaken.

IV. Promotion and management of multi-functional agro-forestry system be undertaken.

V. Promotion of horticulture in Jhalawar, Sri Ganganagar, Hanumangarh, Sawai Madhopur, Bundi, Jalore and Barmer be promoted.

Adaptation to Depleting Forest Cover

The forest cover in the State is about 16036 $k.m.^2$ which is 4.69 per cent of the total geographical area (FSI, 2009). Rajasthan is a home of about 3000 known species of flora and fauna and a large number of undocumented insects, butterflies and micro-organisms. A majority of the native plants are known for their economic importance and are being used by local communities. Many of these species have evolved over the years to survive in the harsh climatic conditions.

Many of the flora and fauna are rare and endangered. Many plant and animal species are endemic to the Thar Desert and the Aravallis. It is estimated that 100 species of plants (61 of the desert region) are threatened in view of their conservation status. All the 19 endemic species are threatened and so are all the 35 wild relatives of cultivated plants. There are 23 species of plant are endemic to the State.

The present and projected changes in temperature and rainfall under climate change scenarios poses risks to structure and functions of the forest eco-systems that are already exposed to multiple stresses. The studies demonstrate that the impacts of forests are under researched in Rajasthan, however, broad inferences from global and national studies point out towards lowered productivity and a shift to a drier forest type. The current projections of the impacts of the climate change suggest increase in the aridity of the existing drylands and sub-humid lands. Hence, the water dependent and the higher and extreme temperature sensitivity species diversity and habitats would face danger of reduction in size or even face extinction. This could further have a cascading affect the crucial ecosystem functions of the forest including carbon sequestration. It could also multiply sound of the current effects such as desertification and land degradation. In view of the above, the state mission on forestry and biodiversity identifies the priority areas based on the Environment Policy and the Climate Action Plan as follows:

- Increasing area under forest cover.
- Defining plantation targets according to land-use.
- Community involvement in afforestation activities.
- Afforestation of private land holdings.
- Explore mitigation potentials of community forest projects.
- Assessment of shifts in forest type.
- Conservation of wildlife outside forest areas.
- Maintaining species diversity under climate change.
- Land degradation and desertification.
- Traditional ecological knowledge and domesticated biodiversity.
- Traditional animal husbandry systems and pastoral nomadism.
- Ecological services of forests.

Knowledge Management

- Create data sharing framework among various data sources to foster climate action.
- Inventorize traditional knowledge systems.
- Dissemination of capacity-building. Television, orientation sessions on climate change for government officials, technical capacity-building for advancement of research on climate change—Advanced

Computering—Regional Climate models etc. Training programme on climate change for school teachers.

Adaptation to Strategic Knowledge on Climate Change

Knowledge Creation

Under this climate change vulnerability assessment and strategies for better preparedness in Rajasthan. The assessment of vulnerability in the following key sectors have been determined:

- Hydrology and water resources.
- Agriculture and food supply.
- Desertification.
- Forestry and biodiversity.
- Livelihood including migration and conflict.
- Human health.
- Understanding low carbon development pathways.
- Increasing the scale and spatial distribution of the data on meteorology, hydrology and land-use.

Enhance the Scope of Renewable Energy—Path Towards Green Economy

Climate change mitigation presents a huge opportunity for Rajasthan to exploit its renewable energy potential. It has a vast potential for power generation from non-convectional energy sources like wind, solar and biomass. The state receives maximum solar radiation intensity in India on accounts of its unique location in sub-tropical region. The average number of rainy days is least in comparison to the other parts of the country. Hence, it is best suited for solar power generation. In order to promote solar energy in Rajasthan, the State government has already initiated policy for promoting Generation of Electricity through Non-Conventional Energy Sources in 2004. Such initiative can result in huge co benefits for the state; about 1656 tonnes of CO_2 can be saved annually by installation of IMW biomass gasifier based power plants (Panwar, N.L. and Rathore, N.S., 2009). It is therefore, suggested that districts of Jaisalmer, Barmer, Bikaner, Jodhpur, Pali, Sri Ganganagar, Hanumangarh, Nagaur, Churu and Jhunjhunu are preferred areas for solar power generation.

Mitigation Opportunities

Renewable Energy Potential: Rajasthan has meagre resources of coal and petroleum, but is richly endowed with renewable energy sources such as wind energy, solar energy, and biomass.

Wind: The Rajasthan State has wind energy potential of 4858 MW, which is around 10 per cent of India's wind energy potential. At present, the state has harnessed only 850 MW of wind energy. Some of the potential sites identified in the State Wind Policy (2003) on the basis of average wind speed, wind density profile and power potential per square metre include Devgarh in Chittorgarh district, Harshnath in Sikar district, Jaiselmer in Jaisalmer district, Khodal in Barmer district, Mohangarh at Jaiselmer disctrict, and Phalodi in Jodhpur district. The State should encourage wind assessment studies in additional locations to identify other potential sites to be able to harness the wind energy potential in the State.

Hydro: Given the water scarce situation, small hydro (up to 25 MW) has only 57.17 MW42 potential out of which 23.85 MW has already been installed.

Solar: Rajasthan is bestowed with significant amount of solar energy potential and the recent interests shown by the developers and investors in addition to the ambitious Solar Policy, Rajasthan is poised to be a leader in the sector. Also, the State transmission utility has developed strong transmission system at Barmer, Jaisalmer, Merta, Jodhpur, Bikaner and Ratangarh along with strong transmission line and sub-station network in desert area. This has led to suitable transmission system to evacuate solar power in the desert area, where the potential of solar energy is maximum. The State can further tap resources and leverage from the Jawaharlal Nehru National Solar Mission. (40 Note on power sector, December 2009. Government of Rajasthan 41 http://www.mnre.gov.in/wpp.htm 42 http://www.mnre.gov.in/145). The arid parts of Rajasthan receive average maximum solar radiation of about 7.5 kWh/m^2 in summers and minimum of about 4.6 kW h/m^2 in winters. In most parts of India, clear sunny weather is experienced from 250 to 300 days a year (Purohit and Michaelowa, 2008). The annual global radiation varies from 1600 to 2200kWh/m^2 and the highest annual global radiation (Z2400 kWh/m^2) is received in Rajasthan and northern Gujarat (Purohit and Garud, 2007). Thar Desert of Rajasthan receives around 2200-2400kWh/m^2 annual direct normal incidences. According to Purohit and Purohit, locations blessed with annual direct solar radiation more than 1800kWh/m^2 are best recommended for installation of concentrating solar power (CSP) systems (Purohit and Purohit, 2010). Further, the authors suggest districts of Jaisalmer, Bikaner, Barmer, Kota, Jodhpur, Udaipur, Jaipur in Rajasthan as preferred locations for CSP systems. Besides this, one of the most widespread uses of solar thermal technology is solar water heating (SWH). Theoretical, annual CO_2 emissions reduction potential of solar water heating systems in Rajasthan through annual electricity saved amounting to 1.3 (TWh) result is 1.0 million carbon equivalents (Purohit and Michaelowa, 2008). SWHs find wide application in large establishments such as hostels, hotels, hospitals, small scale industries such as textile, paper, food processing etc. These applications are also useful for domestic water heating, particularly simple domestic systems. Provision has been made in the building bye-laws of the State for Solar water heating in buildings such as hospitals and nursing homes, hotels, guesthouses, dak bungalows, lodges, government guest houses, hostels, community centres and other public utility buildings, along with panel clauses. Besides, there is also potential for using renewable energy option for autonomous desalination systems and withdrawal pumps as supply of portable water in Rajasthan without any energy input is almost impossible. Even if freshwater is readily accessible under the ground level, energy is required to pump the freshwater from its source. Freshwater drawn from the groundwater source requires 0.14-0.24 kWh/m^2 (0.5-0.9 kJ/kg) for a pumping head of 120-200 ft. Further, treatment of surface waters to potable quality requires 0.36 kWh/m^2 (1.3 kJ/kg) to produce freshwater. The author further estimates energy requirements for different desalination processes and associated green house gas emissions (Gude *et al.*, 2010). Some of the options as discussed by E. Mathioulakis *et al.*, 2007 include PV or wind-powered reverse osmosis, electrolysis or vapour compression; or Solar thermal or geothermal energy and distillation processes. Further solar still can be economically used to provide portable water to remote areas of the State (Khanna, Rathore and Sharma, 2009). Solar photovoltaic technology for the purposes of decentralized electricity generation in remote areas, RREC is already installing SPV Domestic lighting system, street lighting system and solar pump sets.

Biomass: As for biomass energy, the main source of biomass energy in the State of Rajasthan is mustard husk and Julie Flora. Biomass totalling to 72 MW have already been installed on December 2009 and few are for captive generation. Many of the MSME clusters in the state use either diesel or furnace oil or use woody biomass in an inefficient manner and also nearly 65 per cent of the households in the State use fuelwood, that is inefficient and contributes to carbon emissions. A study (Panwar and Rathore, 2009) conducted on production of surplus agricultural biomass from all the major crops grown in Rajasthan which categorizes biomass in two classes, i.e. as essential class and non-essential class of which biomass residues consumed as fodder, thatching and fencing materials and other socially significant nature are grouped under the essential class and the remaining under non-essential class. Currently, non-essential class biomass is being burnt by the farmer itself in the field leading to emissions; this could, however, be used for energy generation purposes. The main type of agricultural biomass in Rajasthan from crop production includes straw from rice, soyabean; stalk from castor, cotton, *moong, moth,* sesame, mustard; and shell from *guar,* groundnut. On an average 1 kg biomass with reasonable moisture content per unit of power generation (per kW) is needed in the biomass gasifier based power plant (Mukhopadhyay, 2004). Thus, annually 7200 tonnes of biomass are required to run the 1 MW capacity power plants round the clock for 300 days. It is estimated that about 1275 MW power could be generated through biomass gasifier based power generation plant through surplus biomass available in Rajasthan leading to savings of around 2 million tonnes of CO_2.

Biogas: Rajasthan possesses the India's second largest livestock population of 49136 thousand (10.13%), with a high degree of diversity in its composition. Revival of community Biogas plants could be a potential activity as besides mitigating emissions, it also supplies energy and manure. This further reduces emissions and local pollutants through substituting firewood for cooking, kerosene for lighting and cooking and chemical fertilizers. A study of Pathak *et al.*, 2005 calculates mitigation potential of a family size biogas plant was 9.7 tons CO_2 equiv. year-1. Same study estimates potential of 39189 thousand tons of CO_2 equivalent in Rajasthan with a potential of 4040 thousand biogas plants. This could be used as energy option in rural areas as access to energy in rural areas is still a challenge for Rajasthan with about 20 per cent of the villages still un-electrified. Further, from above it is clear that state has immense potential for renewable energy particularly solar, wind and biomass. Also, these applications could be used at decentralised levels. In addition, benefits of carbon credits under Clean Development Mechanism (CDM) under the Kyoto Protocol further improve the financial feasibility of the renewable energy applications. 10.2.2 Energy Efficiency Low energy efficiency is one of the challenges faced by the State. In order to improve energy efficiency and address these concerns, Rajasthan Renewable Energy Corporation (RREC) has, under the guidelines of Bureau of Energy Efficiency (BEE), been conducting a survey of government buildings for energy efficiency, energy auditing, energy conservation measures for designated consumers, installing solar water heating systems in private establishments etc. According to the survey conducted by Government of Rajasthan in 2009 the annual electricity sale to agricultural sector is 8.14 BU with major energy consumption in energizing agricultural pumps with energy efficiency varying from 25-35 per cent as compared to the efficiency levels of 50-52 per cent of BEE star labelled agricultural pump sets. Further, it is estimated that by replacement of existing pumps the achievable saving potential is 30-40 per cent and sectoral saving potential works out to be 2.442 BU per year.

Commercial buildings and public works: Also, there are 131 commercial buildings in the state accounting for annual energy consumption of 361 MU (with over 500 kW connected load) which works out to about 23.5 per cent of the sectoral consumption with energy saving potential varying from 20-30 per cent and accounting to be 72 MU. Further, the annual electricity sale to public lighting and public water works and sewage is 1.31 BU. For 169 major municipalities, 5 municipal corporations and 15 municipal councils considered, annual electricity consumption for street lighting is 188 MU and annual consumption for water and sewage pumping is 1.21 BU. The total consumption is 1398 BU. The energy savings potential for street lighting in municipalities and corporations is assessed to be 25 per cent and works out to 47 MU per year, while the energy savings potential for water works and sewage in municipalities and corporations is assessed to be 20 per cent and works out to 242 MU per year.

Industries: The annual electricity sale to the industry sector in Rajasthan is 12.12 BU and works out to 42 per cent of the total electricity sold. While the larger industries segment is covered for energy efficiency under the mandates of EC Act as designated consumers, SME segment is being addressed for energy efficiency through cluster based initiatives by BEE. The electrical energy saving potential in industry sector varies from 7-10 per cent. The energy savings potential for the sector is assessed to be 0.849 BU. According to BEE, four energy intensive clusters have been identified in Rajasthan, namely rolling mills, bearing clusters, marble cluster and textile clusters for energy consumption and savings assessment with estimated total energy consumption in rolling mills (Jaipur) to be 115 MU with 15 per cent energy saving potential; estimated total energy consumption in marble clusters (Kishangarh, Udaipur, Makrana) to be 488.6 MU with 10 per cent energy saving potential; estimated total energy consumption in textile clusters (Pali) to be 46030 TOE with 20 per cent energy saving potential.

Domestic: The annual electricity sale to domestic sector in Rajasthan is 4.46 BU which accounts for 16 per cent of the total electricity sold. The domestic sector electricity consumption varies with respect to rural and urban segments and climatic seasonal variations. In the rural segment major use of electricity is towards lights and fans. In the urban segment the typical energy consumption pattern includes the following AC and refrigeration (56%), lights and fans (24%), coolers, TV, washing machines (16%) and others (4%). The major avenues for energy saving in rural domestic sector include replacement of GLS bulbs with CFLs, adoption of star rated domestic appliances like ceiling fans, refrigerators, AC units, tube lights etc. The savings potential in rural segment by adopting CFLs and BEE star rated products is 40-50 per cent. The savings potential in urban segment by adopting star rated products is 15-20 per cent. On the whole the energy savings potential in domestic sector is estimated 20-25 per cent which accordingly works out to 0.89 BU per year.

Rajasthan State Action Plan on Climate Change: Hence, total saving potential in Rajasthan is 4.542 BU representing 16 per cent of the annual energy sold. Further, from the above background, it is clear that energy efficiency potential is high in commercial buildings and SME sector. Further, analysing the sector-wise electricity consumption in 2007-08 agriculture sectors is the largest consumer at 34.43 per cent followed by large industries (22.73%), domestic (18.86%), commercial (6.47%), medium industries (5.36%), public water works (4.91%), small industries (2.66%) and public lighting (0.65%). From such an analysis, directionally it is clear that demand side measures should be targeted in the agriculture, industries and domestic sectors. However, if further analysed by looking at the past trends the percentage share has been increasing in the agriculture and domestic sector while decreasing in the industries. Within industries since large industries as designated consumers are covered under the national level (perform achieve and trade) PAT scheme, the state could focus on medium and small industries while also ensuring that the implementation of PAT scheme is being facilitated. The Rajasthan Government has already started the pilot programme on e-filling of energy consumption data which is an important first step in this regard.

Mitigation Potential in Forest Sector

The Mission on Afforestation identifies 13 specific districts for the afforestation purpose. The Forest and Environment policies also mention about the programmes to develop the grazing lands and conserve the oran, the traditional sacred woodlands. These provisions provide the scope of undertaking Afforestation/Reforestation CDM possibilities and the REDD+ especially in the context of orans and the Sustainable Forest Management Mechanisms.

State Forest Policy 2010

The objectives of the policy to address the livelihoods of the forest dependent communities and importance of value addition and processing of non-timber forest produce.

- *Area under forest:* To achieve the target additional 45000 sq. km area owned by Government, communities and private persons will be brought under afforestation and pasture development.
- *Forest protection and conservation:* JFM may be used to persuade the encroachers of the forest land, curb forest fires, control illegal grazing. Mining in the forest areas should be discouraged by all means. Adequate infrastructural support and intensified patrolling and interdepartmental co-ordination to reduce poaching and other forest offenses.
- *Afforestation on Government land, Community land and Private land:* Identification of revenue waste land, vacant institutional land, linear patches along roads, railway tracks for plantation of multi-purpose species will be taken up.
- *Demand and Supply of Forest Produce:* Large scale plantations will be raised outside forests to provide wood produce.
- *Enhancement of Productivity:* Selection of Candidate Plus Trees (CPTs), establishment of Seedling seed orchard, modern hi tech nurseries, use of Quality Planting.
- *Checking Land Degradation:* Integrated watershed approach, treatment of catchments, reclamation of waterlogged areas along Indira Gandhi Canal is proposed with suitable tree species.
- *Combating Desertification:* Shelter belt plantations, block plantations of various silvipastoral models and sand dune stabilization programmes and plantations in desert areas with indigenous

xerophytic microphyllus species have been emphasized. 125 Policies Key objectives and goals Strategies/components

- ***Wildlife and Biodiversity Conservation:*** It is targeted to have minimum 5 per cent of the geographical area of the state as protected/ conservation area. Inventory of biodiversity, measures of *in situ* and *ex situ* conservation, rationalization of boundaries of Protected Areas have been suggested for strengthening the wildlife and biodiversity conservation in the State.
- ***Orans/Dev van:*** District-wise database and demarcation on cadastral maps of such areas is proposed. These patches will be declared as deemed forest as per the Forest (Conservation) Act, 1980. Local committee will be fully empowered for the management of these areas.
- ***Financial Provisions:*** Enhancement of sectoral allocation in the state, soft loans, efforts for more central funds, institutional financing such as NABARD etc., NREGA, CAMPA funds, and special funds to be sought for wildlife conservation and forest protection have been suggested. Corporate investment in JFM plantations, donations from Wildlife tourists and other sources for developing biodiversity resources in the state and sponsorships from various sources have been proposed.
- ***Mission on Afforestation:*** Promotion of agro-forestry in order to buttress livelihoods of forest-dependent communities R&D in Monitoring of shifts in forest types and documenting traditional knowledge for adaptation.

State Environment Policy 2010

The strategies are mainly oriented at efficient management of protected areas by Relocation of villages inside protected areas with appropriate package. Conferring heritable but non-transferable rights to the relocated communities to provide defined ecotourism services. Scientific Management Plans with short and long-term strategies for improving biomass productivity, habitat improvement, sustainable tourism and meeting the livelihoods of the local communities. Site-specific, people friendly eco-development programmes in fringe areas of Pas Effective use of Community Reserves and Conservation Reserve Provisions strengthening of the existing PA system by Rajasthan State Action Plan on Climate Change 126 Policies Key objectives and goals Strategies/ components covering unrepresented ecosystems.

- ***Desertification and Land Degradation Scientific Assessment of Desertification:*** Integrated approach for implementation of the schemes to combat the desertification Participatory sand dune stabilization programme shall be encouraged to involve the private land under sand dunes. Biodiversity outside PAs.

- ***The Strategies are Summarised as Follows:*** Proper research to be undertaken before introduction of exotic species. Promote use of traditional crop varieties. Technically, sound programmes for eradication of invasive species. Legal recognition to the traditional entitlement of the local communities under PESA (Panchayat Extension to Scheduled Areas) *ex situ* conservation of genetic resources in form of gene banks. Cultivating bamboo with the help of JFMCs.
- ***Traditional Ecological Knowledge and Domesticated Biodiversity:*** Focus on indigenous domesticated animal diversity for developing the improved breeds having adaptive abilities in deserts. Research on traditional crop varieties shall be promoted so as to understand their adaptation potential against climate change. Traditional knowledge associated with domesticated breeds of animals and traditional cultivars shall be documented in order to 127 Policies Key objectives and goals strategies/components facilitate grant of IPRs for traditional knowledge.
- ***Documentation and Demarcation of Orans:*** Traditional conservation practices of the local communities. The traditional techniques of seed banks, fodder bank, grain bank shall be effectively used for decentralized storage of resources for disaster management. Organic-farming agriculture technologies shall be promoted with proper incentives.

State Climate Change Agenda (2010)

The Climate Change Agenda for the state includes the prioritized issues of forestry and biodiversity as follows—Increasing the forest cover and density; Forest demarcation to facilitate the understanding of likely shifts in forest types under a future climate; Forest development and protection for increasing and maintaining the green cover; and Explore mitigation potential of community forest projects to obtain carbon credits for forest conservation.

Strategy for Combating Desertification

The problem of desertification in Rajasthan is so widespread that massive efforts are needed to improve degraded lands. There are number of institutions and government departments engaged in research, surveys and monitoring activities of arid-zone and drought prone areas of Rajasthan. In the past some sporadic attempts have been made to develop arid zone by former princely states. There is no evidence of any organized and systematic attempt having been made before Independence to tackle the problem of desert and semi-desert areas in a comprehensive manner. In 1951-52, the need to conserve and improve the resources of the desert region was recognized and a symposium was convened by Indian Science Academy. Besides this, an *ad hoc* committee of experts was appointed by the Union Government. Accordingly, Desert

Afforestation Centre was set-up at Jodhpur in pursuance of the recommendations. This programme was renamed in 1957 as the Desert Afforestation and Soil Conservation Station. At the same time, arid zone project was initiated by UNESCO to initiate and intensify research on problem of arid zone in different parts of the world. In the pursuance of UNESCO project recommendation, the Desert Afforestation and Soil Conservation was further reorganized in 1959 as the Central Arid Zone Research Institute. In 1966, a Desert Development Board was set-up. In September 1972, the activities of the Desert Development Board were reviewed and, finally, in Fifth Five Year Plan, it was merged with Drought Prone Area Programme (DPAP). The Desert Development Programme (DDP) was launched in 1977-78. The main objectives of the programme are to check the process of desertification and at the same time to develop and utilize the available resources of the region. The programme covers arid region comprising eleven districts and 84 blocks. The following major activities have been taken up under this programme:

1. Afforestation with special reference to shelter belt plantation, grassland development and sand-dune stabilization.
2. Groundwater development and utilization.
3. Contribution of water harvesting structures.
4. Rural electrification for emerging tubewells and pumpsets.
5. Development of agriculture, horticulture and animal husbandry.

   The Task Force recommended the following programmes in Seventh Five Year Plan:

   (i) Promoting more productive dryland agriculture on the basis of soil-water-climate resources of the area.
   (ii) Development and productive use of water resources of the area.
   (iii) Soil and moisture conservation including production of proper land-use practices.
   (iv) Afforestation including farm forestry.
   (v) Livestock development including development of pasture and fodder resources.
   (vi) Watershed and mini-watershed should be the basis of planning.

Drought Prone Area Programme (DPAP)

A special programme was launched in India in 1970-71 for the development of areas chronically affected by drought. The objectives of this programme are of developing the land, water, livestock and human resources of drought prone areas of the State. In Rajasthan, DPAP was initially started in 1974-75 in 72 blocks of 10 districts, but gradually was extended so as to cover 79 blocks spread over 13 districts. In 9 districts covering 61 blocks, DPAP and Desert Development Programme (DDP) were running simultaneously. However, to rationalize DPAP/

DDP programmes since 1982-83 the Government of India, namely DDP is in operation, while DPAP continued in southern hills and semi-arid districts of Rajasthan. In Seventh Five Year Plan 54 new blocks of some districts of south-eastern part of Rajasthan were included. In Seventh Five Year Plan, it was emphasized that DPAP will continue which would *inter alia* include increased stress on ecological balance and increasing the per capita income through effective utilization of scare water, conservation of scanty rainfall and arresting its runoff in drought prone areas.

Combating Desertification Programme (CDP)

The combating desertification was started in 1998 in which programmes such as dune stabilization, afforestation in Panchayat land, shelter belt plantation and grassland development were taken up.

Despite considerable contribution and prodigious efforts of the Government, significant breakthrough has not yet been achieved. A major reason for this may be the lack of integrated approach, and most of the efforts for controlling desertification have been sectoral in nature without adequate cognizance of collateral issue. In view of changing climate it would be necessary to review all the existing programmes such as DDP, DPAP and CDP in order to bring out new strategies for combating desertification in the State.

Solution to desertification in Rajasthan can be implemented readily if the resources are available to do so. An exploding population will continue to build pressure on land resources. This would hasten the destruction of vegetal degradation by the ever-increasing demand for fire wood and other requirements. The future drought and changing pattern of climate will have the damaging effect on all walk of life. Combating desertification can be done successfully using technology already known if the political will to act is present.

To sum up, it may be stated that the impacts of climate change will be felt across most sections of economy of Rajasthan. Drought and floods could mean higher evaporation which will result in low river flows and the quantity of available runoff may decrease. Crop yields are likely to decrease with rising temperatures. Forest character will change over because of pressing demand for fuel. However, more severe droughts could lead to greater desertification if land resources are not carefully managed. Strategies for drought are also needed. Water supplies have to be maintained.

In view of changing climatic conditions everyone has to contribute slowing climate change and reducing its impacts. People have to take individual responsibility for themselves and families. They have to support the green economy, lower their green house gas emissions, live sustainably and adopt the way of life to the changing climate. At the community level efforts should be made to ensure green economy to make use of products in public buildings such as solar power, lighting control and rain water harvesting.

Questionnaire

General Information

1. Name of Village .... Altitude ....
2. Distance from District HDQ ....
3. Name of the Head of the Households ....
4. Occupation: .... 5. Sex: .... 6. Age: ....
7. Religion: .... 8. Caste ....
9. Number of Members in the Family:
   Total Male .... Total Female .... Adult ....

Climate Change-Impact on Land and Water Resources

1. The world climate is changing:
   Strongly agree ☐ Agree ☐ Abstain ☐ Disagree ☐ Strongly disagree ☐
2. Climate change is the result of human activities:
   Strongly agree ☐ Agree ☐ Abstain ☐ Disagree ☐ Strongly disagree ☐
3. Climate change is natural occurrence:
   Strongly agree ☐ Agree ☐ Abstain ☐ Disagree ☐ Strongly disagree ☐
4. Climate change is due to both (natural and man's activity).
   Strongly agree ☐ Agree ☐ Abstain ☐ Disagree ☐ Strongly disagree ☐
5. I personally can help to limit the effects of climate change:
   Strongly agree ☐ Agree ☐ Abstain ☐ Disagree ☐ Strongly disagree ☐
6. What words or images do you associate with the climate change?
7. Does the temperature have risen in the last 40 years? Yes ☐ No ☐
8. If yes then define High ? Medium ☐ Low ☐

9. Whether the intensity of cold wave increased in this area in the last 20-40 years.
   Yes ☐ No ☐ If yes what are the reasons?..
10. What about the heat waves during the summer, did they increased too in last 20-40 years?
    Yes ☐ No ☐
11. Whether the region suffer physical damages due to extreme:
    Cold wave ☐ Heatwave ☐ Other condition (Freak weather conditions) ☐
12. What are the adaptation measures in the event of cold wave conditions?
    ....
    ....
13. Whether temperature will continue to rise in next 10-15 years.
    Strongly agree ☐ Agree ☐ Abstain ☐ Disagree ☐ Strongly disagree ☐
14. What is the impact of temperature rise on availability of water?
    ....
    ....
15. What are the sources of water including drinking water?
    Pond ☐ Reservoir ☐ Tank ☐ Canal ☐ Other ☐
16. What is the level of undergroundwater (Borewell/open well in feet)
    Now .... 20 Years Back ....
17. Has the undergroundwater level receded? Yes ☐ No ☐
18. Adaptation strategies in the event of temperature rise.
19. What is the intensity of climate hazard assessment?
    Low ☐ Medium ☐ High ☐ Very High ☐
    (a) Drought ☐
    (b) Flood ☐
    (c) Cloud bursts ☐
    (d) Heatwaves ☐
    (e) Cold waves ☐
    (f) Cyclone ☐
    (g) Hailstorm ☐
    (h) Frost ☐
20. Do you think pattern of land-use has changed in last 20-40 years?
    Yes ☐ No ☐
21. Whether land-use changes are due to rainfall variability.
    Yes ☐ No ☐
22. What are reasons for the change in land-use:
    Climate ☐ Institutional factors ☐ Social ☐ Economic return ☐
23. What are change in cropping during last 20-40 years?
    Food crops ☐ Oilseeds ☐ Commercial crops ☐

24. Are there changes in productivity due to climate change (rainfed) land and water resources?
    Yes ☐ No ☐
25. Do you think that land degradation will increase in response to climate change in next 10-15 years?
    Strongly agree ☐ Agree ☐ Abstain ☐ Disagree ☐ Strongly disagree ☐
26. Do you think that wastelands will increase in response to climate change in the next 10-15 years?
    Strongly agree ☐ Agree ☐ Abstain ☐ Disagree ☐ Strongly disagree ☐
27. Do you think that forest will decrease in response to water stress due to climate in the next 10-20 years?
    Strongly agree ☐ Agree ☐ Abstain ☐ Disagree ☐ Strongly disagree ☐
28. Do you think climate change has caused changes in land-use?
    Strongly agree ☐ Agree ☐ Abstain ☐ Disagree ☐ Strongly disagree ☐
29. Productivity has decrease/increase in last 20-40 years:
    Yes ☐ No ☐
30. Do you feel that something can be done on personal basis to reduce the impact of climate change?
    Strongly agree ☐ Agree ☐ Abstain ☐ Disagree ☐ Strongly disagree ☐
31. What is the level of public participation in adaptation/Mitigations of climate change?
    Low ☐ Medium ☐ High ☐
32. What is over all awareness of Government, functionaries about climate change?
    Low ☐ Medium ☐ High ☐
33. What is the level of awareness of elected representatives?
    Low ☐ Medium ☐ High ☐
34. Do you think that people have left village due to climate change, particularly drought?
    Yes ☐ No ☐
35. If they migrate, do they take domestic animals with them?
    Yes ☐ No ☐
36. Do you think the pattern of rainfall has changed during last 20-40 years?
    Yes ☐ No ☐
37. What are the Traditional sources of water conservation?
    Bavari ☐ Tanka ☐ Khadin ☐ Kundi ☐ Village Pond ☐

Appendices

Appendix 1

District Level 112 Years Mean Temperature Data of Rajasthan (1901 to 2012)

Year	*Ajmer*	*Alwar*	*Banswara*	*Baran*	*Barmer*	*Bharatpur*	*Bhilwara*	*Bikaner*	*Bundi*	*Chittor*	*Churu*
1901	26.1	25.6	26.8	26.3	26.3	26.1	25.4	26.5	26.2	25.8	26.3
1902	26.3	25.7	27.2	26.6	26.8	26.2	25.6	26.7	26.3	26.0	26.6
1903	25.3	25.1	26.0	25.6	25.7	25.5	24.5	25.8	25.3	24.9	25.8
1904	25.4	24.9	26.4	25.7	26.2	25.3	24.8	26.3	25.5	25.2	26.0
1905	25.4	25.0	26.1	25.9	25.7	25.4	24.8	25.7	25.8	25.0	25.7
1906	25.4	25.1	26.0	25.8	25.8	25.6	24.6	26.0	25.6	24.9	25.9
1907	25.3	24.9	26.3	25.9	25.7	25.4	24.7	25.6	25.7	25.1	25.5
1908	24.9	24.8	25.9	25.7	25.3	25.3	24.2	25.4	25.3	24.7	25.5
1909	24.7	24.4	25.9	25.5	25.1	24.9	24.1	25.2	25.3	24.6	25.2
1910	24.8	24.7	25.8	25.4	25.2	25.2	24.0	25.4	25.0	24.5	25.4
1911	25.7	25.4	26.5	26.1	25.6	25.9	25.0	26.0	26.0	25.4	26.1
1912	25.6	25.3	26.6	26.1	26.0	25.8	25.0	25.9	25.9	25.4	25.9
1913	25.3	25.1	26.3	25.9	25.6	25.7	24.7	25.7	25.7	25.1	25.6
1914	25.6	25.2	26.7	26.1	26.0	25.8	25.0	25.7	25.9	25.5	25.6
1915	26.3	25.8	27.1	26.6	26.4	26.3	25.7	26.6	26.6	26.1	26.6
1916	25.1	24.8	26.4	25.8	25.7	25.4	24.7	25.6	25.6	25.2	25.5
1917	24.0	23.7	25.6	24.6	25.0	24.2	23.5	24.9	24.2	24.2	24.6
1918	25.4	25.4	26.4	26.1	25.5	26.0	24.7	25.7	25.8	25.2	25.8
1919	25.0	24.8	26.4	25.4	25.5	25.3	24.4	25.5	25.2	25.0	25.5
1920	25.0	24.9	26.4	25.9	25.7	25.5	24.5	25.5	25.5	25.1	25.4
1921	26.2	26.2	27.0	26.7	26.4	26.7	25.5	26.8	26.5	25.9	26.8
1922	25.4	25.1	26.4	25.8	26.0	25.6	24.8	26.0	25.6	25.2	25.8
1923	25.3	25.1	26.5	25.9	25.7	25.7	24.7	25.6	25.5	25.2	25.6
1924	25.0	24.7	26.5	25.6	25.9	25.2	24.5	25.9	25.2	25.2	25.7
1925	25.3	24.6	26.5	25.5	26.1	25.0	24.8	26.1	25.5	25.3	25.7
1926	25.0	24.5	26.6	25.2	25.7	24.9	24.4	25.4	25.0	25.1	25.2
1927	24.5	24.4	25.7	24.9	25.2	24.9	23.8	25.1	24.6	24.4	25.0
1928	25.5	25.6	26.2	25.8	26.0	26.1	24.7	25.9	25.6	25.1	25.9
1929	25.3	25.2	26.3	25.8	25.7	25.8	24.7	25.6	25.6	25.1	25.7
1930	25.3	24.8	26.3	25.9	25.8	25.4	24.9	25.1	25.8	25.2	25.1
1931	25.6	25.1	26.7	26.2	26.2	25.7	25.1	25.5	26.0	25.5	25.2
1932	25.8	25.2	26.5	26.1	26.4	25.8	25.1	26.2	25.9	25.4	26.1
1933	24.6	24.0	26.0	25.1	25.9	24.5	24.0	25.2	24.7	24.6	24.8
1934	25.0	24.6	26.0	25.4	26.0	24.9	24.4	25.7	25.2	24.8	25.5
1935	24.8	24.6	25.9	25.2	25.7	25.2	24.1	25.4	24.8	24.6	25.2
1936	25.1	24.8	26.1	25.2	26.2	25.2	24.4	25.7	25.0	24.9	25.4

(*Contd.*)

(Contd.)

Year	Ajmer	Alwar	Banswara	Baran	Barmer	Bharatpur	Bhilwara	Bikaner	Bundi	Chittor	Churu
1937	25.2	24.7	26.1	25.5	25.9	25.4	24.5	25.4	25.3	24.9	25.1
1938	25.6	25.4	26.3	25.8	26.3	25.9	24.8	26.0	25.6	25.2	25.9
1939	25.6	25.2	26.3	25.7	26.3	25.7	24.8	26.0	25.6	25.2	25.8
1940	25.4	25.0	26.2	25.7	26.4	25.5	24.8	26.0	25.6	25.1	25.8
1941	26.4	26.2	26.9	26.9	26.9	26.8	25.8	26.8	26.8	26.1	26.7
1942	25.5	25.1	26.3	25.9	26.4	25.8	24.8	26.1	25.6	25.1	25.7
1943	25.3	25.0	26.1	25.9	26.3	25.6	24.7	25.9	25.6	25.1	25.7
1944	25.0	24.8	25.9	25.4	25.8	25.4	24.2	25.6	25.0	24.6	25.4
1945	24.6	24.5	25.6	25.1	25.5	25.2	23.9	25.2	24.8	24.4	25.0
1946	25.4	25.3	26.1	25.9	25.9	25.9	24.7	26.3	25.6	25.0	26.1
1947	25.6	25.6	26.3	25.9	26.2	26.1	24.8	26.4	25.7	25.2	26.3
1948	25.6	25.4	26.7	25.9	26.1	25.9	24.9	26.2	25.8	25.4	26.1
1949	25.9	25.5	26.8	26.3	26.1	26.1	25.3	26.0	26.2	25.7	25.9
1950	24.6	24.5	25.9	25.4	25.2	25.2	24.1	24.8	25.0	24.6	24.7
1951	25.5	25.2	26.7	26.2	26.3	25.8	25.1	26.0	26.0	25.6	25.9
1952	25.7	25.6	26.8	26.4	26.2	26.2	25.2	26.1	26.1	25.7	26.0
1953	26.0	25.8	26.9	26.8	26.4	26.4	25.5	26.5	26.5	25.9	26.5
1954	25.6	25.3	26.4	26.2	26.1	25.9	25.1	26.2	26.0	25.4	26.1
1955	25.1	24.8	26.2	25.7	25.8	25.6	24.6	25.8	25.4	25.0	25.5
1956	24.9	24.9	26.0	25.5	25.5	25.5	24.4	25.6	25.3	24.8	25.5
1957	25.0	24.5	26.4	25.6	25.6	25.1	24.6	25.3	25.5	25.1	25.1
1958	26.0	25.5	26.9	26.5	26.8	26.3	25.4	26.7	26.3	25.8	26.4
1959	25.2	25.0	26.3	25.9	26.1	25.8	24.7	26.0	25.6	25.1	25.7
1960	25.2	25.0	26.3	25.9	26.1	25.8	24.6	25.9	25.4	25.1	25.6
1961	24.7	24.6	26.0	25.2	25.4	25.1	24.1	25.3	24.9	24.6	25.0
1962	24.7	25.1	26.1	25.2	25.9	25.7	24.1	25.7	24.8	24.5	25.6
1963	26.2	25.8	26.7	26.0	26.5	26.2	25.6	26.6	26.4	25.9	26.4
1964	25.3	25.0	26.4	26.1	25.8	25.7	24.9	25.6	25.8	25.3	25.3
1965	26.0	25.5	27.1	26.5	25.9	26.1	25.5	25.9	26.5	26.0	25.9
1966	25.7	25.6	27.1	26.5	26.1	26.3	25.3	25.9	26.2	25.8	25.8
1967	24.4	25.0	26.2	25.4	25.7	25.6	23.9	25.5	24.7	24.6	25.4
1968	25.1	25.0	26.3	25.7	25.9	25.5	24.5	25.8	25.3	25.0	25.8
1969	26.0	25.9	27.2	26.3	26.9	26.4	25.5	26.8	26.2	25.9	26.7
1970	25.3	25.3	26.5	26.0	26.1	25.8	24.7	26.2	25.5	25.2	26.0
1971	24.6	24.4	25.9	25.0	25.8	24.9	23.9	25.7	24.6	24.4	25.4
1972	25.3	25.1	26.6	25.9	26.0	25.8	24.9	25.8	25.7	25.4	25.5
1973	25.3	25.3	26.5	25.9	25.9	25.9	24.6	25.9	25.5	25.2	25.7
1974	25.4	25.1	26.7	25.9	26.1	25.7	24.9	25.8	25.6	25.4	25.5
1975	24.6	24.6	25.9	25.3	25.4	25.2	24.0	25.1	24.9	24.6	25.0
1976	25.1	24.6	26.7	26.0	25.6	25.2	24.7	25.6	25.6	25.3	25.4
1977	25.4	24.5	27.1	25.9	26.4	25.0	25.1	26.2	25.7	25.7	25.7
1978	24.7	24.2	26.3	25.4	25.9	24.8	23.9	25.9	24.6	24.5	25.5
1979	25.0	24.3	26.9	25.7	25.6	24.9	24.8	26.0	25.3	25.4	25.6
1980	25.6	24.7	27.1	26.2	26.3	25.2	25.5	26.8	26.1	25.9	26.3
1981	25.3	24.5	26.8	26.0	26.1	24.9	25.0	26.3	25.7	25.5	25.9
1982	24.9	24.3	26.6	25.9	25.9	24.9	24.6	25.5	25.3	25.1	25.3
1983	24.5	24.0	26.0	25.4	25.6	24.6	24.3	25.1	25.0	24.8	24.8
1984	25.1	24.7	26.3	26.0	25.8	25.2	24.8	25.8	25.7	25.2	25.6

(Contd.)

(Contd.)

Year	Ajmer	Alwar	Banswara	Baran	Barmer	Bharatpur	Bhilwara	Bikaner	Bundi	Chittor	Churu
1985	25.6	25.0	26.9	26.6	26.5	25.5	25.4	26.3	26.3	25.9	26.0
1986	25.3	25.1	26.8	26.0	26.3	25.7	25.0	25.7	25.8	25.5	25.4
1987	26.3	26.1	27.7	26.9	27.3	26.7	25.9	26.7	26.6	26.4	26.6
1988	26.1	25.6	27.4	26.8	27.1	26.2	25.6	26.8	26.3	26.1	26.5
1989	25.3	24.8	26.7	26.2	26.0	25.3	24.9	25.7	25.7	25.4	25.5
1990	25.3	24.9	26.7	25.9	26.2	25.5	24.9	26.1	25.7	25.4	25.7
1991	25.3	25.1	26.7	26.4	26.4	25.7	24.8	26.5	25.8	25.4	26.1
1992	25.2	25.0	26.7	26.4	26.1	25.6	24.9	26.0	25.9	25.5	25.7
1993	25.8	25.3	27.2	26.3	27.0	25.9	25.4	26.6	26.1	25.9	26.3
1994	25.4	25.1	26.6	26.1	26.4	25.7	24.9	26.3	25.8	25.4	26.0
1995	25.7	25.3	27.2	26.4	26.6	25.8	25.3	26.4	26.1	25.8	26.1
1996	25.3	24.8	27.2	26.3	26.3	25.2	25.0	26.0	25.9	25.7	25.8
1997	24.5	24.0	26.4	25.3	25.5	24.6	24.1	24.9	24.9	24.8	24.6
1998	26.0	25.3	27.4	26.6	26.7	25.8	25.6	26.3	26.5	26.1	26.0
1999	26.0	25.6	27.0	26.5	26.4	26.1	25.4	26.5	26.3	25.8	26.4
2000	26.1	25.5	27.2	26.7	26.6	26.0	25.5	26.5	26.5	26.0	26.3
2001	25.6	25.3	26.8	26.4	26.4	25.9	25.1	26.2	26.1	25.6	26.0
2002	26.6	25.9	27.3	27.2	27.0	26.5	26.1	26.7	27.1	26.4	26.6
2003								27.0			
2004								—			
2005								—			
2006								27.9			
2007								—			
2008								27.1			
2009								27.6			
2010								—			
2011								26.7			
2012								26.6			
2013								26.9			
2014								—			

Appendix 2

District Level 112 Years Mean Temperature Data of Rajasthan (1901 to 2012)

Year	*Dausa*	*Dholpur*	*Dungarpur*	*G'nagar*	*H'garh*	*Jaipur*	*Jaisalmer*	*Jalore*	*Jhalawar*
1901	26.6	26.3	27.1	26.2	26.0	26.2	26.0	26.9	25.8
1902	26.7	26.3	27.5	26.5	26.3	26.3	26.4	27.4	26.0
1903	26.0	25.6	26.4	25.6	25.6	25.6	25.5	26.3	25.0
1904	25.8	25.5	26.8	26.1	25.9	25.5	26.0	26.6	25.2
1905	26.0	25.6	26.4	25.5	25.5	25.5	25.4	26.2	25.3
1906	26.1	25.6	26.4	25.8	25.7	25.7	25.7	26.2	25.2
1907	25.9	25.6	26.6	25.4	25.3	25.5	25.5	26.3	25.3
1908	25.7	25.4	26.2	25.3	25.4	25.3	25.2	25.8	25.1
1909	25.4	25.1	26.2	25.1	25.0	25.0	25.1	25.5	25.0
1910	25.7	25.4	26.1	25.3	25.2	25.2	25.2	25.8	24.8
1911	26.4	26.0	26.7	25.8	25.8	26.0	25.4	26.2	25.6
1912	26.2	25.9	26.9	25.7	25.7	25.8	25.8	26.5	25.6
1913	26.0	25.9	26.5	25.4	25.3	25.5	25.4	26.1	25.3
1914	26.3	26.0	26.9	25.3	25.3	25.8	25.5	26.5	25.6
1915	26.8	26.4	27.4	26.4	26.4	26.4	26.1	27.0	26.2
1916	25.7	25.6	26.7	25.4	25.4	25.2	25.4	26.3	25.3
1917	24.6	24.4	26.0	24.7	24.4	24.1	25.0	25.5	24.2
1918	26.3	26.1	26.7	25.5	25.7	25.8	25.2	26.1	25.5
1919	25.6	25.4	26.6	25.4	25.4	25.2	25.3	26.1	25.0
1920	25.7	25.8	26.6	25.4	25.3	25.1	25.5	26.2	25.3
1921	27.0	26.8	27.2	26.7	26.8	26.6	26.2	27.0	26.1
1922	26.0	25.8	26.7	25.7	25.7	25.5	25.7	26.4	25.3
1923	26.1	25.8	26.7	25.3	25.4	25.6	25.3	26.2	25.4
1924	25.6	25.4	26.8	25.5	25.5	25.2	25.6	26.4	25.2
1925	25.5	25.2	26.8	25.6	25.4	25.2	25.7	26.6	25.1
1926	25.5	25.0	26.8	25.0	25.0	25.1	25.4	26.2	24.9
1927	25.3	24.9	26.0	24.9	24.8	24.9	25.1	25.5	24.4
1928	26.5	26.1	26.5	25.6	25.7	26.0	25.8	26.3	25.2
1929	26.1	25.9	26.6	25.3	25.6	25.6	25.3	26.3	25.2
1930	25.8	25.7	26.6	24.8	24.9	25.3	25.2	26.4	25.4
1931	26.2	26.0	26.9	25.3	25.1	25.6	25.9	26.6	25.7
1932	26.2	26.0	26.8	26.0	25.8	25.8	26.1	26.7	25.5
1933	25.1	24.7	26.3	25.1	24.7	24.7	25.8	26.0	24.6
1934	25.4	25.0	26.4	25.5	25.4	25.1	25.8	26.1	24.9
1935	25.5	25.3	26.2	25.2	25.1	25.0	25.4	26.0	24.6

(Contd.)

(Contd.)

Year	*Dausa*	*Dholpur*	*Dungarpur*	*G'nagar*	*H'garh*	*Jaipur*	*Jaisalmer*	*Jalore*	*Jhalawar*
1936	25.7	25.3	26.5	25.4	25.2	25.3	26.0	26.3	24.7
1937	25.8	25.6	26.4	25.2	24.9	25.3	25.7	26.3	24.9
1938	26.4	25.9	26.6	25.8	25.8	25.9	26.0	26.5	25.2
1939	26.2	25.7	26.7	25.8	25.6	25.8	26.0	26.5	25.2
1940	26.0	25.7	26.6	25.9	25.6	25.5	26.2	26.6	25.1
1941	27.2	27.1	27.2	26.7	26.5	26.6	26.9	27.1	26.2
1942	26.1	26.0	26.5	26.0	25.5	25.6	26.4	26.5	25.3
1943	25.9	25.8	26.4	25.7	25.5	25.5	26.1	26.4	25.2
1944	25.8	25.5	26.2	24.9	25.2	25.3	25.6	26.1	24.8
1945	25.4	25.4	25.9	25.3	24.8	24.9	25.5	25.7	24.5
1946	26.3	26.2	26.3	26.4	26.0	25.8	26.0	26.2	25.2
1947	26.4	26.1	26.6	26.6	26.3	26.0	26.3	26.4	25.3
1948	26.4	26.1	27.1	26.0	25.9	26.0	25.9	26.7	25.4
1949	26.5	26.2	27.2	25.9	26.0	26.0	25.6	26.7	25.8
1950	25.4	25.4	26.2	24.6	24.8	24.9	24.8	25.8	24.8
1951	26.1	26.0	27.0	25.8	25.9	25.6	26.0	26.7	25.6
1952	26.5	26.4	27.1	26.1	26.1	25.9	26.1	26.7	25.8
1953	26.7	26.7	27.1	26.4	26.6	26.1	26.4	26.8	26.1
1954	26.1	26.1	26.7	25.9	26.0	25.6	25.9	26.4	25.6
1955	25.8	25.8	26.5	25.7	25.5	25.1	25.6	26.3	25.1
1956	25.7	25.6	26.2	25.6	25.7	25.1	25.4	25.9	25.0
1957	25.5	25.4	26.6	25.1	25.1	24.9	25.2	26.2	25.2
1958	26.5	26.6	27.2	26.4	26.4	25.9	26.7	27.1	25.9
1959	25.9	26.1	26.6	25.9	25.8	25.2	26.1	26.4	25.3
1960	25.9	26.1	26.6	25.8	25.8	25.3	25.9	26.5	25.2
1961	25.4	25.3	26.2	25.2	25.1	24.9	25.3	25.9	24.7
1962	25.7	25.8	26.4	25.6	25.6	25.1	25.7	26.3	24.7
1963	26.7	26.1	27.1	26.3	26.0	26.3	26.3	26.8	25.6
1964	26.0	25.9	26.7	25.3	25.0	25.4	25.5	26.2	25.5
1965	26.6	26.2	27.3	25.7	25.6	26.0	25.6	26.5	26.1
1966	26.6	26.5	27.4	25.5	25.6	25.9	25.6	26.7	26.0
1967	25.5	25.8	26.5	25.3	25.4	25.0	25.4	26.2	24.8
1968	25.7	25.6	26.7	25.7	25.7	25.3	25.6	26.4	25.1
1969	26.7	26.5	27.6	26.5	26.5	26.2	26.5	27.4	25.8
1970	26.0	26.1	26.8	26.1	26.0	25.6	26.0	26.5	25.4
1971	25.3	25.0	26.3	25.6	25.3	24.9	25.8	26.1	24.5
1972	26.0	26.0	26.9	25.4	25.3	25.4	25.7	26.5	25.5
1973	26.2	26.1	26.7	25.6	25.6	25.6	25.9	26.3	25.4
1974	25.9	25.8	26.9	25.5	25.4	25.4	25.8	26.6	25.4
1975	25.4	25.4	26.2	24.8	24.8	24.8	25.3	25.8	24.7
1976	25.6	25.7	27.0	25.3	25.2	25.1	25.2	26.3	25.6
1977	25.5	25.4	27.4	25.8	25.4	25.1	26.1	27.0	25.6
1978	25.2	25.3	26.7	25.7	25.4	24.8	25.5	26.5	24.7
1979	25.1	25.4	27.2	25.9	25.6	24.6	25.1	26.5	25.5
1980	25.5	25.6	27.4	26.5	26.2	25.1	26.0	26.9	26.0
1981	25.4	25.5	27.1	26.0	25.7	25.0	25.8	26.7	25.4
1982	25.1	25.3	26.8	25.1	25.2	24.7	25.6	26.4	25.3

(Contd.)

(Contd.)

Year	Dausa	Dholpur	Dungarpur	G'nagar	H'garh	Jaipur	Jaisalmer	Jalore	Jhalawar
1983	24.7	24.8	26.2	24.8	24.9	24.2	25.4	25.9	25.0
1984	25.5	25.7	26.5	25.6	25.5	25.0	25.8	26.1	25.5
1985	25.8	26.0	27.2	26.0	25.9	25.4	26.4	26.8	26.1
1986	25.8	25.7	27.1	25.2	25.2	25.2	26.0	26.7	25.6
1987	26.9	26.8	28.1	26.3	26.3	26.3	26.8	27.8	26.4
1988	26.5	26.4	27.7	26.4	26.2	26.0	26.9	27.5	26.3
1989	25.6	25.5	27.0	25.4	25.3	25.2	25.7	26.5	25.6
1990	25.7	25.7	27.0	25.9	25.6	25.2	26.1	26.6	25.5
1991	26.0	26.0	27.0	26.1	25.9	25.5	26.4	26.7	25.7
1992	25.8	25.8	26.9	25.7	25.5	25.3	26.0	26.4	25.8
1993	26.1	26.1	27.6	26.3	26.2	25.7	26.8	27.4	25.9
1994	26.0	25.9	27.0	26.1	25.9	25.5	26.3	26.7	25.5
1995	26.1	26.0	27.4	26.0	25.9	25.7	26.4	27.0	25.9
1996	25.6	25.4	27.3	25.7	25.6	25.2	26.1	26.7	26.0
1997	25.0	24.9	26.6	24.6	24.5	24.5	25.3	25.9	25.0
1998	26.4	26.0	27.5	26.0	25.8	26.0	26.3	27.0	26.3
1999	26.5	26.3	27.1	26.4	26.3	26.1	26.4	26.7	26.0
2000	26.5	26.2	27.4	26.3	26.1	26.0	26.5	26.9	26.2
2001	26.3	26.2	27.0	26.0	26.0	25.8	26.3	26.6	25.9
2002	27.0	26.7	27.6	26.3	26.3	26.5	26.7	27.3	26.7
2003						25.8	27.1		
2004						26.3	27.9		
2005						—	—		
2006						26.0	27.4		
2007						25.6	27.1		
2008						25.4	27.0		
2009						26.4	27.6		
2010						26.1	27.5		
2011						25.4	26.8		
2012						25.3	26.5		
2013						25.4	27.1		
2014						—	—		

Appendix 3

District Level 112 Years Mean Temperature Data of Rajasthan (1901 to 2012)

Year	*J'jhunu*	*Jodhpur*	*Karoli*	*Kota*	*Nagour*	*Pali*	*R'mand*	*S'Madhopur*	*Sikar*	*Sirohi*	*Tonk*	*Udaipur*
1901	25.5	26.6	26.6	26.2	26.8	26.2	24.4	26.8	25.9	26.6	26.4	25.9
1902	25.7	27.0	26.7	26.3	27.0	26.6	24.7	26.9	26.2	27.0	26.6	26.2
1903	25.1	25.9	25.9	25.3	26.1	25.4	23.5	26.1	25.4	25.9	25.7	25.1
1904	25.0	26.3	25.8	25.5	26.3	25.8	23.9	26.1	25.4	26.3	25.7	25.5
1905	25.0	25.8	26.0	25.7	26.0	25.4	23.5	26.3	25.3	25.9	25.9	25.1
1906	25.2	25.9	26.0	25.5	26.1	25.4	23.5	26.2	25.5	25.9	25.9	25.1
1907	24.8	25.7	26.0	25.7	25.9	25.4	23.6	26.2	25.2	25.9	25.8	25.2
1908	24.9	25.3	25.8	25.4	25.6	24.9	23.1	26.0	25.2	25.5	25.5	24.8
1909	24.4	25.0	25.5	25.3	25.3	24.6	22.9	25.8	24.8	25.3	25.4	24.7
1910	24.6	25.3	25.6	25.1	25.5	24.8	23.0	25.8	25.0	25.4	25.3	24.7
1911	25.4	26.0	26.4	26.0	26.4	25.6	23.8	26.6	25.8	26.0	26.3	25.4
1912	25.2	26.1	26.3	25.9	26.2	25.7	23.9	26.5	25.6	26.2	26.1	25.5
1913	24.8	25.8	26.2	25.7	26.0	25.4	23.6	26.4	25.3	25.8	25.9	25.2
1914	25.0	26.1	26.3	25.9	26.2	25.8	24.0	26.5	25.4	26.3	26.1	25.6
1915	25.9	26.7	26.9	26.5	26.9	26.4	24.6	27.1	26.2	26.7	26.7	26.0
1916	24.8	25.8	25.9	25.6	25.8	25.4	23.7	26.1	25.1	26.0	25.6	25.3
1917	23.6	24.9	24.7	24.3	24.8	24.4	22.7	24.9	24.0	25.2	24.4	24.5
1918	25.3	25.7	26.4	25.8	26.0	25.4	23.6	26.5	25.6	25.8	26.0	25.3
1919	24.8	25.6	25.7	25.2	25.7	25.3	23.5	25.9	25.1	25.8	25.4	25.2
1920	24.7	25.6	26.0	25.6	25.6	25.3	23.5	26.1	25.0	25.9	25.5	25.2
1921	26.2	26.7	27.1	26.5	27.0	26.2	24.4	27.3	26.5	26.7	26.8	26.0
1922	25.1	26.1	26.1	25.6	26.1	25.6	23.8	26.2	25.4	26.1	25.8	25.4
1923	25.0	25.8	26.1	25.6	26.0	25.5	23.7	26.3	25.3	25.9	25.8	25.3
1924	24.7	26.0	25.7	25.3	25.9	25.5	23.8	25.9	25.1	26.1	25.4	25.4
1925	24.6	26.3	25.6	25.4	26.2	25.8	24.0	25.9	25.1	26.2	25.6	25.5
1926	24.5	25.7	25.4	25.0	25.7	25.4	23.7	25.6	24.9	25.9	25.3	25.4
1927	24.3	25.2	25.2	24.6	25.3	24.7	22.9	25.4	24.7	25.2	25.0	24.5
1928	25.4	26.1	26.4	25.5	26.3	25.6	23.7	26.5	25.8	25.9	26.0	25.2
1929	25.1	25.8	26.2	25.6	26.0	25.5	23.7	26.3	25.4	26.0	25.9	25.2
1930	24.6	25.8	26.1	25.8	25.8	25.6	23.8	26.3	25.0	26.1	25.8	25.3
1931	24.7	26.2	26.3	26.0	26.1	25.9	24.0	26.5	25.2	26.3	26.1	25.6
1932	25.2	26.5	26.3	25.9	26.5	26.0	24.1	26.5	25.6	26.3	26.1	25.6
1933	24.0	25.6	25.1	24.8	25.3	25.0	23.2	25.3	24.4	25.7	24.9	24.9
1934	24.7	25.8	25.4	25.1	25.8	25.3	23.4	25.7	25.1	25.8	25.4	25.0
1935	24.5	25.6	25.5	24.9	25.6	25.2	23.3	25.6	24.8	25.7	25.2	24.9
1936	24.6	26.1	25.6	24.9	26.0	25.5	23.7	25.8	25.1	26.0	25.4	25.2

(*Contd.*)

(Contd.)

Year	J'jhunu	Jodhpur	Karoli	Kota	Nagour	Pali	R'mand	S'Madhopur	Sikar	Sirohi	Tonk	Udaipur
1937	24.4	25.9	25.8	25.2	25.8	25.4	23.6	26.0	24.9	25.9	25.6	25.1
1938	25.2	26.2	26.2	25.5	26.4	25.7	23.9	26.4	25.6	26.1	26.0	25.3
1939	25.1	26.3	26.1	25.5	26.3	25.8	23.9	26.3	25.6	26.2	25.9	25.4
1940	25.0	26.3	26.0	25.5	26.2	25.7	23.8	26.2	25.4	26.2	25.8	25.3
1941	26.0	27.0	27.3	26.7	27.1	26.4	24.6	27.5	26.4	26.6	27.0	26.0
1942	25.0	26.4	26.2	25.6	26.2	25.7	23.8	26.4	25.3	26.1	25.9	25.2
1943	24.9	26.2	26.1	25.6	26.1	25.6	23.7	26.2	25.2	25.9	25.8	25.2
1944	24.7	25.8	25.7	25.0	25.9	25.3	23.4	25.9	25.1	25.7	25.4	24.9
1945	24.3	25.4	25.6	24.8	25.3	24.8	23.0	25.6	24.6	25.3	25.0	24.5
1946	25.2	26.1	26.3	25.6	26.3	25.5	23.6	26.4	25.6	25.8	25.9	25.0
1947	25.5	26.3	26.4	25.7	26.4	25.6	23.8	26.5	25.8	26.0	26.0	25.3
1948	25.3	26.2	26.4	25.7	26.4	25.8	24.0	26.5	25.7	26.4	26.1	25.7
1949	25.2	26.4	26.6	26.1	26.5	26.1	24.3	26.8	25.6	26.5	26.4	25.9
1950	24.1	25.2	25.6	25.0	25.2	24.9	23.2	25.7	24.5	25.5	25.2	24.9
1951	25.2	26.3	26.3	26.0	26.2	25.9	24.1	26.5	25.4	26.4	26.0	25.7
1952	25.4	26.3	26.7	26.1	26.3	25.9	24.1	26.8	25.7	26.4	26.2	25.7
1953	25.7	26.6	27.0	26.5	26.6	26.1	24.4	27.1	26.0	26.4	26.5	25.9
1954	25.3	26.3	26.4	26.0	26.3	25.8	24.0	26.6	25.5	26.1	26.0	25.4
1955	24.7	26.0	26.0	25.4	25.8	25.6	23.7	26.1	25.0	26.0	25.5	25.2
1956	24.8	25.7	25.9	25.3	25.6	25.2	23.3	26.0	25.0	25.6	25.4	24.9
1957	24.4	25.7	25.7	25.4	25.6	25.5	23.7	25.9	24.7	25.9	25.4	25.3
1958	25.5	26.9	26.8	26.2	26.7	26.4	24.5	27.0	25.8	26.7	26.3	25.9
1959	25.0	26.2	26.2	25.6	25.9	25.6	23.7	26.3	25.2	26.0	25.6	25.2
1960	24.9	26.2	26.2	25.5	26.0	25.7	23.8	26.2	25.1	26.1	25.6	25.2
1961	24.4	25.6	25.6	25.0	25.4	25.1	23.2	25.7	24.7	25.6	25.1	24.8
1962	25.0	25.9	25.9	24.9	25.7	25.4	23.3	25.9	25.1	25.9	25.1	24.9
1963	25.6	26.6	26.7	26.2	26.8	26.2	24.4	27.0	26.1	26.5	26.6	25.9
1964	24.7	26.0	26.2	25.8	25.9	25.6	23.8	26.4	25.1	26.0	25.8	25.3
1965	25.3	26.0	26.7	26.4	26.4	25.9	24.3	27.0	25.7	26.3	26.6	26.0
1966	25.3	26.2	26.8	26.2	26.3	26.0	24.3	26.9	25.7	26.5	26.3	26.0
1967	24.8	25.6	25.8	24.8	25.5	25.2	23.2	25.7	25.0	25.9	24.9	25.0
1968	25.0	25.9	25.9	25.3	25.9	25.6	23.7	26.0	25.3	26.2	25.5	25.4
1969	25.9	27.0	26.8	26.1	26.9	26.6	24.7	26.9	26.2	27.1	26.4	26.2
1970	25.3	26.1	26.3	25.6	26.1	25.5	23.7	26.3	25.5	26.1	25.7	25.4
1971	24.4	25.6	25.3	24.7	25.6	25.0	23.1	25.5	24.8	25.7	25.0	24.8
1972	24.8	26.1	26.2	25.7	26.0	25.7	23.9	26.3	25.1	26.2	25.8	25.5
1973	25.1	25.9	26.3	25.6	26.0	25.4	23.7	26.4	25.4	25.9	25.8	25.2
1974	24.8	26.3	26.1	25.6	26.1	25.9	24.0	26.2	25.2	26.3	25.7	25.6
1975	24.3	25.3	25.6	25.0	25.3	24.9	23.1	25.7	24.6	25.4	25.1	24.7
1976	24.6	25.8	26.0	25.7	25.9	25.5	23.7	26.1	25.0	26.2	25.6	25.4
1977	24.6	26.4	25.8	25.7	26.2	26.1	24.3	26.0	25.0	26.7	25.7	26.0
1978	24.4	26.0	25.6	24.8	25.9	25.6	23.4	25.7	24.8	26.3	25.0	25.1
1979	24.4	26.0	25.6	25.5	25.9	25.9	24.0	25.6	24.7	26.5	25.1	25.7
1980	24.9	26.7	25.9	26.1	26.5	26.3	24.5	26.1	25.3	26.7	25.8	26.0
1981	24.8	26.3	25.8	25.7	26.1	25.9	24.1	26.0	25.1	26.4	25.6	25.7
1982	24.5	25.9	25.5	25.5	25.6	25.6	23.7	25.7	24.7	26.1	25.2	25.4
1983	24.0	25.6	25.0	25.1	25.2	25.1	23.3	25.2	24.2	25.5	24.7	24.8
1984	24.7	25.9	25.9	25.8	25.8	25.3	23.6	26.0	25.0	25.7	25.5	25.1

(Contd.)

(*Contd.*)

Year	*J'jhunu*	*Jodhpur*	*Karoli*	*Kota*	*Nagour*	*Pali*	*R'mand*	*S'Madhopur*	*Sikar*	*Sirohi*	*Tonk*	*Udaipur*
1985	25.1	26.4	26.2	26.4	26.2	25.9	24.3	26.5	25.4	26.4	26.0	25.8
1986	24.7	26.2	26.0	25.8	25.9	25.8	24.1	26.2	25.0	26.4	25.7	25.7
1987	25.9	27.2	27.0	26.6	27.0	26.9	25.1	27.1	26.1	27.6	26.6	26.8
1988	25.6	27.1	26.7	26.4	26.9	26.6	24.7	26.9	25.9	27.1	26.4	26.3
1989	24.7	26.1	25.9	25.8	26.0	25.8	24.0	26.1	25.1	26.2	25.7	25.6
1990	24.8	26.3	26.0	25.7	26.1	25.8	24.0	26.1	25.1	26.3	25.6	25.6
1991	25.1	26.3	26.3	25.9	26.3	25.7	23.9	26.4	25.5	26.3	25.9	25.5
1992	24.8	26.0	26.1	26.0	25.9	25.5	23.8	26.2	25.1	26.0	25.7	25.5
1993	25.3	26.9	26.4	26.1	26.6	26.4	24.5	26.5	25.6	27.0	26.0	26.2
1994	25.0	26.3	26.1	25.8	26.2	25.8	24.0	26.3	25.4	26.3	25.8	25.6
1995	25.2	26.6	26.3	26.1	26.4	26.1	24.3	26.5	25.5	26.7	26.1	26.0
1996	24.9	26.1	25.8	26.0	25.9	25.6	24.0	26.0	25.2	26.3	25.7	25.8
1997	23.9	25.2	25.2	25.0	25.1	24.8	23.2	25.4	24.3	25.6	25.0	25.0
1998	25.2	26.6	26.5	26.5	26.6	26.2	24.5	26.8	25.7	26.6	26.5	26.1
1999	25.6	26.5	26.7	26.3	26.6	25.9	24.2	26.9	25.9	26.3	26.4	25.7
2000	25.5	26.6	26.6	26.5	26.7	26.1	24.4	26.9	25.9	26.6	26.5	26.0
2001	25.3	26.3	26.5	26.1	26.3	25.8	24.0	26.6	25.6	26.2	26.2	25.6
2002	25.8	27.0	27.1	27.0	27.1	26.7	24.9	27.4	26.2	26.9	27.1	26.4
2003		26.9										
2004		—										
2005		—										
2006		27.6										
2007		27.0										
2008		26.5										
2009		27.8										
2010		27.5										
2011		26.9										
2012		26.4										
2013		26.5										
2014		—										

Appendix 4

District Level 112 Years Mean Rainfall Data of Rajasthan (1901 to 2012)

District	Ajmer	Alwar	Banswara	Baran	Barmer	Bharatpur	Bhilwara	Bikaner	Bundi	Chittorgarh	Churu
1901	343.59	400.57	741.12	704.76	107.89	437.82	448.88	146.18	558.74	656.96	234.33
1902	336.76	491.50	744.84	708.85	228.20	567.98	424.34	151.16	558.16	602.43	226.62
1903	640.19	492.03	997.46	760.60	308.66	560.96	742.83	245.04	765.53	922.69	313.77
1904	388.67	606.86	490.68	715.14	95.81	776.05	418.66	164.70	577.58	480.81	268.22
1905	216.56	222.62	782.04	589.24	114.55	267.27	334.03	83.06	438.81	571.46	121.81
1906	483.71	586.26	1214.02	847.17	278.29	593.82	581.53	286.13	695.49	840.46	382.10
1907	406.73	430.55	823.89	666.43	294.22	572.74	474.04	280.98	577.42	643.18	324.77
1908	1084.82	1076.43	1244.49	985.44	464.87	1136.28	1148.87	439.29	1145.57	1294.30	622.89
1909	733.81	779.04	1014.43	947.64	377.22	808.07	832.10	509.26	912.41	1036.74	656.51
1910	423.09	491.19	1030.18	756.69	346.55	563.66	530.72	327.62	634.95	778.55	385.96
1911	287.89	387.62	520.27	606.06	115.70	442.67	372.46	127.42	476.09	487.36	187.88
1912	639.16	674.27	899.88	762.23	305.48	720.37	743.58	165.59	781.36	877.46	284.56
1913	298.14	309.11	1013.51	595.96	451.85	312.57	423.15	168.52	500.64	691.09	195.02
1914	531.51	642.46	1011.52	1035.16	261.82	767.84	699.96	290.20	870.95	974.64	397.78
1915	282.92	340.35	560.29	710.15	92.25	420.43	406.48	123.19	526.98	579.76	190.81
1916	643.56	707.21	1070.13	927.72	485.09	834.62	750.36	465.37	879.78	950.67	540.38
1917	1204.66	1113.88	1136.26	990.20	674.10	1251.60	1224.04	547.85	1181.66	1239.28	694.37
1918	130.93	205.09	526.64	412.33	91.44	275.78	197.61	57.79	288.62	338.17	91.28
1919	505.07	545.42	1170.77	900.93	362.57	739.16	587.94	420.73	730.48	827.47	476.54
1920	373.68	506.63	814.19	517.64	141.31	557.72	386.87	296.49	450.90	488.68	379.33
1921	325.68	440.78	1064.77	764.64	369.85	505.14	435.62	356.32	578.48	700.07	430.50
1922	429.39	525.06	921.43	786.00	203.12	673.02	501.33	207.69	633.77	683.93	290.09
1923	575.46	625.79	706.69	943.06	230.76	678.67	682.97	263.42	801.15	859.16	381.44
1924	498.89	620.40	993.72	763.78	265.82	710.35	567.35	218.44	660.99	741.90	327.94
1925	401.65	584.56	690.76	808.14	118.25	785.65	445.65	144.49	641.31	567.65	233.61
1926	704.38	723.99	1171.48	824.24	508.96	820.67	785.40	306.83	840.90	944.05	403.39
1927	702.32	669.11	1231.91	848.95	499.00	780.90	790.84	255.16	854.67	1009.23	329.22
1928	516.21	446.60	873.12	700.74	257.86	544.57	609.79	267.10	671.18	764.60	335.18
1929	444.26	365.89	842.08	520.72	435.49	390.06	505.01	260.00	509.73	646.56	297.13
1930	425.49	583.18	967.78	771.30	163.46	749.54	492.11	215.21	629.28	723.79	301.02
1931	605.39	588.10	1048.97	1047.03	258.12	682.57	777.09	275.20	918.37	1049.46	372.38
1932	589.84	551.14	901.48	858.01	252.01	658.03	702.40	206.72	787.19	876.89	287.47
1933	829.83	993.75	1391.86	1118.16	458.88	1003.35	948.94	335.64	1064.00	1191.89	501.42
1934	586.86	584.95	1486.14	1176.49	300.27	605.80	818.87	231.16	953.70	1223.16	338.57
1935	466.14	563.65	923.71	820.84	279.00	640.35	544.27	338.62	666.92	744.34	412.40
1936	516.53	702.62	720.55	877.19	228.61	837.56	577.63	179.50	728.42	691.97	305.46

(Contd.)

(*Contd.*)

District	Ajmer	Alwar	Banswara	Baran	Barmer	Bharatpur	Bhilwara	Bikaner	Bundi	Chittorgarh	Churu
1937	413.73	436.64	1061.51	624.58	217.05	435.80	496.06	239.68	537.58	723.24	338.29
1938	470.08	378.41	1066.66	918.46	178.06	506.52	650.61	130.55	767.71	952.95	182.41
1939	326.20	388.72	773.93	756.52	82.84	504.76	432.34	89.09	589.60	643.81	141.70
1940	452.35	481.78	804.29	732.35	241.92	518.13	564.57	206.16	641.81	739.24	282.54
1941	334.45	324.94	1097.16	567.39	264.15	405.58	415.56	163.78	492.29	644.69	181.20
1942	599.41	785.76	1121.15	1167.39	321.94	995.71	687.47	517.32	926.32	935.31	649.84
1943	827.29	715.30	846.47	781.11	291.30	853.84	849.35	230.64	852.98	914.64	322.30
1944	733.07	671.88	1524.32	1055.54	422.53	561.42	888.04	337.25	921.99	1095.02	492.49
1945	677.49	711.60	1152.55	968.43	414.82	796.30	755.44	418.58	860.01	957.48	509.36
1946	423.77	520.61	1092.94	895.32	247.28	622.80	574.64	180.77	725.01	867.71	266.93
1947	626.90	637.10	1064.76	1063.81	269.43	916.26	714.80	170.80	883.95	898.81	244.80
1948	546.42	695.19	867.52	1008.63	200.42	807.92	658.28	147.97	833.49	871.81	255.67
1949	477.85	674.36	1034.94	1062.09	241.13	888.26	575.70	304.59	811.91	819.41	380.79
1950	473.92	652.54	974.95	772.97	246.99	705.02	540.94	363.71	658.14	735.55	541.97
1951	501.17	516.08	599.80	718.18	173.49	669.18	568.92	155.61	651.23	633.42	235.41
1952	592.52	651.65	804.73	738.90	262.09	859.68	601.41	207.64	692.64	680.97	300.40
1953	521.10	557.34	887.72	640.50	324.98	565.54	599.75	199.58	636.60	748.38	296.27
1954	637.90	581.45	1406.23	975.20	322.16	729.81	773.52	277.40	875.80	1047.05	356.20
1955	614.49	741.27	1280.86	1112.90	327.65	989.77	716.99	284.87	933.04	963.57	380.42
1956	727.07	750.55	1271.82	957.56	457.79	869.32	795.97	276.52	887.64	989.46	380.16
1957	470.55	642.10	770.25	793.53	195.14	782.46	538.32	181.89	682.03	671.48	291.82
1958	719.80	847.79	1043.58	1134.64	327.97	1011.28	870.55	195.96	1011.67	1061.27	361.76
1959	601.76	555.36	1412.69	1058.51	525.82	647.52	790.29	469.29	902.44	1133.48	545.74
1960	344.00	564.65	826.91	922.65	112.00	774.32	431.82	174.72	653.45	643.93	270.85
1961	613.18	808.09	1095.07	1006.70	599.83	913.62	709.35	363.40	856.94	905.70	463.05
1962	499.88	544.45	819.28	854.64	262.28	672.29	592.64	245.67	698.73	755.91	326.52
1963	408.17	570.73	1196.05	830.65	180.01	723.74	500.54	113.50	664.12	735.28	213.46
1964	594.00	902.73	871.78	1048.61	362.82	1014.62	674.25	279.65	882.30	851.30	449.22
1965	466.53	542.16	707.35	702.51	219.81	638.27	516.01	166.84	625.47	655.11	257.68
1966	599.26	634.00	785.06	802.45	225.44	747.88	686.95	281.65	789.08	821.19	395.90
1967	842.14	925.06	875.02	961.27	421.24	1046.91	915.92	295.56	978.02	987.38	427.99
1968	332.06	456.42	806.82	828.11	136.00	544.23	442.71	65.24	619.23	677.11	154.54
1969	373.99	511.61	793.22	986.79	80.27	652.25	512.66	127.90	711.97	762.31	223.04
1970	727.93	717.61	1192.91	955.08	372.42	852.09	849.66	275.37	931.65	1065.56	367.40
1971	467.92	666.30	814.33	748.64	250.90	735.52	508.05	269.84	629.13	625.33	379.96
1972	443.44	554.35	711.03	692.12	200.85	560.11	542.56	254.11	623.18	707.16	361.63
1973	831.27	775.58	1380.01	945.85	393.85	898.47	920.84	225.99	966.52	1095.70	340.42
1974	486.80	441.50	713.69	758.81	115.06	382.13	629.35	288.48	677.86	831.05	403.95
1975	998.22	985.03	1282.23	958.43	532.90	928.46	1088.07	348.94	1057.35	1238.51	514.47
1976	681.64	786.75	821.48	792.81	430.14	772.45	762.33	368.70	795.10	865.76	525.08
1977	594.52	767.63	1091.15	936.24	376.15	871.15	679.99	484.64	807.76	881.47	630.14
1978	643.42	762.73	862.24	870.84	478.26	761.09	739.75	382.15	804.95	870.85	514.37
1979	761.75	582.96	850.44	657.80	341.53	539.00	850.52	212.49	792.30	977.84	304.83
1980	386.04	524.57	1021.45	713.48	155.69	594.08	457.95	173.97	573.89	661.69	299.57
1981	529.66	626.08	1007.05	961.12	242.39	809.09	652.64	302.08	855.70	860.16	405.22
1982	509.29	698.28	775.73	739.76	224.37	749.07	535.24	533.18	634.56	635.32	727.30
1983	786.19	750.09	1036.87	1012.20	403.42	710.03	888.21	478.66	889.92	1043.65	644.58
1984	452.64	564.99	858.45	575.69	275.67	635.84	488.75	145.43	555.55	593.10	251.11

(*Contd.*)

(Contd.)

District	Ajmer	Alwar	Banswara	Baran	Barmer	Bharatpur	Bhilwara	Bikaner	Bundi	Chittorgarh	Churu
1985	413.43	780.71	630.48	843.21	210.91	952.07	401.88	208.30	622.52	527.53	342.86
1986	370.06	436.49	774.14	515.86	217.35	499.14	423.61	283.99	469.37	619.36	355.63
1987	229.43	286.10	565.93	512.62	104.36	351.81	306.96	110.09	376.82	459.57	152.13
1988	548.84	682.33	981.12	645.19	381.48	691.68	670.55	235.25	659.75	869.87	349.42
1989	399.70	447.25	717.75	610.47	203.28	598.70	449.86	246.83	544.93	578.82	303.12
1990	587.13	733.70	1107.43	1180.05	285.01	904.72	708.21	213.95	917.71	921.76	329.17
1991	375.41	565.90	676.87	652.45	169.97	670.41	451.03	138.50	599.80	591.42	215.05
1992	403.26	464.99	774.43	582.45	237.24	525.38	466.83	199.66	529.32	634.80	291.56
1993	568.83	755.17	943.98	745.57	288.60	812.76	635.96	210.96	731.72	830.58	355.54
1994	725.04	859.03	1254.82	1118.23	544.33	995.49	817.61	293.20	995.98	1014.04	397.52
1995	604.91	707.68	861.26	746.77	280.66	717.01	674.01	341.57	727.60	849.33	480.49
1996	928.27	898.08	1250.92	1002.84	388.97	959.68	1038.31	317.21	1070.18	1250.06	464.50
1997	685.22	647.58	1054.00	925.15	366.48	692.04	820.00	289.22	906.18	1020.18	393.06
1998	612.44	664.63	958.88	814.65	558.66	635.72	720.78	338.98	774.93	858.29	441.27
1999	500.93	550.91	781.78	954.79	169.69	788.63	568.04	176.57	758.18	708.23	238.64
2000	415.87	538.40	632.61	672.55	180.76	540.22	467.33	175.03	580.08	555.95	307.09
2001	573.77	627.44	702.20	718.05	300.50	653.87	627.51	236.81	698.51	692.90	331.55
2002	155.71	285.37	539.85	417.88	73.85	381.65	206.75	76.94	288.91	360.57	117.05
2003								288.79			
2004								—			
2005								—			
2006								—			
2007								—			
2008								—			
2009								—			
2010								—			
2011								—			
2012								319.26			
2013								441.98			
2014								—			

Appendix 5

District Level 112 Years Mean Rainfall Data of Rajasthan (1901 to 2012)

District	Dausa	Dhaulpur	Dungarpur	G'nagar	H'garh	Jaipur	Jaisalmer	Jalor	Jhalawar	Jhunjhunu
1901	421.92	529.81	722.40	151.24	211.64	365.38	74.50	163.63	818.86	320.46
1902	505.60	663.70	716.25	122.44	192.02	412.78	165.88	289.48	775.15	349.15
1903	631.82	618.01	976.36	198.17	236.53	587.62	174.22	444.52	903.15	374.82
1904	654.04	889.56	454.81	162.76	221.87	515.51	84.65	139.16	646.43	389.87
1905	266.68	375.37	733.04	99.29	136.46	207.19	59.23	285.62	730.51	164.45
1906	591.77	677.13	1213.82	277.39	343.99	544.18	226.43	444.86	976.85	488.75
1907	537.48	655.52	855.94	238.18	272.93	427.81	223.92	495.47	726.11	329.26
1908	1182.87	1102.61	1347.17	378.70	498.79	1120.62	253.87	739.14	1060.00	806.84
1909	831.03	856.71	989.78	460.37	557.54	775.62	283.69	567.30	1108.49	734.35
1910	535.42	656.81	1049.88	277.01	322.85	453.61	297.80	491.48	898.67	436.79
1911	384.87	512.93	480.13	146.67	184.40	321.21	72.98	189.55	656.99	273.11
1912	712.49	749.78	899.01	135.64	212.96	654.91	130.26	482.22	847.39	452.38
1913	332.85	374.14	1005.42	160.32	214.04	292.56	259.03	663.00	780.39	254.24
1914	719.28	919.89	932.71	290.48	363.30	578.51	190.90	474.80	1196.86	498.29
1915	380.88	546.31	406.67	134.30	179.39	299.76	67.61	158.34	828.87	243.00
1916	809.62	900.39	1027.16	398.43	458.68	681.91	378.89	586.39	1035.28	584.53
1917	1315.38	1213.30	1213.01	433.39	531.26	1244.48	346.80	1039.93	1012.22	805.89
1918	223.89	357.95	542.32	77.43	104.49	156.74	56.60	216.88	456.54	131.95
1919	681.45	875.42	1222.82	364.20	410.28	543.23	254.63	693.88	978.09	464.27
1920	527.19	598.98	909.91	251.42	298.43	469.86	121.16	285.90	488.62	432.34
1921	448.80	613.89	1079.73	312.75	385.41	363.75	297.06	588.76	877.58	454.24
1922	610.05	775.57	920.09	207.67	259.53	493.16	110.76	436.16	839.59	359.81
1923	677.48	770.94	540.67	243.90	313.55	606.35	167.57	260.24	1072.35	497.98
1924	644.38	786.91	1049.08	216.90	289.98	566.52	138.34	529.47	798.54	446.93
1925	710.70	915.78	702.72	162.61	221.55	536.21	76.09	246.63	775.33	336.37
1926	804.09	853.93	1343.62	250.78	326.18	723.99	291.84	798.51	879.24	525.77
1927	801.17	816.61	1373.75	195.15	247.17	712.80	280.86	816.55	949.35	442.98
1928	581.47	606.59	827.53	236.73	285.61	499.96	156.97	357.21	808.23	355.55
1929	450.99	408.80	909.30	251.98	280.02	425.37	274.26	467.70	604.62	316.09
1930	678.45	870.94	878.20	203.96	257.10	538.01	125.96	259.31	836.66	389.63
1931	693.67	793.38	937.45	249.71	329.24	587.71	189.58	431.45	1254.73	455.49
1932	664.98	731.23	819.09	183.52	220.13	575.06	142.60	407.93	980.73	373.58
1933	987.66	1011.51	1381.26	267.59	384.75	908.67	272.77	727.91	1292.81	735.26
1934	632.01	727.47	1348.54	191.56	267.64	560.03	193.44	541.18	1488.44	470.48
1935	608.83	737.71	899.13	295.03	337.00	523.85	203.16	386.04	902.02	471.15
1936	745.17	930.78	637.42	175.82	269.93	624.96	151.12	313.37	905.96	466.78

(Contd.)

(Contd.)

District	Dausa	Dhaulpur	Dungarpur	G'nagar	H'garh	Jaipur	Jaisalmer	Jalor	Jhalawar	Jhunjhunu
1937	449.71	475.92	1115.23	242.69	294.08	425.75	137.87	530.51	738.82	403.25
1938	541.69	669.63	914.84	120.30	152.18	434.17	102.00	350.89	1141.88	233.23
1939	481.61	623.65	681.85	98.49	140.34	363.69	79.86	171.75	856.50	216.52
1940	519.08	579.76	688.34	169.54	227.93	462.57	164.30	339.36	860.94	376.78
1941	406.99	480.36	1198.73	134.00	170.42	335.35	198.38	423.52	643.23	220.49
1942	898.12	1177.80	1109.06	456.53	533.64	717.79	258.57	352.86	1219.80	697.20
1943	902.23	861.40	872.05	174.60	212.70	836.65	150.81	602.46	829.50	433.99
1944	616.23	706.77	1616.82	301.49	391.26	675.29	248.35	727.96	1183.46	656.06
1945	795.00	872.69	1152.51	350.79	411.20	723.45	271.45	649.49	1080.61	584.49
1946	578.06	738.55	968.99	162.31	237.72	465.11	165.16	459.52	1088.70	373.93
1947	849.29	1084.35	1051.15	131.54	187.27	661.31	132.12	447.18	1086.05	329.63
1948	747.24	933.96	720.33	164.10	240.17	627.01	147.71	245.66	1101.07	428.75
1949	769.89	1080.40	984.17	250.38	339.83	592.59	326.64	329.26	1114.67	490.68
1950	638.26	778.54	1004.60	382.06	498.85	564.07	128.66	522.18	826.94	630.13
1951	630.67	753.10	569.60	143.01	194.83	523.33	101.09	258.68	714.70	319.68
1952	799.57	930.86	868.59	160.35	224.31	671.22	135.04	443.78	699.70	393.18
1953	569.16	583.68	900.90	182.80	254.60	536.55	192.72	454.29	730.97	425.80
1954	734.44	852.88	1447.85	250.01	315.20	622.23	181.64	526.49	1123.29	418.79
1955	894.58	1140.63	1283.84	259.11	329.13	695.94	212.48	422.21	1173.26	472.87
1956	889.00	950.40	1392.95	244.31	310.98	792.54	268.87	714.52	1019.24	508.92
1957	695.94	882.55	764.27	183.86	267.05	567.51	114.70	308.41	798.18	422.03
1958	924.81	1156.25	893.12	184.10	304.04	777.48	148.66	415.68	1217.05	564.63
1959	667.02	770.95	1312.19	393.17	472.96	572.27	350.60	783.14	1315.70	550.48
1960	625.81	967.12	781.50	176.08	231.13	457.75	96.63	208.17	947.30	373.37
1961	811.41	1011.09	1043.37	286.95	398.71	697.38	421.57	688.67	1065.31	617.99
1962	637.42	784.78	722.62	224.98	291.63	536.83	189.02	307.06	936.03	401.35
1963	616.74	842.82	1264.60	112.51	179.80	481.27	71.45	358.02	859.44	345.63
1964	888.46	1127.62	752.06	258.63	388.32	753.06	251.37	437.99	1093.78	667.66
1965	617.83	718.74	724.79	158.86	212.91	531.36	116.66	329.07	738.66	363.17
1966	735.97	802.20	741.56	278.56	352.92	633.55	136.91	393.36	893.92	477.51
1967	992.12	1089.08	836.84	266.52	347.68	896.77	250.04	679.88	999.53	622.03
1968	490.29	692.25	704.16	86.60	146.43	392.30	61.68	235.87	934.40	288.25
1969	572.48	821.25	569.64	161.98	210.11	436.46	44.64	176.22	1129.35	333.65
1970	843.70	910.02	1196.01	234.29	295.19	733.41	208.81	486.46	1104.07	480.02
1971	672.86	799.04	812.61	251.48	338.52	581.53	182.53	341.91	744.66	501.06
1972	526.17	591.29	671.42	217.75	295.71	477.03	136.19	296.97	813.56	471.59
1973	920.16	938.13	1362.92	211.43	259.88	829.86	154.83	692.08	1033.31	486.25
1974	454.77	458.52	542.45	256.99	315.79	459.81	93.44	153.41	968.98	470.54
1975	1002.80	905.15	1311.10	287.89	389.29	1019.76	303.63	862.62	1087.79	759.80
1976	763.24	771.23	762.54	356.79	479.66	726.85	309.03	583.33	860.30	679.56
1977	805.00	970.35	1103.50	404.36	528.54	699.99	276.46	670.89	1002.35	708.27
1978	752.38	793.04	807.20	360.84	464.16	701.61	355.59	629.70	947.78	648.23
1979	689.97	486.43	804.65	189.09	232.48	712.42	160.14	431.63	854.30	417.24
1980	535.56	689.33	1016.25	241.06	329.83	463.73	86.98	317.55	786.11	409.17
1981	767.23	937.28	948.10	316.00	388.61	594.39	165.64	463.74	1055.80	464.90
1982	682.24	777.51	802.05	501.37	613.84	618.73	143.57	438.94	745.61	745.05
1983	731.76	849.47	951.69	431.69	536.93	771.63	304.51	485.67	1141.75	749.10
1984	585.83	659.39	923.64	191.91	273.04	513.76	119.73	442.64	567.90	371.23

(Contd.)

(Contd.)

District	*Dausa*	*Dhaulpur*	*Dungarpur*	*G'nagar*	*H'garh*	*Jaipur*	*Jaisalmer*	*Jalor*	*Jhalawar*	*Jhunjhunu*
1985	809.04	1076.26	591.05	212.12	310.25	651.53	182.79	300.92	759.44	514.31
1986	491.21	539.15	680.41	293.42	346.72	427.83	182.04	302.22	623.10	376.49
1987	313.85	423.87	463.61	126.66	158.82	253.22	71.52	167.32	605.58	198.71
1988	608.55	699.08	933.41	226.68	318.26	554.49	211.82	548.41	828.91	516.31
1989	548.71	678.33	714.38	237.15	283.57	439.92	173.20	355.06	650.50	328.25
1990	816.91	1078.15	1096.13	239.17	317.84	664.34	136.76	531.38	1237.14	473.98
1991	601.19	668.08	640.66	144.60	201.09	477.16	91.97	301.71	716.74	348.42
1992	499.78	571.70	727.71	200.08	267.52	436.98	163.44	359.20	679.34	359.77
1993	776.93	795.56	881.77	186.86	278.93	687.75	182.05	437.09	867.79	535.84
1994	963.08	1102.58	1299.58	224.01	323.00	822.17	432.84	797.67	1178.36	579.74
1995	703.31	737.69	752.17	336.21	417.65	664.41	193.74	389.21	841.40	594.49
1996	1006.69	987.77	1193.71	278.91	366.57	944.77	184.08	640.24	1146.94	631.75
1997	735.14	737.84	992.82	294.84	389.09	673.69	249.45	542.20	1085.89	492.60
1998	646.85	661.16	991.59	274.04	349.76	624.38	294.10	722.75	938.10	560.29
1999	737.87	950.51	712.45	155.31	201.93	566.94	120.52	304.01	960.86	300.04
2000	531.60	600.05	657.86	175.68	266.94	485.02	95.85	333.44	679.27	443.20
2001	666.51	683.22	724.99	203.60	271.41	622.60	201.06	468.08	751.67	459.66
2002	295.70	455.95	473.65	105.75	137.04	218.36	55.50	142.84	477.76	171.64
2003						522.46	202.72			
2004						862.09	41.67			
2005						—	—			
2006						366.52	325.37			
2007						—	—			
2008						516.88	144.27			
2009						—	—			
2010						—	327.15			
2011						748.29	—			
2012						770.89	254.53			
2013						725.62	126.49			
2014						—	—			

Appendix 6

District Level 112 Years Mean Rainfall Data of Rajasthan (1901 to 2012)

District	*Jodhpur*	*Karauli*	*Kota*	*Nagaur*	*Pali*	*Rajsam.*	*S'madho-pur*	*Sikar*	*Sirohi*	*Tonk*	*Udaipur*
1901	166.47	520.06	661.23	251.67	314.41	426.05	499.57	319.57	269.94	436.69	573.56
1902	193.63	614.42	653.70	236.95	303.82	394.68	564.97	338.91	376.99	456.23	576.43
1903	383.52	673.52	799.07	486.13	661.24	759.18	686.49	448.45	611.24	680.09	894.98
1904	160.24	785.61	640.69	275.83	260.24	338.44	689.96	384.45	212.84	528.74	386.87
1905	96.21	371.14	553.54	127.01	244.02	329.12	358.99	163.69	453.93	307.25	544.27
1906	306.78	667.60	797.51	388.98	498.93	630.83	647.90	488.59	649.26	579.85	936.13
1907	300.23	632.64	645.73	324.41	435.26	492.93	587.04	332.94	625.12	493.82	709.15
1908	635.25	1179.15	1099.27	874.45	1131.44	1244.46	1173.06	900.70	1089.58	1126.62	1449.94
1909	494.69	889.29	983.00	642.23	712.71	810.50	875.89	721.82	758.23	811.98	938.37
1910	326.49	633.36	732.12	346.14	447.20	558.66	600.74	407.21	605.03	521.49	815.95
1911	143.47	478.41	553.93	200.02	255.58	366.75	448.60	269.90	259.77	379.88	422.52
1912	340.26	765.83	800.75	475.96	629.59	711.02	750.57	516.41	649.56	701.06	813.74
1913	257.09	403.53	599.33	207.83	397.43	456.34	404.99	254.67	804.17	374.01	778.52
1914	280.76	882.56	1012.22	370.24	477.38	600.18	828.44	475.82	661.44	698.50	797.91
1915	111.23	498.30	645.78	176.80	207.19	310.99	463.92	233.49	217.93	390.73	352.98
1916	465.18	909.18	950.66	514.41	615.36	712.48	873.51	567.82	683.75	770.16	922.11
1917	785.86	1262.68	1096.85	1002.11	1226.55	1232.40	1247.34	949.43	1360.16	1209.12	1326.10
1918	51.58	313.74	357.89	68.77	149.07	202.15	278.80	116.86	327.33	212.84	391.93
1919	377.34	813.09	835.04	424.51	583.00	659.20	749.02	441.73	982.45	621.17	1004.50
1920	236.41	566.05	476.06	337.98	382.56	450.56	523.15	419.68	459.20	436.41	678.85
1921	281.67	570.23	700.56	276.52	397.20	469.23	536.93	373.91	777.82	448.95	820.92
1922	215.48	713.55	721.29	305.56	438.52	508.52	659.52	365.28	661.59	547.91	756.38
1923	318.74	770.88	896.32	448.52	466.86	553.49	753.35	517.70	301.18	680.96	552.30
1924	288.15	726.93	726.17	387.18	562.34	633.02	679.35	460.77	794.62	590.08	900.87
1925	149.72	845.72	724.30	252.56	293.96	367.42	753.66	346.83	364.70	569.96	534.23
1926	460.49	863.59	854.17	539.88	813.21	911.39	836.92	568.56	1089.65	776.56	1201.85
1927	443.31	850.98	879.89	519.63	810.47	887.90	834.96	521.52	1118.80	778.09	1257.01
1928	313.87	628.64	715.78	390.28	507.59	609.88	621.80	377.66	483.53	581.26	730.62
1929	358.77	456.48	517.37	357.93	477.77	553.86	469.19	348.90	497.60	467.26	737.69
1930	211.41	797.66	715.86	309.77	367.17	492.23	718.26	394.86	385.21	561.72	683.60
1931	354.14	807.87	1039.20	435.19	584.11	682.21	799.41	469.84	664.74	735.68	859.72
1932	305.90	746.54	847.79	422.04	547.83	638.92	733.38	426.07	494.80	680.92	731.91
1933	491.61	1045.68	1143.37	661.06	828.18	933.47	1025.43	770.66	969.07	941.90	1208.45
1934	320.62	760.93	1122.86	403.02	607.32	763.97	763.97	477.67	815.83	719.14	1090.65

(Contd.)

(Contd.)

District	Jodhpur	Karauli	Kota	Nagaur	Pali	Rajsam.	S'madho-pur	Sikar	Sirohi	Tonk	Udaipur
1935	308.53	706.05	761.34	387.92	454.10	550.23	667.37	459.02	540.05	568.17	742.02
1936	248.97	867.20	816.62	380.02	408.34	509.80	787.98	486.39	374.74	638.23	574.65
1937	240.50	486.03	594.15	340.68	546.31	612.03	484.88	400.34	938.96	459.36	977.64
1938	191.69	676.18	903.60	273.94	418.95	549.50	657.40	275.29	496.30	592.01	739.34
1939	116.51	604.30	706.01	184.56	269.67	381.09	566.58	232.92	306.04	461.74	544.28
1940	284.47	592.76	712.84	342.26	415.71	517.40	580.05	391.50	443.75	532.74	607.21
1941	233.70	483.55	554.56	229.13	442.84	554.08	464.81	239.60	672.20	409.81	931.76
1942	436.87	1083.40	1074.39	512.23	515.12	645.73	988.85	632.43	527.17	791.83	883.93
1943	411.89	895.33	828.97	628.10	818.05	852.96	886.04	578.56	816.04	850.86	921.55
1944	448.92	694.91	994.02	603.76	799.60	900.59	711.26	685.18	1059.24	766.68	1229.28
1945	452.34	869.05	939.53	568.08	693.43	747.90	843.37	614.35	901.19	767.95	977.23
1946	233.33	713.25	854.80	286.49	432.75	518.54	669.65	370.99	729.27	564.36	793.00
1947	296.38	994.70	975.25	398.39	602.12	702.60	922.21	391.31	680.97	779.13	921.77
1948	258.72	891.11	954.66	374.32	423.85	556.12	835.02	468.92	322.38	700.20	638.28
1949	300.93	966.39	965.67	343.34	395.34	518.12	867.91	466.41	449.98	673.27	759.68
1950	265.99	732.00	733.59	423.12	517.08	586.98	684.94	564.86	786.47	581.59	878.19
1951	248.14	715.19	687.85	350.61	425.46	506.65	671.44	363.96	357.71	586.82	544.53
1952	322.63	873.43	707.94	448.07	568.07	604.99	802.61	459.10	623.17	681.22	763.86
1953	324.90	608.54	664.88	410.84	542.37	615.28	602.82	456.85	578.37	570.18	774.93
1954	364.27	841.28	972.07	454.76	682.83	842.97	814.16	457.56	778.57	739.88	1171.43
1955	340.20	1058.95	1058.99	420.91	565.27	721.66	975.07	481.65	581.69	799.80	1028.60
1956	432.60	947.59	932.78	557.00	756.08	842.52	914.88	588.75	944.99	820.54	1145.56
1957	234.24	815.87	751.38	334.50	402.61	499.03	745.45	430.00	425.69	606.92	633.75
1958	342.14	1080.97	1108.50	492.90	619.73	784.44	1010.59	594.46	523.24	870.62	853.53
1959	464.96	785.80	1040.83	474.47	667.95	786.26	773.25	510.43	1005.11	715.07	1105.46
1960	143.42	823.47	801.25	228.00	261.55	365.77	717.03	346.14	325.67	526.36	570.44
1961	484.45	942.05	957.22	494.93	602.09	704.92	874.65	605.28	738.68	740.74	909.51
1962	289.48	738.75	778.47	370.14	438.03	555.70	696.81	417.04	372.47	606.06	643.99
1963	173.08	758.83	763.37	262.98	432.07	575.63	689.95	349.41	589.56	553.59	936.76
1964	351.81	1041.35	998.91	468.70	477.71	553.77	953.60	649.22	492.32	772.00	673.90
1965	248.02	697.02	686.98	350.63	430.95	501.42	657.74	401.58	450.33	560.77	631.37
1966	324.05	819.28	835.82	452.15	544.00	613.41	784.69	497.41	565.43	701.85	713.42
1967	482.15	1064.68	994.63	651.59	819.95	867.61	1014.61	694.96	881.75	923.15	892.58
1968	122.53	641.60	762.18	204.89	273.58	377.48	589.83	300.57	342.88	473.76	553.72
1969	126.83	749.63	877.63	227.60	266.77	374.70	683.51	327.44	276.04	544.15	473.06
1970	408.14	922.94	992.41	521.13	699.42	837.84	894.13	535.53	636.56	825.92	1032.22
1971	289.75	749.63	689.27	382.20	415.04	487.18	692.53	490.45	441.15	572.86	650.68
1972	274.33	593.49	687.32	367.42	416.63	493.72	576.43	454.59	402.22	523.03	604.36
1973	454.97	964.67	980.79	616.53	880.94	964.05	948.90	594.72	984.66	896.40	1222.56
1974	263.42	510.78	769.82	409.68	401.80	516.15	533.30	470.77	226.00	536.79	521.87
1975	607.60	982.47	1046.94	825.63	1059.79	1100.86	1003.29	872.23	1141.97	1007.32	1237.19
1976	465.20	795.35	807.27	575.89	670.64	734.33	781.95	673.35	707.67	737.33	792.60
1977	433.56	907.18	891.17	527.87	591.71	665.52	845.04	655.09	859.08	714.86	922.24
1978	442.47	805.67	847.86	544.38	616.77	692.53	786.57	648.20	719.45	719.41	783.77
1979	437.57	628.63	774.47	606.86	744.03	816.55	683.98	551.71	556.17	736.51	832.25
1980	191.59	635.40	662.27	299.84	379.20	477.18	583.81	404.39	513.92	486.45	747.73
1981	264.59	922.71	976.73	357.15	471.21	583.98	860.60	442.53	666.77	708.08	786.70
1982	331.15	743.69	676.17	526.87	491.50	543.39	691.88	657.48	628.30	588.76	695.25

(Contd.)

Bibliography

Abhichandani, C.T. and Roy, B.B. (1966). *Rajasthan Desert—Its Origin and Amelioration.*

Agrawal, D.P. and Kursumgar, S. (1975). Radio Carbon dates of some Late Quaternary Samples, *Current Science,* Vol. 44, pp. 149-51.

Agrawal, D.P.; Datta, P.S.; Husain, Z.; *et al.,* (1980). Palaeoclimate, Straigraphy and Pre-history in north and west Rajasthan, proceedings of the *Indian Academy of Sciences,* Vol. 89 (1), pp. 51-66.

Agrawal, D.P.; Dhir, R.P. and Krishnamurthy, R.V. (1976). *The Relevance of Kankar Studies of the Archaeology and Palaeo-ecology of Rajasthan,* Annual Archaeological Congress, University of Rajasthan, Jaipur.

Ahmed, E., (1969) *Origin and Geomorphology of the Thar Desert,* Annals of Arid Zone, pp. 171-80.

Ajay, A.S., (2007) *Desertification and Land Degradation Atlas of India,* SAC, Ahmedabad. Indian Space Research Organization, GoI.

Ajay, A.S., Arya, P.S.; Dhinwa, S.K.; Pathan and K.G. Raj (2009). "Desertification/ land degradation status mapping of India". *Current Science,* 97 (10). 1478-1483.

Alka Gadgil and Amit Dorde (2005). Temperature trends in Twentieth Century at Pune, India, *Atmospheric Environment,* 39: 6550-6556.

Allchin, B. and Allchin, R. (1982). *The Rise of Civilization in India and Pakistan.* Cambridge University Press, London.

Allchin, B., (1959) *The Indian Middle Stone Age.* Bulletin of the Institute of Archaeology, London. Vol. 2, pp. 1-35.

Allchin, B. and Goudie, A. (1971). Dunes, Aridity and Early Man in Gujarat, Western India, *Man,* Vol. 6, pp. 248-65.

Allchin, B.; Goudie, A.; and Hegde, K., (1978). *The Pre-history and Palaeogeography of the Great Indian Desert,* Academic Press, London.

Allchin, B.; Hegde, K.T.M., and Goudie, A. (1972). Pre-history and Environmental Change in Western India, A note on the Budha-Pushkar Basin, Rajasthan, *Man,* Vol. 74, pp. 542-64.

Allchin, Band; Goudie, A., (1974). Pushkar: Pre-history and Climatic Change in Western India, *World Archaeology,* Vol. 5, pp. 358-68.

(Contd.)

District	Jodhpur	Karauli	Kota	Nagaur	Pali	Rajsam.	S'madho-pur	Sikar	Sirohi	Tonk	Udaipur
1983	511.64	783.29	939.72	704.01	714.47	792.38	768.91	765.51	594.02	795.03	850.23
1984	239.33	635.87	573.12	335.74	480.06	547.57	600.10	398.42	612.04	523.61	770.84
1985	200.58	947.95	712.09	334.75	278.60	338.72	838.08	512.78	363.08	613.27	470.23
1986	250.99	532.64	510.83	319.56	344.20	420.74	504.10	364.96	371.10	434.09	552.51
1987	142.58	392.61	458.45	164.60	220.04	291.06	359.46	194.81	237.21	295.53	368.55
1988	378.11	663.51	699.59	437.73	617.04	701.53	617.90	504.08	692.88	559.20	850.21
1989	242.40	638.06	593.82	303.83	401.11	447.48	587.73	326.88	507.68	491.51	608.33
1990	307.86	985.83	1058.45	404.43	583.65	678.72	908.82	494.21	795.08	760.45	939.38
1991	184.36	696.65	669.91	246.19	311.26	390.50	651.80	350.97	401.18	518.05	534.55
1992	247.26	559.27	578.12	318.21	409.92	479.14	533.82	364.63	494.68	472.31	631.37
1993	301.28	826.71	792.45	443.01	524.27	610.72	792.86	559.74	590.17	676.23	771.02
1994	476.05	1089.28	1087.56	529.98	733.82	809.49	1029.61	624.75	1023.60	882.34	1118.22
1995	366.71	746.73	764.46	517.13	548.60	637.05	730.23	601.79	496.12	672.17	712.17
1996	504.43	1038.99	1091.03	720.48	893.60	999.72	1033.58	737.82	868.88	992.24	1146.54
1997	406.82	796.18	971.60	521.22	674.62	773.66	800.23	545.98	707.97	765.74	915.56
1998	458.56	680.60	835.95	514.74	651.34	715.14	681.24	576.06	838.15	656.55	894.26
1999	225.14	877.83	867.28	324.22	411.66	479.67	804.86	344.35	445.04	653.46	603.67
2000	217.42	603.64	637.11	336.96	389.45	423.51	581.87	444.21	485.79	505.74	558.30
2001	338.40	701.25	729.83	453.58	534.01	567.38	685.71	505.53	591.81	636.96	668.20
2002	63.72	381.17	354.05	96.56	130.13	198.36	327.43	160.31	216.90	238.65	333.66
2003	361.69										
2004	—										
2005	—										
2006	161.81										
2007	—										
2008	—										
2009	—										
2010	—										
2011	359.67										
2012	406.38										
2013	372.87										
2014	—										

Allchin, R. and Allchin, B. (1997). *Origins of a Civilization.* Viking-Penguin India, New Delhi.

Amal Kar (2012). GCM-Derived Future Climate of Arid Western India and Implication for Land Degradation, *Annals of Arid Zone*, 51 (3 & 4): 147-169.

Amal Kar (2014). Agricultural land-use in arid Western Rajasthan: Resource exploitation and emerging issues, *Agropedology*, 24 (2): 179-196.

Amal Kar (2014). The Than or the Great Indian Sand Desert, in *Landscapes and Landforms of India* (ed., V. S. Kala), 79-90, Springer, Dordrecht.

Amal Kar, (2009). "Wind erosion potentials in the Thar-Megathar Region from AD 1951 to 2100: Patterns and Implications, Abstract in International Conference on Nurturing Arid Zones for People and the Environment: Issues and Agenda for the twenty-first Century." *CAZRI*, Jodhpur, Rajasthan, India, pp. 24-25.

Amit Dhoda and Anargha Wakhare (2004). Evidence of Long-Term Climate Change at Major Cities of India during the Twentieth Century, the *International Journal of Climate Change, Impacts and Responses* 1(3): 15-41.

Amit Dhorda, Anargha and Alka S. Gadgil (2009). Long-term Temperature Trends at Four Largest Cities of India during Twentieth Century, *Indi. Geophys Union,* 13(2): 85-97.

Anderson, D.M., J.T. Overpeck and A.K. Gupta (2002). "Increase in the Asian southwest monsoon during the past four Centuries." *Science*, 2,97 (5581): 596-599.

Arnell, N.W., (2006) Climate change and water resources: A global perspective. In: H.J. Schellnhuber, W. Cramer, N. Nakicenovic, T.M.L. Wigley and G. Yohe (Editors), *Avoiding Dangerous Climate Change.* Cambridge University Press, Cambridge, pp. 167-175.

Bakliwal, P.C. and Grover, A.K. (1999), Report, *Geol. Soc. India, Mem.*, 42: 113-119.

Bandyopadhyay, A., A. Bhadra, N.S. Raghuwanshi and R. Singh (2009). "Temporal trends in estimates of reference evapo-transpiration over India." *Journal of Hydrologic Engineering*, 14(5). 508-515.

Basistha, A., Goel, N.K., Arya, D.S., Gangwar, S.K. (2007). Spatial pattern of trends in Indian sub-divisional rainfall, *Jalavigyan Sameeksha*: It is also a Journal of National Institute of Hydrology, 27: 47-57.

Bhaskar, R., (2004) Re-evaluating the sustainability of irrigated agriculture in arid zones—A case study of Indira Gandhi Nahar Pariyojana. *Indian Cartographer*, 24: 132-141.

Bhalla, G.S. and Gurmail Singh (2009). Economic Liberaliztion and Indian Agriculture: A State-wise Analysis, *Economic and Political Weekly*, 46(52): 34-44.

Bhardwaj, V. and Geeta Sharma (2008), Flood Calamity: Challenge to Sustainable Development in the Thar Desert, *Geographer*, Vol. V, No. 1.

Bhaskaran, B., Mitchell, J.F.B., Lavery, J.R. and Lal, M., (1995) Report, *Int. J. Climatol.*, 15: 873-892.

Bhatacharya, S., C. Sharma, R.C. Dhiman and A.P. Mitra (2006). "Climate change and Malaria in India." *Current Science*, 90(3): 369-375.

Bhuiyan, C., W.A. Flugel and R.P. Singh (2009). "Erratic monsoon, growing water demand, and declining water table." *Journal of Spatial Hydrology*, 9(1): 1-19.

Billett, S. (2010). "Dividing climate change: Global warming in the Indian mass media." *Climate Change*, 99(1): 1-16.

Blaikie, P. and Brookfield, H. (1987). *Land Degradation and Society*. Methuen, London.

Bokil, M. (2000). "Drought in Rajasthan: In search of a perspective." *Economic and Political Weekly*, 35(48): 4171-4175.

Breankert, A. and E. Malone (2005). Modelling Vulnerability and Resilience to Climate Change: A Case Study of India and Indian State, *Climate Change*, 72 (1-2): 57-102.

Bruce Battrick (2006). *The Changing Earth* (ed.), ESA Publication Division, The Nitherlands.

Bryson, R.A. and Barries, D.A. (1967), Possibilities of Major Climatic Modifications and their Implications: North-west India. A Case for Study. *Bulletin of American Meteorological Society*, Vol. 48, pp. 136-42.

Chase, T.N., Pielke Sr. R.A., Kittel, T.G., Running, S.W. (2000). Simulated Impacts of Historical Land Cover Changes on Global Climate in Northern Winter, *Climate Dynamics* 16: 93-105.

Chatterjee, K., A. Chatterjee, and S. Das (2005). "Community Adaptation to Drought in Rajasthan." *IDS Bulletin* 36 (4): 33-52.

Chaudhury, S.K., J.M. Gore and K.C.S. Ray (2000). "Impact of heat waves over India."*Current Science*, 79(2): 153-157.

Climate of Rajasthan, Climatological Summaries of State series No. 16, India Meteorology Department (20).

C.A. Rama Rao, B.M.K. Raju, A.V.M. Subba Rao, K.V. Rao, V.U.M. Rao, Kausalya Ramchandran, B. Venkateswarlu, A.K. Sikka, M. Srinivas Rao, M. Maheswari and Ch. Srinivasa Rao (2016): A. District-level assessment of Vulnerability of Indian Agriculture to climate change, *Current Science*, Vol. 110, No. 10, 1939 to 1946.

Dash, S.K., M.A. Kulkarni, U.C. Mohanty and K. Prasad (2009). "Changes in the characteristics of rain events in India." *J. Geophys. Res.*, 114: D10109, doi:10110.11029/12008JD010572.

Dash, S.K., R.K. Jenamani, S.R. Kalsi and S.K. Panda (2007). "Some Evidence of Climate Change in twentieth-century India." *Climate Change*, 85(3): 299-321.

Desertification and Land Degradation Atlas of India (2016), Space Application Centre, Indian Space Research Organisation, Government of India, Ahmedabad, pp. 165-170.

Dhabria, S.S. (1984). Intensification of Desertification Hazard Through Sambhar Gap: A Problem of Environmental Degradation in India Arid Zone", *18th International Symposium on Remote Sensing of Environment*, Paris.

Dhir, R.P., A.K. Singhvi, J.E. Andrews, A. Kar, B.K. Sareen, S.K. Tondon, A. Kailath and J.V. Tondon (2010). Multiple Episodes of Aggradation and Calcrete Formation in Late Quarternary Eolian Sands, Central. Thar Desert, Rajasthan, India, *Journal of Asian Earth Sciences*, 37(1): 10-16.

Doi, R.D. (1993). *Analysis of Rainfall Variability and Drought Occurrences in Rajasthan*, Readings of Geographical Paper, University of Reading, U.K.

Doi, R.D. (2000). *Rainfall Pattern and Agricultural Response to its Occurrence in Rajasthan, Studies in Geography*, 18: 56-64.

Dregne, H.E. (1983). *Desertification of Arid Lands*, Harwood Academic Publishers, London.

Dynamic Water Resource of India as on 31st March, 2004 (2006), CGWB, Government of India, New Delhi.

Enzel, Y., L.L. Ely, S. Mishra, R. Ramesh, R. Amit, B. Lazar, S.N. Rajaguru, V.R. Baker, A. Sandler. "High-resolution Holocene Environmental changes in the Thar Desert, Northwestern India." *Science*, 284: 125-128.

Foley, J.A., Costa, M.H., Delise, C., Ramakutty, N., Snyder, P. (2003). Green Surprise? How terrestrial ecosystems could affect earth's climate, *Frontiers in Ecology and the Environment*, 1: 38-44.

Garg, Shukla, *Emission Inventory* Book, Annexure-II.

Garg, Shukla, *Emission Inventory* Book, Annexure-III.

Ghose, B. and Singh, S. (1965) Geomorphological Studies in Rajasthan—Analysis of Landforms of Siwana Block from Aerial Photographs. *Annals of Arid Zone*. Vol. 4: (2). pp. 207-216.

Ghose, B., (1964). Geomorphological Aspects of the Formation of Salt Basin the Lower Luni Basin, *Proceedings of Symposium of Problems of India Arid Zone*, Jaipur, pp. 169-78.

Gleick, P.H., R.M. Adams, *et al.*, (2010). "Climate change and the Integrity of Science." *Science*, 328(5979): 689-690.

Government of India (1998). *Water Resources Planning for the State of Rajasthan*, Tahal and WAPCoS.

Government of Rajasthan, December (2009). *Note on Power Castor.*

Government of India (2011). *Report on Ground Water Resources of Rajasthan*, Central Ground Water Board, New Delhi.

Government of Rajasthan (2012-a). *Draft State Agricultural Policy*, Government of Rajasthan, Jaipur.

Government of Rajasthan (2012-b). *Rajasthan Agriculture Competitiveness Project*, Department of Agriculture, Government of Rajasthan, Jaipur.

Government of Rajasthan (2013-14). *Draft Annual Plan*, Planning Department, Government of Rajasthan.

Government of Rajasthan (2010). *State Water Policy Document*, State Water Resource Planning Department.

Government of Rajasthan (2010-11). *Agricultural Statistics of Rajasthan*, Directorate of Economics and Statistics, Rajasthan.

Gore, P.G., B.A. Roy and H.R. Hatwar (2011). Impact of Climate Change on Land Degradation, *National Climate Centre Report* 1/2011, Pune: 1-12.

Gosain, A.K. and Rao, S., (2003) *Climate Change and India: Vulnerability Assessment and Adaptation* (eds. Shukla, P.R., *et al.*), Universities Press (India) Pvt. Ltd., Hyderabad, p. 462.

Gosain, A.K., Rao, S. and Basuray, D., (2006). Climate Change Impact Assessment on Hydrology of Indian River Basins. *Current Science*, 90(3): 346-353.

Goswami, B.N., V. Venugopal, D. Sengupta, M.S. Madhusoodanan and P.K. Xavier (2006). "Increase trend of extreme rain events over India in a warming environment." *Science*, 314(5804): 1442-1445.

Goswmi, P. and K.V. Ramesh (2008). The Expanding Indian Desert: Assessment through Weighted Epochal Trend Ensemble." *Current Science*, 94: 476-480.

Goyal, R.K. (2004), "Sensitivity of Evapotranspiration to Global Warming: A Case Study of Arid Zone of Rajasthan (India)." *Agricultural Water Management*, 69: 1-11.

Gude, V.G., *et al.* (2010). Renewable and Sustainable Energy, *Energy Reviews*, 14: 2641-2654.

Guhartha, K.P., M. Rajeevan (2006). Trends of Rainfall over India, National Climate Centre, *Research Report*, 2: 1-22.

Gupta, R. (1996). Correlation of rainfall with upsurge of Malaria in Rajasthan, *Journal of the Association of Physicians of India*, 44 (6): 385-389.

Gupta, A., (1995) *In American Geophysical Union, Geophysical Monograph* (eds. Costa, J.E., *et al.*), Vol. 89, pp. 125-136.

Gupta, A.K. (2004) "Origin of Agriculture and Domestication of Plants and Animals linked to early Holocene Climate Amelioration." *Current Science*, 87: 54-59.

Gupta, D.M., D.M. Anderson, D.N. Pandey, A.K. Singhvi (2006). "Adaptation and Human Migration, and Evidence of Agriculture coincident with changes in the Indian Summer Monsoon during the Holocene". *Current Science*, 90: 1082-1090.

Gupta, R.K. (1968). "Anthropogenic Influences on the Vegetation of Western Rajasthan." *Vegetation*, 16(1): 79-94.

Gupta, S.K. and R.D. Deshpande (2004). "Water for India in 2050: First-order assessment of available options." *Current Science*, 86(9). 1216-1224.

Hashimi, N. H., Nigam, R., Nair, R.R., and Rajagopalan, G. (1995). Holocene sea level fluctuations on western India continental margin: An update. *Jour. Geol. Soc. India*, 46: 157-162.

Hema Malini and Nageshwar Rao (2004) Coastal Erosion and Habitat Loss along the Godavari Delta Front—A fallout of dam construction, *Current Science*, Vol. 87, No. 9, pp. 1232-36.

Hingane, L. S., Rupa Kumar, K. and Ramana Murthy, (1985) Bh. V., *J. Climatol.*, 5: 521-528.

Hooja, B. (1985). Reforming Desert Development Administration and Strategies in Rajasthan —An Extensive Approach for Desert Development and Environmental Preservation, *Indian Journal of Public Administration*, 31(3): 910-32.

Ian Douglas (2013). Climate Change and Green Economy with reference to India: Strategies Options at Different Scales in H.S. Sharma *et al.* (Ed.) *Climate Change Biodiversity and Green Economy*, Concept Publishing Company Pvt. Ltd., New Delhi: 33-42.

IPCC (1992) *Global climate change and the rising challenge of the sea. Supporting Document for the IPCC*, WMO and UNEP, Geneva.

IPCC (1998) *Special Report of IPCC Working Group II* (eds. Watson, R.T., *et al.*), Intergovernmental Panel on Climate Change, Cambridge University Press, Cambridge, UK, p. 517.

IPCC (2007) *Climate Change, Synthesis Report.*

IPCC (2014). *Summary for Policy-makers. Climate Change 2007: The Physical Science Basis*. Contribution of Working Group I to the Fourth Assessment Report of the Inter-governmental Panel on Climate Change. Cambridge University Press, Cambridge.

ISRO (2007). *Desertification and Land Degradation Atlas of India Space Application Centre*, Government of India, Ahmedabad.

ISRO (2007). *Desertification Monitoring and Assessment using Remote Sensing and GIS: A Pilot Project under TPN-1 UNCC*, Space Application Centre, Government of India, Ahmedabad.

Jain, S.K., R. Keshri, A. Goswami, A. Sarkar and A. Chaudhry (2009) "Identification of drought-vulnerable areas using NOAA AVHRR data." *International Journal of Remote Sensing*, 30(10): 2653-2668.

Jaglan, M.S. and M.H. Qureshi (1996). Irrigation development and its environmental consequences in arid regions of India, *Environmental Management*, 20: 323-336.

Jodha, N.S. (1985). "Population growth and the decline of common property resources in Rajasthan, India." *Population & Development Review*, 11(2): 247-264.

Jodha, N.S. (1988). "Farmers adjustment strategies against rainfall variability in dry regions of India." *Fertiliser News*, 33(4): 67-72.

Joseph D'Aleo and D. Easterbrook (2010). Multi-decadal Tendencies in ENSO and Global Temperature related to Multidecadal Oscillalions, *Energy and Environment*, Vol. 21, No. 5, pp. 437-460.

Joyce, Terrence (2008). *Are We on the Brink of a 'New Little Ice Age'?* Woods Hole Oceanographic Institute website.

Kale, V.S. (ed.), (1998). Flood Studies in India, Geological Society of India, Bangalore, *Memoir*, 41, pp. 229-256.

Kale, V.S., (2003). *Nat. Hazards*, 28: 65-84.

Kale, V.S. and Rajaguru, S.N. (1985). Neogene and Quaternary Transgressional and Regressional History of the West Coast of India. *Bull. Deccan Coll. Res. Inst.*, 44: 153-165.

Kale, Vishwas S., Ashok K. Singhvi, Praveen Mishra and Debabrata Banerjee (2000). Sedimentary Records and Luminescence Chronology of Late Holocene Palaeofloods on the Luni River, Thar Desert, North-West India, *Catena*, 40: 337-358.

Kaul, M., V.K. Dadhwal and G.M.J. Mohren (2009). "Land-use change and net C flux in Indian Forests." *Forest Ecology and Management* 258(20): 100-108.

Khanna, R.K., R.S. Rathore and C. Sharma (2009). Solar Still and Appropriate Technology for portable water need of the remote villages of desert state of India, Rajashtan (*A Report*).

Kolli and Rupa Kumar (Nov., 2006), Modelling of Regional Climate in South Asia Region and Future Climatic Scenario. *Global Change Conference*, Islamabad, Pakistan, 16-17.

Koteswaram, P., (1958), *Monsoons of the World*, India Meteorology Department, New Delhi, 105.

Kothawala, D.R., Munot, A.A. and Krishna Kumar, K. (2010). Surface Air Temperature Variability over India during 1901-2007 and its association with ENSO, *Climate Research*, 42: 89-104.

Kothyari, U.C. and Singh, V.P., (1996) *Hydrol. Process.*, 10, 357-372.

Kripalani, R.H., Kulkarni, A., Sabade, S.S. and Chaudhari, H.S., (2007), South Asian Summar Monsoon Precipitation Variability: Coupled Climate model Simulations and Projection under IPCC ARY, *Theoretical and Applied Climatology*, 90: 133-159. *Hazards*, 29: 189-206.

Kulkarni, A.V., Rathore, B.P., Mahajan, S. and Mathur, P., (2005) Alarming Retreat of Parbati Glacier, Beas Basin, Himachal Pradesh. *Current Science*, 88(11): 1844-1849.

Kumar, K.K., K.R. Kumar, R.G. Ashrit, N.R. Deshpande, J.W. Hansen. (2004). "Climate impacts on Indian agriculture." *International Journal of Climatology*, 24(11): 1375-1393.

Kumar, K.R., A.K. Sahai, K.K. Kumar, S.K. Patwardhan, P.K. Mishra, J.V. Revadekar, K. Kamala and G.B. Pant (2006). High-resolution Climate change scenarios for India for the twenty-first Century." *Current Science*, 90(3): 334-345.

Kumar, K.R., G.B. Pant, B. Parthasarathy and N.A. Sontakke (1992). "Spatial and Subseasonal Patterns of the long-term trends of Indian Summer Monsoon Rainfall." *International Journal of Climatology*, 12(3): 257-268.

Lal, M. and Singh, S. (2001). *Mausam*, 52: 1.

Laul, V.P. and Vivek Laul (2007). Floods in Barmer and other Tertiary terrains in Rajasthan. *Current Science*, 92 (5): 558.

Leshnik, L.S., (1968). Pre-historic Explorations in North Gujarat and Parts of Rajasthan. *East and West* (N.S.), Vol. 1183 and 1184, pp. 295-310.

Liverman, D. (Editor), (1998). *People and Pixels: Linking Remote Sensing and Social Science*. National Academy Press, Washington, DC.

Lonergan, S. (1998). *Climate Warming and India* (eds. Dinar, *et al.*), World Bank Technical Paper No. 402, Washington, DC.

Mall, R.K., R. Singh, A. Gupta, G. Srinivasan and L.S. Rathore (2006). "Impact of climate change on Indian Agriculture: A review." *Climate Change*, 78(2): 445-478.

Mall, R.K., A. Gupta, R. Singh, R.S. Singh and L.S. Rathore (2006). "Water Resources and Climate Change: An Indian Perspective." *Current Science*, 90(12): 1610-1626.

Mall, R.K., Akhilesh Gupta, Ranjeet Singh, R.S. Singh and L.S. Rathore (2006). Water Resources and Climate Change—An Indian Perspective, *Current Science*, Vol. 90, No. 12, pp. 1610-125.

Mathur, U.B. (2008) Are We Prepared for Past Climate Change, *Souvenir CHEC India*, Jaipur, pp. 29-32.

Mathur, U.B. (2010). *Climate Change: Past, Present and Future*, Geological Society of India, Bangalore.

Mathur, U.B., Pandey, D.K. and Tej Bahadur (2004). Falling Late Holocene sea-level along the Indian Coast. *Current Science*, 87 (4): 439-440.

Meher-Homji, V.M., (1973) Is the Sind-Rajasthan Desert the result of a recent climatic change, *Geogforum*, Vol. 15, pp. 47-57.

Mishra, V.N. (1961) Palaeolithic Cultures of Western Rajputana. *Bulletin of the Deccan College*, Research *Institute*, Vol. 21, pp. 85-156.

Mishra, V.N. (1973) Bagor—A late Mesolithic Settlement in North-West India, *World Archaeology*, Vol. 5, No. 1, pp. 92-110.

Mishra, V.N.; Rajaguru, S.N.; Agrawal, D.P.; Thomas, P.K.; Husain, Z. and Dutta, P.S., (1980) Pre-history and Palaeoenvironment of Jayal, Western Rajasthan. *Man and Environment*, Vol. IV, pp. 18-30.

Misra, R.P. (2013). "Facing Environmental Challenges: The Gandhian Way" in H.S. Sharma, *et al.* (Ed.). *Climate Change, Biodiversity and Green Economy*, Concept Publishing Company Pvt. Ltd., New Delhi, pp. 1-32.

Mooley, D.A. and Parthasarathy, B. (1984). *Climatic Change*, 6: 287-301.

Mrutyunjay Swain, S.S. Kalamkar and Manish Kant Ojha (2012). *State of Rajasthan Agriculture-2011-12*, Agro-Economic Research Centre, Sardar Patel University, Anand, Gujarat, Project Report: 1-35.

Mukhopadhyay, K. (2004). An Assessment of a Biomass Gasification based Power plant in Sunderbans, *Biomass and Bioenergy*, 27: 253-264.

O'Brien, K.R. Leichenko, U. Kelkar, H. Venema, G. Aandahl, H. Tompkins, A. Javed, S. Bhadwal, S. Barg, L. Nygaard, J. West (2004). "Mapping Vulnerability to Multiple Stressors: Climate Change and Globalization in India." *Global Environmental Change*, Part A 14: 303-313.

Panda, A. (2009). "Assessing vulnerability to Climate Change in India." *Economic & Political Weekly*, 105-107.

Pant, G.B. and Hingane, L.S. (1988). Climatic Change in and around the Rajasthan desert during the Twentieth Century, *J. Climatol.*, 8: 391-401.

Panwar, N.L. and Rathore, N.S. (2009). Potential of surplus biomass gasifier based power generation: A Case study of an Indian State Rajasthan, *Mitigation and Adaptation Strategies for Global Change*; 14 (8): 1381-2386.

PHED, Government of Rajasthan (2010) *Proposal for PPP Projects—Water Supply System* as accessed in April 2013.

Poonia, S.A.S. Rao (2013). Climate Change and its impact on Thar Desert Ecosystem, *Journal of Agricultural Physics*, 13(1): 71-79.

Prabhakar, S.V.R.K. and R. Shaw (2008). "Climate change adaptation implications for drought risk mitigation: A perspective for India." *Climatic Change*, 88(2): 113-130.

R.K. Mall, R. Singh, G. Srinivasan, L.S. Rathore (2007). "Impact of climate change on Indian agriculture: A Review (*erratum*)." *Climate Change*, 82: 225-231.

Ragab, R. and C. Prudhomme (2002). "Climate change and Water Resources management in arid and semi-arid regions: Perspective and challenges for the twenty-first century." *Biosystems Engineering*, 81(1): 3-34.

Raghav, K.S. (1992a) Geoenvironmental Appraisal of the Mendha and Bandi River Catchment area of Jaipur upland, *Jaipur District Record*, GSI29 (7): 134-135.

Raikes, R.L. (1965). The Mahenjo-Daro Floods, *Antiquity*, Vol. 39, pp. 196-203.

Raikes, R.L. and Dyson, R.H. (1961). The Pre-historic Climate of Baluchistan and the Indus Valley. *American Anthropologist*, Vol. 63, pp. 265-282.

Rajasthan State Action Plan on Climate Change (2010) Prepared by Government of Rajasthan, Teri the Energy and Resources Institute, New Delhi: 1-171.

Rajendran, K. and A. Kitoh (2008). "Indian summer monsoon in future climate projection by a super high-resolution global model." *Current Science*, 95(11): 1560-1569.

Ratnapriya and Padmini Pani (2014); *Understanding Cause to Land Degradation: Is it Climate or Deforestation?*, Abstract Volume 27th IGI Conference on Geomorphology and Geo-spatial Technology in Environment and Societal Development, 14 to 17 November, 2014, Vidya Sagar University, Midanapur, West Bengal.

Rathore, M.S. (2004). State Level Analysis of Drought Policies and Impacts in Rajasthan, India International Water Management Institute (IWMI), *Working Paper* 93: 1-29.

Rathore, M.S. (2005). *Ground Water Exploration and Augmentation Efforts in Rajasthan—A Review.* Institute of Development Studies: 1-33.

Rao, A.S., Gupta, J.P., Faroda, A.S. (1997). Changing climate scenarios and strategies for Drought Management in the Indian arid region, *Drought Network News*, 9(1): 8-11.

Ravindranath, N.H., N.V. Joshi, R. Sukumar and A. Saxena (2006). Impact of Climate Change on Forest in India, *Current Science*, 90 (354): 16.

Reddy and Mahendra (2003). Drinking Water and Sanitation in India: Need for Demand Management Structures, presented at *Seminar on Drinking Water and Sanitation*, Delft, The Netherlands.

Reddy, V.R. (2010). *Water Sector Performance under scarcity conditions: A case study of Rajasthan*, India.

Report, National Commission for Integrated Water Resources Development, Ministry of Water Resources, September 1999, Vol. 1, p. 542.

Romani, S. (2004). *Groundwater Management.*

Roy, B.B. and Pandey, S. (1970) Expansion or Contraction of the Great Indian Desert. *Proceedings of the Indian National Science Academy*, Vol.36, B. No. 6, pp. 331-344.

Roy, S.S. (2009). "A spatial analysis of extreme hourly precipitation patterns in India." *International Journal of Climatology*, 29(3): 345-355.

Rupa Kumar, K., *et al.* (2006) High-resolution Climate change scenarios for India for the twenty-first Century. *Current Science*, 90(3): 334.

Rupa Kumar, K., Kumar, K., Prasanna, V., Kamala, K., Deshpande, N.R., Patwardhan, S. K. and Pant, G.B. (2003). *Climate Change and India: Vulnerability Assessment and Adaptation* (eds. Shukla, P.R., *et al.)*, Universities Press (India) Pvt. Ltd., Hyderabad, p. 462.

Rupa Kumar, K., Pant, G.B., Parthasarathy, B. and Sontakke, N.A. (1992). *Int. J. Climatol.*, 12: 257-268.

Rupa Kumar, *et al.* (2005): High Resolution Climate change Scenario for India for twenty-first Century, Current Science, 90: 334-345.

Sankaran, A.V. (1999). Saraswati—The ancient river lost in the desert. *Current Science*, 77(8).

Shailesh Kumar Kharol, D.G. Kaskawtis, K.V.S. Badrinath, Annu Rani Sharma, R.P. Singh (2013). Influence of land-use/land cover (LULC) Changes on Atmospheric Dynamics over the arid region of Rajasthan State, India, *Journal of Arid Environments*, 38: 90-101.

Sharma, H.S. (1976) Morphology of Ravines of the Morel Basin, India, *International Geography*, Moscow, Vol. 1, pp. 219-223.

Sharma, H.S. (1987) *Tropical Geomorphology*, Concept Publishing Company, Pvt. Ltd., New Delhi.

Sharma, H.S. (1988). The Problem of Desertification in Rajasthan, in V. Vidyanath and R.R. Rao, (Eds.). *Development of India's Resource Base; Patterns, Problems and Prospects*, Gyan Books, New Delhi, pp. 358-66.

Sharma, H.S. (1993). Geomorphological Aspects of the Processes of Desertification in the Northeastern Rajasthan (India), *Z. Geomorph N.F. Suppl. Bd.*, 87: 61-69, Berlin-Stuttgart.

Sharma, H.S. (2013). Climate Change, Desertification and Water Resource Management in India in H.S. Sharma, *et al.*, (Ed.) *Climate Change, Biodiversity and Green Economy*, Concept Publishing Company Pvt. Ltd., New Delhi: 43-66.

Sharma, H.S., (1968) Genesis of Ravines of the Lower Chambal Valley, India, *Selected Papers, 21st IGU*, Vol. I, pp. 114-118.

Sharma, H.S., (1983) Inland Drainage of North-eastern Rajasthan and Its Possible Link with the Ghaggar Yamuna System. *Transaction's of Institute of Indian Geographers*, Vol. 5, No.1, pp. 19-22.

Sharma, V.K. (2007) Global Warming its impact on India. *Theme paper (4)* the member Annual Conference, 118A, New Delhi.

Singh, G.; Joshi, R.D., Chopra, S.K. and Singh, A.B. (1974) Late Quaternary History of Vegetation and Climate of the Rajasthan Desert, India. *Philosophical Transactions of the Royal Society*, 267B, pp. 467-501.

Singh, G. (1971) The Indus Valley Culture seen in the Context of Post-glacial Climate and Ecological Studies in North-west India. *Archaeology and Physical Anthropology in Oceania*, Vol. 42, pp. 178-89.

Singh, G. (1977) Stratigraphical and Palynological Evidence for Desertification the Great Indian Desert, *Annals of Arid Zone*, Vol. 16, No.3, pp. 310-320.

Singh, N. and Sontakke, N.A., (2002). *Climatic Change*, 52: 287-313.

Singh, P. and Jain, S.K., (2002) *Snow and Glacier melt in the Satluj River.*

Singh, R.L., (1971). *India: A Regional Geography.* National Geographical Society of India, Varanasi, p. 992.

Singh, R.S., K.D. Sharma, Faroda, A.S. (1999). "Climate change and its impact on drought and floods in Luni River basin of north-west arid India." *Journal of Agrometeorology,* 1(2): 99-111.

Singh, R.S., P. Narain and Sharma, K.D. (2001). "Climate Changes in Luni river basin of arid western Rajasthan (India)." *Vayu Mandal*, It is Journal hence, there is no another Indian Meteorology Society. New Delhi, 31: 103-106.

Singh, S. (1977). Geomorphology and Climatic Changes During Quaternary period in Rajasthan Desert, *Recent Researches in Geology*, Vol. 9, pp. 121-130.

Singh, S. (1977). Sand Dunes and Palaeoclimate in Jodhpur District, Western Rajasthan, *Man and Environment*, Vol. 1, pp. 7-15.

Singhvi, A.K. (1995). Luminescence Dating: An Appraisal, IGCP, 349, I meeting at Jaipur, 8-11 April, 1995.

Sood, R.K. and Pathan, S.K. (1982). Palaeoenvironment of Rajasthan—Observation from Landsat Imagery, Abstracts of paper, Fourth Indian Geography Congress, Bombay.

Srivastava, H.N. and Balakrishnan, (1993). Synthesis of meteorological and tide gauge observations on a decadal scale near the Indian Coast. *TOGA* Notes, pp. 15-17.

Staubwaser, M., Sirocko, F., Grootes, P.M. and Segl, M. (2003). "Climate change at the 4.2 ka BP termination of the Indus Valley Civilization and Holocene South Asian Monsoon variability." *Geophys. Res. Lett.*, 3002. 30: 1425, doi:10.1029/2002GL016822.

Stein, (1942). A Survey of Ancient Cities along the 'Lost' Saraswati River, *Geographical Journal,* Vol. 99, pp. 1 73-82.

Sweta M. Huda (2013). Rajasthan Water Assessment: Potential for Private Sector Intervention International Finance Corporation (World Bank Group). *Report:* 1-140.

Thapliyal, V. and Kulshrestha, S.M., (1991). *Mausam,* 42: 333-338.

Theme Paper on Management of Floods and Droughts, Indian Water Resources Society, New Delhi, 22 March, 2001, p. 53.

UNCOD (1977). *Plan of Action*, Nairobi.

UNDP/UNEP (1992). *Draft Report on Expert Meeting on Desertification/Land Degradation and the Global Environment Facility*, Nairobi, November.

UNEP (1992). Status of Desertification and Implementation of the United Nations Plan of Action to Combat Desertification, *Report of the Executive Director*, February.

Verstappen, H.T. (1970). Aeolian Geomorphology of the Thar Desert and Palaeo-climates, *Zeitschrift fur Geomorphische Abhandlungen*, Vol. 60, pp. 341-353,

Verstappen, H.T.; Ghose, B. and Pandey, S. (1969). Landforms and resources in Central Rajasthan (India) Result of Jalore Pilot Survey, *Publi. Inst. Aerial Survey and Earth Sc.*, Vol. 51, pp. 1-20.

Vijay Shanker Singh, Deep Narayan Pandey, Anil K. Gupta and N.H. Ravindranath (2010). *Climate Change Impacts, Mitigation and Adaptation, Science for Generating Policy Options in Rajasthan, India*, Rajasthan State Pollution Control Board, Jaipur: 1-150.

Vulnerability and Adaptation Experiences from Rajasthan and Andhra Pradesh (2009). SDCV and A Programme.

Wadhawan, S.K. (1995). Quaternary Event Stratigraphy of Rajasthan Desert and its Palaeoenvironmental Implications, *Abstract IGCP 349*, I meeting at Jaipur from 8-11 April, 1995.

Wilhelmy, H. (1969). Das Vrstromtal am Ostrand der Indusebene und dar Sarasvati-problem, Zeitochrift fur Geomorphologie, *Supplement band*, Vol. 8, pp. 76-83.

Yadav, R.K., K. Rupa Kumar and M. Rajeevan (2010). "Climate change scenarios for Northwest India winter season." *Quarterly International*, 213 (1-2): 12-19.

Yash Pal, Baldev Sahai, Sood, R.K. and Agrawal, D.P. (1980). *Remote Sensing of the 'Lost' Sarasvati/Space Application Centre* (ISRO), Ahmedabad: 53.

Index